T0234919

CAMBRIDGE LIBRARY COLLECTION

Books of enduring scholarly value

Technology

The focus of this series is engineering, broadly construed. It covers technological innovation from a range of periods and cultures, but centres on the technological achievements of the industrial era in the West, particularly in the nineteenth century, as understood by their contemporaries. Infrastructure is one major focus, covering the building of railways and canals, bridges and tunnels, land drainage, the laying of submarine cables, and the construction of docks and lighthouses. Other key topics include developments in industrial and manufacturing fields such as mining technology, the production of iron and steel, the use of steam power, and chemical processes such as photography and textile dyes.

Underground Life

Born in Marseilles, Louis Simonin (1830–86) became a leading mining engineer of his age. He travelled widely on government and private commissions, particularly around the United States, where he was held in very high esteem. His posthumous renown rests primarily on this substantial work on mining, first published in 1867. The book is divided into three parts, dealing with coal mining, metal mining, and the mining of precious stones. It covers metallurgy and mineralogy, the history of mining, and techniques, methods and equipment. Bringing the struggles of miners to life, and enhanced by numerous illustrations by some of the leading engravers of the day, the book is regarded as having inspired and informed Émile Zola, whose great novel *Germinal* (1885) depicts coal miners' lives during a strike. Simonin's work reached a wide readership in his native France, and this English translation appeared in 1869.

Cambridge University Press has long been a pioneer in the reissuing of out-of-print titles from its own backlist, producing digital reprints of books that are still sought after by scholars and students but could not be reprinted economically using traditional technology. The Cambridge Library Collection extends this activity to a wider range of books which are still of importance to researchers and professionals, either for the source material they contain, or as landmarks in the history of their academic discipline.

Drawing from the world-renowned collections in the Cambridge University Library and other partner libraries, and guided by the advice of experts in each subject area, Cambridge University Press is using state-of-the-art scanning machines in its own Printing House to capture the content of each book selected for inclusion. The files are processed to give a consistently clear, crisp image, and the books finished to the high quality standard for which the Press is recognised around the world. The latest print-on-demand technology ensures that the books will remain available indefinitely, and that orders for single or multiple copies can quickly be supplied.

The Cambridge Library Collection brings back to life books of enduring scholarly value (including out-of-copyright works originally issued by other publishers) across a wide range of disciplines in the humanities and social sciences and in science and technology.

Underground Life

Or, Mines and Miners

LOUIS SIMONIN

EDITED AND TRANSLATED BY

H.W. BRISTOW

CAMBRIDGE
UNIVERSITY PRESS

University Printing House, Cambridge, CB2 8BS, United Kingdom

Cambridge University Press is part of the University of Cambridge.
It furthers the University's mission by disseminating knowledge in the pursuit of
education, learning and research at the highest international levels of excellence.

www.cambridge.org
Information on this title: www.cambridge.org/9781108072014

© in this compilation Cambridge University Press 2014

This edition first published 1869
This digitally printed version 2014

ISBN 978-1-108-07201-4 Paperback

This book reproduces the text of the original edition. The content and language reflect
the beliefs, practices and terminology of their time, and have not been updated.

Cambridge University Press wishes to make clear that the book, unless originally published
by Cambridge, is not being republished by, in association or collaboration with,
or with the endorsement or approval of, the original publisher or its successors in title.

The original edition of this book contains a number of colour plates,
which have been reproduced in black and white. Colour versions of these
images can be found online at www.cambridge.org/9781108072014

A Faguet pinxt.

G. Regamey, Chromolith.

1 Native Copper, *Corocora,(Bolivia)*.

2. Cuprite, or Red Oxide of Copper, *Kurmski. (Siberia)*

3. Blue carbonate of Copper or Azurite, *Chessy*

4. Malachite, Green Carbonate of Copper, *Ural.*

5. Copper Pyrites, *Tenez.(Algeria)*.

6. Peacock Copper, *Cornwall.*

CHAPMAN & HALL, London.

Imp Lemercier & Cie Paris.

UNDERGROUND LIFE;

OR.

MINES AND MINERS.

BY L. SIMONIN.

TRANSLATED, ADAPTED TO THE PRESENT STATE OF BRITISH
MINING, AND EDITED

BY H. W. BRISTOW, F.R.S.,

OF THE GEOLOGICAL SURVEY; HONORARY FELLOW OF KING'S COLLEGE, LONDON.

ILLUSTRATED WITH 160 ENGRAVINGS ON WOOD, 20 MAPS GEOLOGICALLY COLOURED,
AND 10 PLATES OF METALS AND MINERALS IN CHROMOLITHOGRAPHY.

———

LONDON:
CHAPMAN AND HALL, 193, PICCADILLY.
1869.

PRINTED BY WILLIAM MACKENZIE, 45 & 47 HOWARD STREET, GLASGOW

AUTHOR'S PREFACE.

VICTOR HUGO has recently described the struggles of the sailor in "The Toilers of the Sea." What he so happily calls the άναγκη, or irrepressible power of the Elements, addresses itself alike to the Mariner and the Miner, for each is the soldier of the deep, against whom the powers of nature wage at times their utmost fury.

In the following pages we purpose to describe the struggle of the miner in its reality, without exaggeration of any sort. We shall follow him to the field of his labours, observe him in his subterranean life, and describe his habits in various countries; and as we would not only amuse, but instruct, we shall speak of the countries he inhabits, and of the substances he digs from the earth: in short, we shall endeavour to explain the social position of this pioneer of civilization. We ourselves have long sojourned with him both in Europe and in America, and have made ourselves acquainted with the manly qualities which so eminently distinguish him.

The first part of the present work is devoted to Coal, a substance indispensable to all civilized nations; the second to Metals, the origin of all progress; the last to the Precious Stones, which play so important a part in the decorative arts. The intrepid coal-miner, whose advent is but of recent date; the veteran of the mineral world, whose origin dates from

a

the dawn of history; and the patient seeker after gems— are the types of industry we have to consider.

It will be thought, perhaps, that a chapter ought to have been dedicated to Petroleum, Sulphur, and Salt Mines, but these do not happen to belong to any of the divisions we have under consideration. They are touched upon incidentally, and the processes by which they are worked differ but little from those described.

Fiction has been strictly excluded, but the artist's crayon has been employed whenever it was thought necessary as a forcible appeal to the eye, but in no instance has any merely fanciful design been admitted. All the fossils and minerals, all the implements and apparatus, have been drawn from models; and the specimens so well represented in the coloured plates, whether chosen from the national collections of the Museum of the School of Mines (of Paris) generously placed at our disposal, or from the private collections of our friends or ourselves, have been selected as the best within reach.

Among those artists who have seconded the author with so much zeal, we may mention MM. de Neuville, Faguet, Dumas-Vorzet, Bonnafoux, Lançon, and Bonhommé, but we return to all without exception our most sincere thanks.

L. SIMONIN.

NOTE BY THE TRANSLATOR.

M. SIMONIN, in "La Vie Souterraine," looks at Mining from a French point of view, and consequently he does not give any special prominence to the Mineral Industries of this country.

It would have been impossible, therefore, to give a more elaborate account of British Mines and Miners, without re-modelling M. Simonin's book. Under the circumstances, I have chosen to preserve his text entire when it was possible to do so, making only some few interpolations where they appeared to me to be absolutely necessary.

Amongst the friends to whom I am indebted for advice and assistance, my thanks are due to Mr. R. Etheridge, Mr. H. Bauerman, and Mr. W. Whitaker, and more especially to my esteemed colleague, Mr. Robert Hunt, of the Mining Record Office, for much valuable aid kindly rendered during the progress of my undertaking.

The Geology of the maps has either been revised by Mr. James B. Jordan, of the Mining Record Office, or new maps constructed by him to replace those in the original work when considered necessary. In this task Mr. Jordan has had the benefit of the advice of Mr. John Arthur Phillips with respect to the map of California; and for the map of Chile I am indebted to my friend Mr. David Forbes, whose personal observations in that part of South America extend over several years.

HENRY W. BRISTOW.

LONDON, *October*, 1868.

a 2

ERRATA:

Page 33, third line from bottom, for Plate I. read Plate VIII.
Plate V., fig. 2, for Sulphate of Arsenic read Sulphide of Arsenic.
Plate VII., fig. 2, for Bisulphate of Iron read Bisulphide of Iron.
Plate X., fig. 7, for Grenada read Granada.

COLOURED ILLUSTRATIONS.

GEOLOGICAL MAPS.

COAL-FIELDS OF THE WORLD.

METALLIFEROUS REGIONS.

TABLE OF ILLUSTRATIONS.

COAL-FIELDS.

TABLE OF CONTENTS.

FIRST PART.

COAL MINES.

CHAPTER V.

HOW COAL IS DISCOVERED.

CHAPTER VI.

SHAFTS AND LEVELS.

CHAPTER VII.

HOW THE COAL IS WORKED.

CHAPTER VIII.

THE FIELD OF BATTLE.

CHAPTER IX.

FALLS OF ROOF AND INUNDATIONS.

CHAPTER X.

THE PERILOUS PASSAGE.

SECOND PART.

METALLIFEROUS MINES.

CHAPTER I.

THE STAGES OF THE HUMAN RACE.

CHAPTER II.

THE LABORATORY OF NATURE.

CHAPTER III.

THE PRINCES OF THE MINERAL KINGDOM.

CHAPTER IV.

THE METALLIFEROUS WORLD.

EUROPE AND ASIA.

CHAPTER V.

THE METALLIFEROUS WORLD.

AMERICA, POLYNESIA, AFRICA.

CHAPTER VI.

EUREKA.

CHAPTER VII.

HIDDEN TREASURES.

CHAPTER VIII.

THE ATTACK ON THE GROUND.

CHAPTER IX.

THE SISTERS OF THE CATACOMBS.

CHAPTER X.

CRUSHING AND WASHING.

CHAPTER XI.

THE PHALANX OF MINERS.

THIRD PART.

MINES OF PRECIOUS STONES.

CHAPTER I.

THE FAMILY OF GEMS.

CHAPTER II.

IN THE EAST AND UNDER THE TROPICS.

CHAPTER III.

THE SEEKERS.

PART I.

COAL MINES.

UNDERGROUND LIFE;

OR

MINES AND MINERS.

CHAPTER I.

COAL MINES—PAST AND PRESENT.

Coal in Paris and London.—Uses of Mineral Fuel.

In the year 1769 wood was very dear in Paris, as it is now. Some English merchants conceived the idea of sending cargoes of coal from the English collieries to supply the scarcity of fuel in France. Vessels were dispatched with coal from Newcastle, which was sent up the Seine, and soon reached Paris.

The coal which was sent from England was at first used by the common people, and then by some of the higher classes in the stoves and fireplaces of antechambers.

There was a general outcry. The unfortunate mineral fuel was accused of vitiating the air, of soiling the linen when drying, of causing affections of the chest, and, worst of all, impairing the delicacy of the female complexion. Complaints being unceasing, the Academies of Science and Medicine were appealed to for counsel in the matter: they declared in favour of the British fuel; but in such a case it is not the Academies, but the public, whose decision is final.

Many years before this the black mineral had not been better received by the Parisians. In 1714 they expelled it for the first time. Under Henri II. the Doctors of the Sorbonne had condemned it for its noxious sulphurous vapours, and a royal edict prohibited the iron-merchants, on pain of imprisonment and fine, from making use of coal. Some time afterwards the interdict was repealed, and Henri IV. even exempted coal from the duty which importers paid to the crown in virtue of the regal right.

The combustible mineral had been as ill received in London, on its first introduction, as at Paris. The doctors objected to it on account of its smoke; and the proprietors of forests, because it injured their trade, kept the gates of the city long closed against it. Royal ordinances rejected it, as was the case afterwards in France, and it was only by degrees that these restrictions became relaxed.

London now receives annually more than six millions of tons of coal (6,013,265 tons in 1866), upwards of three million tons of which quantity are brought by sea, and two millions nine hundred and eighty thousand tons by railway and canal. Nearly six thousand ships of five hundred tons burthen are engaged in this trade for the great metropolis. Six thousand ships of five hundred tons equals the whole annual marine trade, on the average, of a large sea-port like Marseilles! It is more than all the traffic circulating yearly throughout the French coasts. All the coal brought is not, however, consumed in London; a large portion is used in the steam-vessels leaving the Thames, and a considerable quantity is passed on by the railways to the south-eastern counties. Between three and four million tons of coal are actually consumed annually in London alone; and yet people are astonished to see that a constant atmosphere of smoke envelopes the capital of the three kingdoms, and that a thick layer of black dust covers its edifices.

Paris burns much less coal than London, but still it consumes a million of tons annually. The mineral hitherto proscribed is now admitted everywhere ; and if stopped by the customs or tolls, it has only to pay the duty. It matters not that the employment of coal provokes many complaints, causing men to dream of smoke-consuming furnaces, many of the inventions being ingenious, but none of them entirely complete ; the reign of mechanical labour is come, and the nineteenth century has inaugurated the era of industry, which finds in coal—what has been so well termed—its daily bread.

In the present day coal makes the fortune of bold and patient explorers, of numerous mining companies, and even of entire countries ; England herself owes no small part of her industrial and maritime power to her coal. Coal is the motive power of all machines, whether of mills, manufactories, or workshops, and also of marine and locomotive engines. Being heavy, it answers the purpose of ballast in merchant vessels, in the form of a most profitable cargo, and supplies half the traffic of the canals and railways. Now that the naval service is transformed by means of steam, coal not only conduces materially to the prosperity of a state, but also contributes to its defence—so much so, that it has been declared to be contraband of war. Towns are lighted, and furnaces—whether for manufacturing or domestic purposes—are heated, by means of coal ; it is also the agent employed in reducing metallic minerals. And as if nothing might be wanting to its many and varied uses, it is from coal that skilful chemists have recently produced the most brilliant colours, such as render famous the names of Magenta, Solferino, and mauve, which have made the tour of the world with the novelties of Lyons and Paris and the textile fabrics of England. Finally it is from coal that the volatile organic compound phenic (or carbolic) acid has been extracted, which possesses very remarkable disinfecting or rather antiseptic properties, and has been

used for curing gangrene and accelerating the healing of
wounds.

It is necessary, before proceeding further, to describe the
origin of this useful mineral—

"King COAL, the mighty hero of the mine"—

and to explain how it has been deposited in the earth which
contains it.

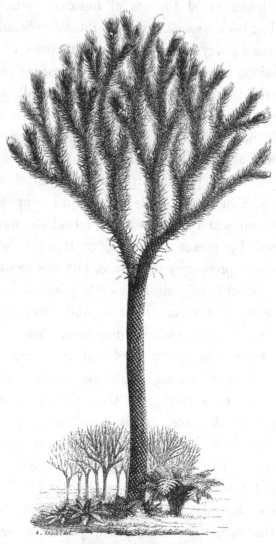

Lepidodendron Sternbergii restored.

CHAPTER II.

ORIGIN OF COAL.

Carboniferous Period.—Mineral Fuel most abundant in rocks of Carboniferous age.—Climate.—Dr. Daubeny's estimate of its temperature.—Theory of the formation of Coal.—Limestones, how formed.—Elimination of carbonic acid from the atmosphere.—Carboniferous formation of Scotland and Ireland.—Volcanic rocks associated with Carboniferous rocks of Devon and Cornwall, Derbyshire, North of England.—Fauna of Carboniferous period; Foramenifera, Corals, Crinoids, Crustaceans, Insects, Molluscs, Fishes, Reptiles. Coprolites, Foot-prints, Rain-drops, Ripple-marks. Coal Measures. Flora.—Enormous time required.—Chemical changes undergone by vegetation during the formation of Coal.—Theories as to mode of formation of Coal; Elie de Beaumont; the ancient Geologists; Adolphe Brongniart.—Baroulier's experiments.—Analogy between Coal and Peat.—Coal of any geological age, but most abundant in the Carboniferous formation.—Lignite and Brown Coal; of Germany and Russia; of Bovey Tracey.—Coal of Mackenzie River, China, Scarborough.—Classification of Coals.—Anecdote connected with the discovery of Coal in Etruria.

At the very remote geological period when the coal was formed, immense jungles and swamps covered a large portion of the earth's surface. The atmosphere was charged or saturated with vapour, and possibly contained a much larger proportion of carbonic acid than it does now. The gas in question being that which especially promotes the growth of plants, while it is and was probably unfavourable to the existence of animal life, it has been plausibly suggested that the gradual withdrawal of the carbonic acid by the growth of the vegetation of the Carboniferous period slowly purified the atmosphere, and fitted it for the existence of those higher forms of terrestrial and vertebrate life which made their appearance at later dates. A more extended knowledge of the Carboniferous formation, and the now well-ascertained fact that although mineral fuel is more abundant in the Carboniferous rocks than in any other, numerous accumulations and deposits of a precisely similar nature, though perhaps represented by higher forms of vegetable life, also occur in rocks of various and very different geological ages—especially in those of the Oolitic, Cretaceous, Eocene, and Miocene periods—tend to render the above hypothesis somewhat questionable.

Neither is it necessary to suppose that the climate of the Carboniferous period should have been such as is understood by our term tropical, although it certainly must have been warm, moist, and equable—nowhere colder than temperate. This is evidenced by the corals and mollusca, which must have inhabited seas as warm as those of the present epoch, and illustrated by the existing coral-reef seas which lie between parallels of 28° on either side of the equator, the corals forming which flourish only in waters having a temperature not lower than 82°.

Dr. Daubeny* arrived at the conclusion that a knowledge of the temperature of this and other periods of the ancient world may be inferred by ascertaining the mean between the extremes of heat and cold, within which each one of the great families of plants predominating in the vegetation of each period is now capable of maintaining itself unassisted by man.

Taking the arborescent ferns, of which a large portion of the coal is believed to be composed, Dr. Daubeny inferred that the temperature prevailing at the time when they flourished in such abundance as appears to have been the case during the period of the coal-formation, could not have been less than 62·5° Fahrenheit.

The most plausible and reasonable theory of the formation of coal, seems to be that it is for the most part the remains of vegetable matter which became decomposed and mineralized on the spot where it grew and is now found. The fibrous tissues of the aquatic vegetation, which flourished as a thick carpet at the feet of the larger plants, mingled and became matted together. as is the case now with the turf and peat of

* "On the Temperature of the Ancient World."

Fig. 1.—Ideal Landscape of the Coal-measure Period.

most peat-bogs, forming swamps and marshy plains; only at the period in question the phenomenon was more general than at the present day, and occurred on an enormous scale—on the borders of great lakes and estuaries, that is to say, where the fresh water of rivers first mingles with the salt water of the sea. Vast lagoons were thus formed, as is still seen, only on a smaller scale, beneath the tropics, in Senegal and Madagascar, &c. In these lagoons vegetation must have been developed with the greatest luxuriance and profusion (fig. 1).

These wide, swampy, or peaty plains occupied portions of the earth's surface which, owing to subterranean movements, were undergoing slow depression. Hence by degrees the layer of matted vegetation was carried down beneath the level of the sea, which then flowed over it, and deposited above it those layers of sand, silt, and mud which we now find among and alternating with our coal-seams as beds of sandstone and shale. The downward movement, however, was not constant. It seems to have been intermittent, or interrupted by minor elevations or by long pauses, during which the accumulating sediment filled up the lagoons, and allowed new jungles to spring up and marshy tracts to be formed. As the submergence recommenced these were, in like manner, entombed beneath the silt of the encroaching waters. Thus, stage by stage, the vast pile of strata known as the Coal Measures was built up. Each coal-seam, then, represents the vegetation of an old land-surface, while the intervening strata of sandstone, shale, fire-clay, &c., mark the varying sediments which were brought together by the combined action of the sea and estuarine waters, during the slow subsidence of their beds.

The limestones of the Carboniferous period were derived from the secretions of the coral polypes, from the shells of the mollusca, and the remains of other organisms which lived in the waters in which they were formed.

" The gradual removal, in the form of carbonate of lime, of

the carbonic acid from the primeval atmosphere, has been connected with great changes in the organic life of the globe. The
air was doubtless at first unfit for the respiration of warm-
blooded animals, and we find the higher forms of life coming
gradually into existence as we approach the present period of
a purer air. Calculations lead us to conclude that the amount
of carbon thus removed in the form of carbonic acid has been
so enormous, that we must suppose the earlier forms of air-
breathing animals to have been peculiarly adapted to live in
an atmosphere which would probably be too impure to support
modern reptilian life. The agency of plants in purifying the
primitive atmosphere was long since pointed out by Brongniart;
and our great stores of fossil fuel have been derived from the
decomposition, by the ancient vegetation, of the excess of carbonic acid of the early atmosphere, which through this agency
was exchanged for oxygen gas. In this connection, the vegetation of former periods presents the curious phenomenon of
plants allied to those now growing beneath the tropics flourishing within the polar circles. Many ingenious hypotheses have
been proposed to account for the warmer climate of earlier
times, but are at best unsatisfactory; and it appears to me that
the true solution of the problem may be found in the constitution of the early atmosphere, when considered in the light of
Dr. Tyndall's beautiful researches on radiant heat. He has
found that the presence of a few hundredths of carbonic acid
gas in the atmosphere, while offering almost no obstacle to the
passage of the solar rays, would suffice to prevent almost
entirely the loss by radiation of obscure heat, so that the surface of the land beneath such an atmosphere would become
like a vast orchard-house, in which the conditions of climate
necessary to a luxuriant vegetation would be extended even to
the polar regions." *

At certain intervals during the deposition of the Coal

* "On the Chemistry of the Primeval Earth," by T. Sterry Hunt, pp. 5, 6.

Measures, carbonate of iron (clay-ironstone) was formed, derived from ferruginous substances and carbonic acid ; and occurs in layers of nodules interstratified with the shales.

Transported fragments of other rocks, the various components of which, consisting mostly of quartz and schists, became cemented together by argillaceous, arenaceous, or ferruginous matter, were deposited chiefly at the base, and towards the upper part of the Carboniferous formation—as though the Old Red Sandstone or Devonian rocks, as well as those of the Permian series, marked the beginning and end of periods of greater relative quietness and repose than those which prevailed before and after the Carboniferous period.

The Coal Measure period, as has been described, was one of comparative tranquillity, favourable to the growth of a rank and luxuriant vegetation in the moist and equable climate which then prevailed. The crust of the earth over well-marked and widely-spread areas, as we have just observed, was subject to gradual but oft-repeated oscillations of level, due to subsidence and elevation. Nor were there absent active volcanic vents, throwing out streams of lava and showers of ash, as volcanoes still do.

Mr. Archibald Geikie, in his presidential address to the Geological Section of the British Association, at the meeting held at Dundee in 1867, has shown that the Carboniferous formation of Scotland displays striking evidences of volcanic activity; volcanic rocks of several varieties abounding throughout the Carboniferous Limestone group of that country (in which its lower coals occur), down to the base of the Carboniferous system.

The great Carboniferous Limestone series of Ireland is also stated by Mr. Jukes to contain evidence that, at various intervals during its deposition, volcanic vents were active on different parts of the sea-bottom. In the county of Limerick great masses of trap, eight hundred or a thousand feet thick, with

well-marked interstratifications of volcanic ash, are found to
lie among the limestones—Pallas Hill forming a conspicuous
example of this occurrence.*

In Cornwall and Devon Sir Henry De la Beche† has shown
that the base of the Carboniferous series is characterized by
the occurrence in it of sheets of trappean ash and amygdaloidal
greenstone, which were probably erupted over what was then
the bottom of the Carboniferous sea, where they became interstratified, and this about the time of the deposition of the
middle portion of the Carboniferous Limestone, and also near
its summit, with the ordinary marine sediment of the period.‡

It has also been pointed out by Professor J. Phillips that
the well-known bands of igneous rock in Derbyshire, called
toadstone, which are interstratified with the Carboniferous
Limestone of that county for many miles, sometimes with a
thickness of fifty feet, are contemporaneous trap rocks, indicating volcanic action proceeding from different vents.

In the north of England, also, in Yorkshire, Durham, and
Northumberland, the great beds of basalt, one of which, commonly known as the Great Whin Sill, alternating with those
of the Carboniferous Limestone, and extending for many miles,
and varying in thickness from twenty to three hundred feet,
are believed by Professor Sedgwick to have been masses of
molten matter horizontally injected between the ordinary sedimentary deposits; while Professor John Phillips supposes them
to have been possibly submarine lavas erupted from several
centres or lines. §

We are able to restore, in a great degree, the extinct fauna
and flora of this period, a short summary of which may not be
out of place here, and may be described as follows.

<hr>

* "Student's Manual of Geology," 2nd edition, page 326.
† "Report on the Geology of Cornwall and Devon," page 119.
‡ Jukes' "Student's Manual of Geology," 2nd edition, p. 523.
§ Phillips' "Manual of Geology," p. 522; also Jukes' "Student's Manual of Geology," 2nd
edition, p 524.

The great mass of limestones, &c., known as the Carboniferous or Mountain Limestone, forming the base of the series, of which it is the lowest and oldest member, is a truly marine deposit, rich in remains of corals, crinoids, brachiopods, &c., and has usually no seams of coal mixed with it, except in Scotland.

As in the case of all limestones that have not been metamorphosed into pure crystalline carbonate of lime, the Carboniferous Limestone consists largely of Foramenifera, which are minute shelled animals having a very simple organization, and abounding in the open seas where shell-banks and coral reefs flourished. Sometimes these Foramenifera of the Carboniferous Limestone occur in such abundance and so loosely compacted that they can be easily separated, the limestone being friable : this is the case with the white Carboniferous Limestone of Russia. The Foramenifera forming the mass of this rock is the *Fusulina*, so called from its distaff-like shape. The limestone seems to be nearly wholly composed of this and other smaller forms, with many corals ; and is found in typical Coal Measure regions all over the globe. In more solid varieties of the Carboniferous Limestone, polished slices of the rock also show abundant evidences of Foramenifera as constituents of the mass.

Corals are extremely abundant in the Carboniferous Limestone, and resembled in every particular those now existing in the warm sea of the equatorial region. Both single and compound forms were numerous : amongst the former the genera Cyathophyllum, Zaphrentis, and Amplexus abounded ; of the latter group, Lithostrotion, Michelinia, fig. 2, and Syringopora may be mentioned.

Numerous species of Crinoids, of very varied forms, are met with in the Carboniferous Limestone, in such numbers as frequently to cause the stone in which the remains are met with to be known as crinoidal or encrinital limestone; familiar examples of which are afforded by the so-called marbles of Clifton, Derbyshire, and elsewhere. The occurrence of animals of this class denotes that the water of the sea in which they flourished in such profusion was clear and free from muddy

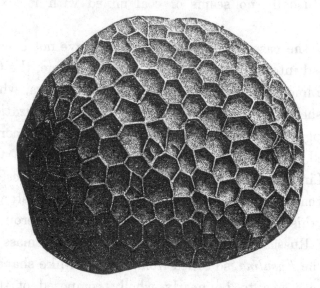

Fig. 2.—Fossil Coral (*Michelinia favosa.*) Belgian Coal Measures. Scale, ⅔.

sediment, which would be unfavourable or altogether fatal to their existence. Besides the true Crinoids, many forms of Star-fishes occur, of which that shown in fig. 4 is a good example. Amongst the Polyzoans, which were a numerous class during this epoch, the peculiar fossil Archimedipora

Fig. 3.—Fossil Polyzoan (*Archimedipora Archimedis*). Coal Measures of Illinois, U.S. Nat size.

Archimedis (fig. 3) may illustrate one of the forms of this singular group.

Of crustaceans the Carboniferous Limestone and Coal Measures yield Trilobites, and afford proofs of the existence of centipedes and spiders, two species of scorpion, Eurypteri, at least six species of king-crabs, and three species of lobster-like forms.

A climate like that of the Coal Measure period, when the surface of the earth was covered with marshes and swamps, and clad with a thick mantle of vegetation, must necessarily have been favourable to the existence of insect life; and it is not so surprising that the remains of insects are of comparatively rare occurrence (not more than twelve species having been found in the Coal Measures), as it is that so many fragile organisms have been preserved.

Insect remains have been found by Mr. Prestwich in the Coal Measures of Coalbrook Dale, and portions of an orthopterous insect, nearly allied to Blatta, or the cockroach, have lately been discovered by Mr. James Kirkby in the Coal Measures of Claxheugh, near Sunderland.*

Fig. 4.—Cast of a Star-fish (*Asterias constellata?*) in a nodule of Clay Ironstone. Scale, ⅔.

Of the Mollusca, or shell-bearing animals, the Brachiopoda are the most abundant throughout the Carboniferous Limestone age, and especially species of the genera Productus and Spirifer (fig. 5). The Nautili (fig. 6) and ammonite-like Goniatites, as well as many of the numerous Gasteropoda and Conchifera, which peopled the sea of the Carboniferous period, have now entirely disappeared; and the species have been replaced by others altogether different, even when many of the genera have not themselves become extinct.

* Geological Magazine, vol. iv. p. 388.

The remains of Fishes occur in great numbers both in the Carboniferous Limestone and Coal Measures, especially the

Fig. 5.—*Spirifera striata.*　Yorkshire Coal Measures.　Scale, ⅔.

Ganoid (fig. 7) and Placoid orders, allied to our modern sharks.

Fig. 6.—*Nautilus cariniferus?*　Coal Measures of Indiana, U.S.　Scale, ½.

One hundred and fifty species of fishes have been discovered, some of which were of great size, as is proved by their remains, especially the Ichthyodorulites, or fin-spines, which are in some instances as much as from eight to ten inches long.

The highest forms of animal hitherto met with in the Carboniferous rocks are those of swimming reptiles.

The remains of no less than twenty-two species of land-dwelling or amphibious reptiles have been brought to light, which must have lived on the borders of estuaries or in the marshes and muddy waters of their shores. Most likely the smaller reptiles of the coal period were insectivorous, the head and some other fragments of a large insect, probably Neuropterous, having been found by Principal Dawson in the coprolites, or petrified

Fig. 7.—Fossil Fish (*Amblypterus macropterus*). Coal Measures of Saarbrück, in Rhenish Prussia. Scale, ⅔.

excrement, of a fossil reptile, inclosed in the trunk of an erect Sigillaria, at the Joggins Mines of Nova Scotia, along with the shells of a small land-snail (Pupa), and other animal remains.

C

Numerous amphibian Labyrinthodonts (fig. 8) have also been found in the coal-beds of Germany and Ireland.*

In the United States footprints have been met with in the

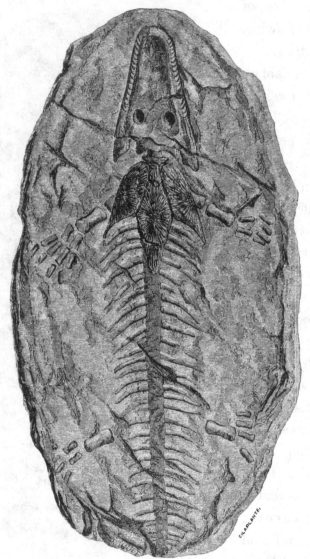

Fig. 8.—Archegosaurus, or the most Ancient Lizard (*Archegosaurus Decheni*). Coal Measures of Saarbrück. Scale, ½.

Coal Measure shales, produced by animals in walking over what was then the soft and moist mud of the shore. The impressions of rain-drops have likewise been preserved, as

* Geological Magazine, vol. iv. p. 385.

well as ripple- and current-markings, represented by roughly parallel, tortuous, and curved irregularities, produced by the changing level and current of the water in the soft sediment

Fig. 9.—Impression of a Fern (*Odontopteris Schlotheimii*). Coal Measures of Saarbrück.   Scale, ⅔.

over which it flowed—marks which remain still uneffaced after so many subsequent revolutions of the globe.

The Carboniferous period, as we have just shown, abounded in animal life. It is, however, pre-eminently remarkable for

c 2

the luxuriance of its flora, both in number of species and individuals, which then flourished to such an extent that the upper or Coal Measure series may be truly considered as the age of plants.

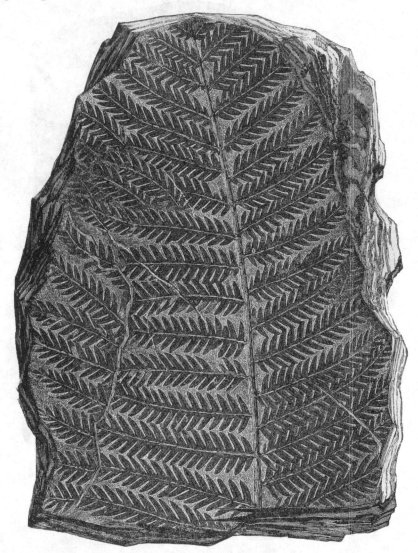

Fig. 10.—Impression of a Fern (*Pecopteris dentata*).   Scale, ⅔.

Jointed Calamites of gigantic proportions with fluted stems, forming the early representatives of the Equisetaceæ; Sigillaria, large trees with slender stems marked with long, oval, or pentangular scars exhibiting a more or less spiral arrangement, and by some considered to be a Cycad, while others regard it as a highly-developed Cryptogam; Cycadeæ, which botanists have long believed to be the early representatives of existing

Fig. 11.—Impression of a Fern (*Neuropteris speciosa*, Ad. Brongniart, MS.). Coal Measures of Blanzy (Saône-et-Loire). Scale, ⅓.

bamboos and palms; Annularia, Asterophyllites with star-shaped leaves, and arborescent (but mostly herbaceous) ferns, grew in thick forests. With them were associated Lepido-

dendra, or scaly-barked trees of no less lofty stature and of
equally irregular appearance, of which there are now no living
representatives, but which in their structure bear the nearest
resemblance to the Lycopods or club-mosses of the present
flora; and Conifers of the Araucanian type, being like them in
fructification as well as foliation.   The Sigillariæ, so named

Fig. 12.—Impression of *Annularia longifolia.*   Coal Measures of Saarbrück.   Scale, ⅔.

from the seal-like (*sigillum,* "a seal") marks made by the leaf-
scars on their stems, appear to have been marsh-plants; while
the Lepidodendra and Conifers may have occupied higher and
drier situations: they are, however, found associated.   The
Stigmariæ are the roots and stems of Sigillariæ and perhaps
of Lepidodendra, whose long cylindrical rootlets are found
penetrating the clay (underclay) which is of almost invariable

occurrence beneath each seam of coal, and the soil in which the vegetation grew out of which the coal has been formed.

Remains of plants are usually very abundant in the sand-stones and roof-shales above the coal-seams. These consist of

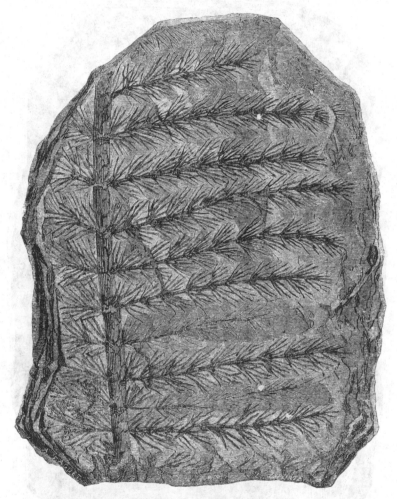

Fig. 13.—Impression of Asterophyllites (*Asterophyllites equisetiformis*). Coal Measures of Saarbrück. Scale, ⅔.

the impression of many species of ferns, some of which were arboreal in their growth, like the tree-ferns of New Zealand, mingled together with the stems of Sigillaria, Lepidodendron, Dadoxylon, and Calamites. These plants have, no doubt, contributed very largely to form the coal. But although about

three hundred species of plants have been described from the
Coal Measures, we only possess a very incomplete list of the
great mass of the vegetation of that period.   Experiments

Fig. 14.—Impression of a Coal-plant (*Nöggerathia lactuca*).   Scale, ⅓.

made by the late Dr. Lindley have proved that after two years'
submersion beneath fresh water, one hundred and twenty-one
out of one hundred and seventy-seven species of plants entirely
disappeared, and of the fifty-six remaining the most perfect

were Coniferæ, ferns, and Lycopodiaceæ. There are no palms found in the Coal Measures, neither grasses, nor flowering-plants

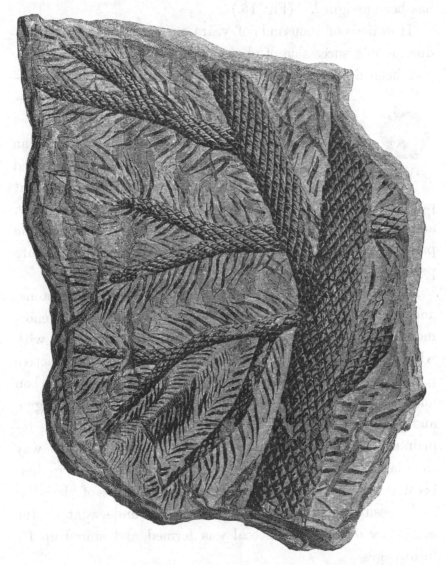

Fig. 15.—Impression of *Lepidodendron gracile*. Coal Measures of Eschweiler, near Aix-la-Chapelle. Scale, ½.

(so far, at least, as is at present known), but remains of the other three classes are abundantly preserved. (Figs. 8 to 18.)

There are also fruits (Carpolites) which have been referred to Conifers, the ancient representatives of our modern pines and firs; and others in the form of angular, nut-like fruits,

which are met with in certain sandstones (as in the Peel Delph
Quarry near Bolton), and to which the name of Trigonocarpum
has been assigned.   (Fig. 16.)

Hundreds of thousands of years are necessary for the production of a succession of phenomena such as those which have
just been described; but nature does not reckon by years—

time for her has no existence.
Otherwise, if needful, some sort
of approximation towards an
estimate of the period required
for the epoch under notice might

Fig. 16.—*Trigonocarpum Nöggerathii.*

possibly be arrived at; for fossils are the medals of geology,
and the layers of shale and rock in which they occur form the
pages on which are inscribed the history of the Coal Measure
period.

The layers of peat, after being covered by shales, sandstone,
and limestone, were strongly compressed beneath the enormous weight of the overlying strata, simultaneously with
which a slow distillation and an almost insensible putrefactive
fermentation took place, during which a portion of the carbon
and hydrogen escaping in the form of carburetted hydrogen,
and part of the oxygen and carbon as carbonic acid, the remaining carbon became more concentrated; and in this way
the plants and peaty tissue, originally loosely matted together,
became more and more compact, and by means of the slow
decomposition which the vegetation thus underwent in the
laboratory of nature, the coal was formed and stored up for
future ages.

The explanations which have just been given of these grand
changes are those generally accepted by modern science.   M.
Elie de Beaumont, the father of French geology, suggested a
theory to clear up the confused ideas which formerly prevailed
with regard to the origin of coal.   Before his time it had been
freely conceded that coal was formed out of the vast forests

which had been transported by running water, as are at the present day the great trees of the virgin forests which flourish along the banks of the Mississippi. Rafts of this kind, after having been buried in the soil, were supposed to have become converted into coal; but M. Elie de Beaumont has demonstrated by figures that such a mode of explanation is inadmissible, and that all the carbon contained in the immense surfaces of forests would scarcely furnish a very thin bed of coal.

Neither can the crude hypotheses of the ancient geologists be any longer accepted. They saw in the coal-deposits either streams of bitumen which had become petrified or else had impregnated certain very porous kinds of rock, or forests carbonized on the spot where they once grew, or traversed by streams of sulphuric acid (oil of vitriol), which possesses the property of hardening and carbonizing wood. It is easy to imagine the agency of bitumen, fire, and acids, to account for the phenomena in question, but it is not so easy to tell from whence they were derived. However, it was not by such means that coal has been formed; it was out of a succession of peaty accumulations, which were subsequently covered up, compressed, heated, distilled, and mineralized, that coal or fossil carbon originated.

In addition to the mathematical proofs of this furnished by M. Elie de Beaumont, there are also physical proofs advanced for the first time by M. Adolphe Brongniart. It is known that this learned naturalist, who so ably carries out, at the Museum of Natural History in Paris, the traditions left by his illustrious father, has been able to depict the forms of the fossil flora of the Coal Measures. A. Brongniart having observed, at the mine of Treuil, near Saint Etienne, that trunks of prostrate Sigillariæ, parts of a true forest petrified in place, were met with in the middle of the coal itself, as well as of the sandstone near the surface, this proved a revelation to him (fig. 17). The coal, then, had been formed at the foot of these trees in

the same way as peat; and the geological study of the ground
afforded a confirmation of the truth of the hypothesis to which
figures had already furnished a clue.

Fig. 17.—Trunks of Sigillaria in the mine of Treuil, at Saint-Étienne.   (From a photograph and an
original drawing by M. Brunet de Boyer, communicated by M. L. Gruner, Inspector General of Mines.)

Since M. Brongniart published his discovery, plants standing erect have been discovered at the collieries of Saint
Etienne, the trunks of the trees being converted into stone in

place. The same has been noticed also in the mines in the north of France, especially at Anzin, and in many of the English coal-mines, as well as in those of the United States, where the casts of perished trees have been counted by hundreds.

This local burial of the vegetation which gives rise to coal, affords an explanation of certain facts which might otherwise be, perhaps, unintelligible. In a mine the quality of the combustible varies more according to the particular spot from which it is taken, than with the position of the bed from which it is extracted. It is clear, in fact, that the quality of the coal must in great measure depend on the special nature of the plants from which it is derived, and, consequently, on the height and situation of growth, as we learn from geographical botany, rather than on the precise geological period when the phenomenon takes place. The oldest strata and the lowest, which have undergone the greatest amount of pressure and heat, are those most productive of anthracite.

In many mines the experiments of the laboratory have confirmed these conclusions. Chemists, and amongst others M. Baroulier, in prosecuting their researches, have been able to convert sawdust into coal by the aid of suitable temperature and pressure. With sands and clays they have reproduced sandstones and shales, and have obtained impressions artificially with the leaves of vegetables.

The analogy between coal and peat accounts for other facts. Mineral fuel is found in many formations, but it is, as a general rule, less pure and compact, and is, besides, spread over areas of more limited extent in proportion as it ascends or descends in the geological scale—above or below the true Coal Measures. This last formation is the only one in which the botanical and climatal conditions have allowed a grand accumulation of vegetables for the production of coal. These plants have subsequently disappeared, or little by little changed their nature, until they have assumed the forms which they now display ; nevertheless, owing to particular circumstances, true compact bituminous coal, caking in the fire, has been formed

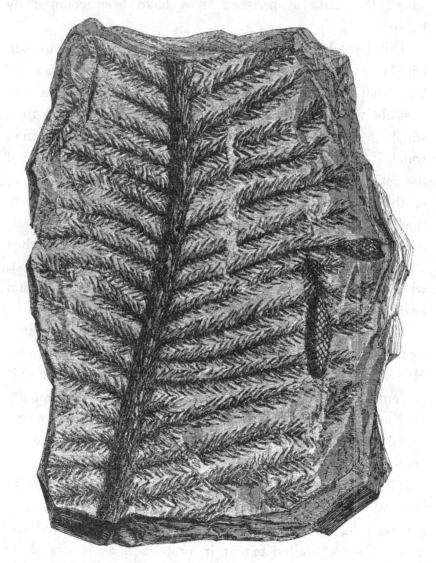

Fig. 18.—Impression of *Walchia piniformis*. L'Hérault. Scale, ⅔.

at several periods, and is not confined exclusively to one geological horizon, the Carboniferous formation, as has been supposed by certain learned persons. It must, therefore, be allowed that coal may be of any geological age, because to deny this

would be to shut the eyes to facts. Still less proper is it to
give the name lignite, or brown coal (which only reminds us

Fig. 19.—Fossil Palm (*Flabellaria raphifolia*).   Tertiary formation of the Somme.
Scale, ⅓.

of wood—*lignum*), to coals belonging to formations of an age
later than the Carboniferous, which were often the scene, as in
the Tertiary period, of a luxuriant vegetation, the Palmacites,

or fossil palms, which were the precursors of the palms of the present equatorial regions.*   (Figs. 9–20.)

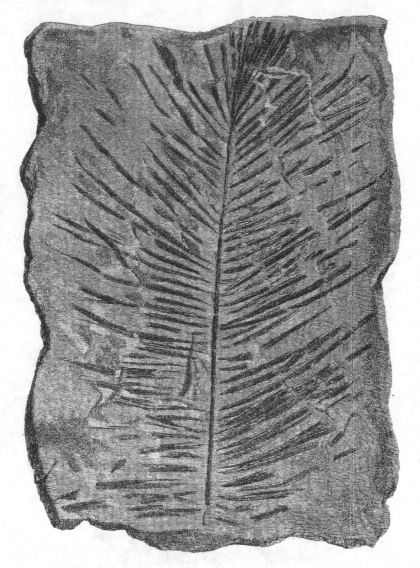

Fig. 20.—Impression of a Date (*Phœnicites Italica*).   Tertiary formation of Lombardy.   Scale, $\frac{1}{10}$.

---

* A notable example of beds of coal, with true underclays, occurs in the Bracklesham (Middle Eocene) beds of Alum Bay, in the Isle of Wight.   See "Memoirs on the Geology of the Isle of Wight," by H. W. Bristow, in the "Memoirs of the Geological Survey;" also Vertical Sections, sheet 25.   True black bituminous coal also occurs associated with the volcanic rocks (Miocene) of the Hebrides.   See A. Geikie, "On the Tertiary Volcanic Rocks of the British Islands," in "Proceedings of the Royal Society of Edinburgh," vol. vi. p. 72.

CARBON

A. Faguet pinx.

G. Regamey Chromolith.

1. Crystallized Diamond (Carbon).
2. Graphite or Plumbago.
3. Anthracite (Stone Coal).
4. Iridescent or Peacock Coal.

5. Schistose or Slate Coal.
6. Perfect Lignite or Jet.
7. Imperfect Lignite or Fossil-wood.
8. Turf or Peat.

CHAPMAN & HALL. London

Imp. Lemercier & Cie. Paris

As we have explained, these great accumulations of coal were the slow results of growth and decay as they go on at the present day; and although this mineral fuel is more abundant in the Carboniferous series than in any other, yet it may be also met with in rocks of Secondary and even of Tertiary age. Indeed, such is the case with the Tertiary Brown coal of Germany and Russia; the Bovey Tracey lignites, and the Miocene coal-beds of the Mackenzie River; and there is every reason to believe that the great coal-deposits of China are of Secondary age, the accumulated growth of Cycadeæ and plants corresponding in character with those found in the Oolitic plant-beds of Scarborough—such as Neuropteris and Pecopteris, of peculiar species.

The best classification of coals, satisfying both the requirements of science and commerce, is that which does not take into consideration the geological age, but only the quality of the combustible. By means of such an arrangement, coals may be conveniently divided into two classes; 1st, Non-bituminous, comprising anthracite or stone-coal; and 2nd, the Bituminous, common or pit-coal—included under four heads:—viz., 1, Caking coal; 2, Splint or hard coal; 3, Cherry or soft coal; and 4, Cannel or parrot coal.*

From anthracite the passage is into graphite,† and even diamond, which is pure crystallized carbon; while cannel coal (of which jet‡ is an extreme variety) passes into brown coal, imperfect earthy lignite, into fossil wood with its fibrous tissue preserved, and finally into peat and turf: so that the complete series would be, diamond, graphite, anthracite, common coal, lignite fossil wood, peat, turf. (Plate I.)

As a commentary upon the above, the story may be told of a coal formation of Tertiary age. The ancient Palæozoic

---

* Jukes' "Manual of Geology," p. 152.

† Graphite or Plumbago, called also Black Lead, consists of nearly pure carbon, with the occasional admixture of a variable quantity of iron.

‡ Bristow's "Glossary of Mineralogy," p. 219.

D

Carboniferous formation does not exist in Tuscany. The assertion by geologists of the non-existence of coal in Etruria having reached the ears of a certain large farmer of the Maremma, named Lenzi, who had discovered the outcrop of a seam of coal in a ravine one day, he informed some capitalists of Leghorn of his discovery. They, without troubling themselves about the opinions of geological authorities, took shares in the affair. The Grand Duke Leopold was astonished, and having himself but little confidence in Italian savants, he sent a Saxon engineer to examine into the matter. The result was a great dispute amongst the geologists as to whether the mineral was true bituminous coal, or only lignite; but the public, who did not at all understand the discussion, only wanted to know whether there really was any coal, and if so, whether it burned well. The remarkable part of the story is that the mineral in question possessed all the properties of the best coal, even to that of affording coke. In reply to this the orthodox geologists asserted that because it furnished coke it could not be lignite, but must rather be English coal bought at Leghorn, and thrown purposely to the bottom of the pit. Thence arose lawsuits; for something always comes of slander, as Basil says. The discussion lasted a long while, and perhaps the matter is not settled even now. But in any case, it is very certain that the unfortunate shareholders never recovered from the heavy blow and great discouragement inflicted upon them by the geologists.

CHAPTER III.

HISTORY AND TRADITION.

Coal amongst the Greeks, Romans, and Chinese; in England and Belgium.—The blacksmith of Plénevaux.—Coal in France.—How mineral fuel has been produced. —Quantity of Coal annually brought to London.—Quantity of Coal annually raised in the United Kingdom.—Quantities of Coal annually raised at some of the largest Collieries of this country.

LIKE most mineral substances, coal has a history with which fable is largely combined. The Greeks and Romans were probably acquainted with fossil fuel; and it is supposed by some to be mentioned by ancient authors under the name of *lithanthrax* ("stone-coal"), which is handed down to our days in the Italian word *litantrace*. Aristotle's favourite pupil, Theophrastus, in his *Treatise on Stones*, does not omit to speak of coal as being found in Liguria and in Elis (Chapter xxviii.).

The ancients made little or no use of this combustible; probably they did not know the proper way to burn it. The forests of those early times afforded ample supplies for all the wants of industry, which was then in its infancy. Possibly some miners who smelted and worked the metals, or some blacksmiths who made and tempered arms, might have made occasional use of this fuel; but this is exceedingly doubtful. In the houses wood and charcoal were the only fuel; and as the polished people of those ages inhabited countries favoured by heaven, such as Italy, Greece, Egypt, and Asia Minor, where, besides, coal is of rare occurrence, they warmed themselves in the sunshine during the winter, even while discussing public affairs in the forum. Perhaps it may also be allowable to suppose that the mean temperature of the air was in those days slightly in excess of that of the present day. However

this may be, the laws of heat were unknown to the ancients;
the pressure of vapour was not even suspected then, and
mechanical force was solely derived from living beings. When
the wind-bags of Eolus were empty, convicts rowed the galleys.
When the absence of a stream of water did not allow of a bad
water-wheel being put up, animals, and even men, turned the
mill—the slave Plautus having executed this laborious task.
We read on the walls of Pompeii, beneath a caricature representing an ass in a mill, this line, which might have been
written by the great comedian himself:—

  Labora, aselle, quomodo laboravi, et proderit tibi.

  "Work away, little donkey, as I have worked, and you will profit by it."

Thus, amongst the ancients there was little need of coal, either
for industrial purposes or in common life. We see with what
indifference the masters of the world passed by mineral fuel,
coal. In Provence, the aqueduct of Fréjus (*Forum Julii*), passing through the middle of the coal formation at the foot of
l'Esterel, lays bare beds of mineral fuel, traversing them uninterruptedly. In the Lyonnaise, the aqueduct which carried the
rapid waters of the Gier to the capital beloved by Claudius,
likewise traverses the Carboniferous formation; one of the subterranean branches of the channel is even cut through the coal:
and yet nobody troubled his head about this substance. This
incident, which in the present day would cause no little excitement amongst the people, not only finds the populace of Rome
indifferent, but the Senate equally so, since they had no real
knowledge of its value.

Matters advanced somewhat better in the extreme East,
whose civilization had preceded that of Italy and Greece. The
Chinese, to whom all the great discoveries, except that of
America, have been rightly or wrongly accorded by some persons, were acquainted with mineral fuel from a very remote
antiquity. They knew how to work it, to apply it to various

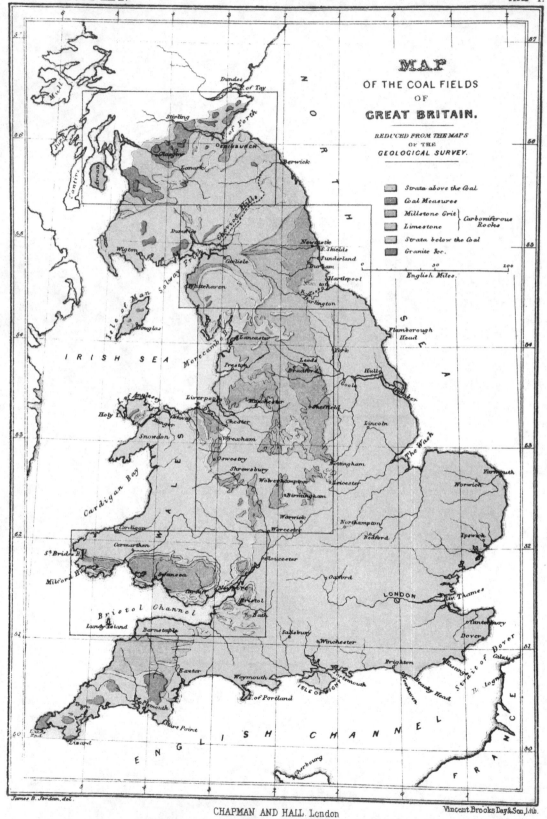

MAP
OF THE COAL FIELDS
OF
GREAT BRITAIN.

REDUCED FROM THE MAPS
OF THE
GEOLOGICAL SURVEY.

Strata above the Coal
Coal Measures
Millstone Grit
Limestone
Strata below the Coal
Granite &c.

Carboniferous Rocks

0   50   100
English Miles.

James B. Jordan, del.

industrial uses, such as (for example) the baking of porcelain;* they even knew how to collect the inflammable gases which exude spontaneously from coal, and to use them for illumination. The accounts of the early missionaries state that, from time immemorial, the Chinese sought for this gas beneath the soil with a boring-rod (a true Chinese borer, worked slowly with a rope); then that they conveyed it in pipes to the places where it was wanted. Here is the invention of lighting with gas and boring attributed to the Chinese, together with that of the compass, gunpowder, printing, macadamized roads, and several others.

The missionaries also inform us that even prior to our epoch mines of coal were worked in the Celestial empire; but that the mode of working was after a very barbarous fashion. No care was taken to support the underground ways, or to provide a proper outlet for the water; still less was it considered necessary to avoid explosions of fire-damp, or the inflammable gas of the Coal Measures. The Chinese have remained in this primitive state in working their mines up to the present time; and it was scarcely worth while to begin so early if so little progress were made afterwards.

Let us return to our own country, where more useful lessons await our notice. The coal-mines which seem to have been worked the earliest in Europe, are those of Britain. There are many evidences which show that at a very early period the uses of coal were known in Great Britain. Stone implements have been found in ancient workings in Derbyshire and other places, and amongst the ruins of the Roman Uriconium —the modern Wroxeter—coal has been found. There are other evidences which establish the fact that coal was known to the Romans, and probably to the Britons previously to the Roman invasion. During the Anglo-Saxon period coal was

* The substitution of coal for charcoal in the baking of porcelain, which has lately been made in France, but long used in England, has been found to be attended with a considerable saving of expense.

worked, but in all probability only at the outcrops of the coal-seams. It is exceedingly doubtful if coal was worked by the Norman invaders. No mention is made of coal until the time of Henry II. In 1259 Henry III. granted a charter to the freemen of Newcastle, by which they obtained liberty to " dig for cole;" and subsequently we read of " sea-cole" being carried to London. In 1306 the Lords and Commons in Parliament assembled petitioned Edward I. to prevent the importation of coal, and that monarch issued a proclamation forbidding the use of that fuel. In the fourteenth century coal was worked to some extent in Northumberland, Durham, Yorkshire, and Derbyshire. Towards the close of the thirteenth century there is mention made of the mines of Wales and Scotland, countries whose vigorous people, not then under English rule, long fought for independence. In the seventeenth century the British coal-mines were in full work. The coal-trade contributed to the support of the English mercantile marine by giving employment, in 1615, to four thousand ships, which, as well as foreign vessels, carried coal to the coasts of France, Germany, Holland, &c., returning with cargoes of corn. In 1612 Simon Sturtevant obtained letters patent from James I. for the making of iron with pit-cole and sea-cole, "for the preservation of wood and timber so greatly consumed by the iron-works." Simon Sturtevant failed in his experiments; his patent was withdrawn, and a new patent granted to John Rovenson. Doctor Jordanie and others were also encouraged to make experiments on the use of coal for the smelting of iron ore. In 1619 we find Dud Dudley securing, through the interest of his father, a patent from King James for the same purpose. After numerous failures and some persecution Dud Dudley succeeded, and established in South Staffordshire the use of coal in the manufacture of iron. From this period may be dated that activity in coal-mining which has distinguished Great Britain beyond any other country.

Belgium began to work its coal-mines about the same time as England; the earliest workings seeming to have been those begun towards the twelfth century at the village of Plénevaux, near Liége. Here fable is mixed up with history in a very ingenious manner, the story told by the chroniclers being as follows:—

Houillos, a farrier at Plénevaux, was so poor as not to be able to earn enough for his wants; not having, very often, bread to give to his wife and children. One day, being without work, he had almost made up his mind to put an end to his life, when an old man with a white beard entered his shop. They entered into conversation. Houillos told him his troubles; that being a disciple of Saint Eloi he worked in iron, blowing the bellows himself to save the expense of an assistant. He could easily realize some advantages if charcoal were not so dear, for it was that which ruined him. In short, like the poor wood-cutter in the fable—

> "Sa femme, ses enfants, les soldats, les impôts,
> Le créancier et la corvée,
> Firent d'un malheureux la peinture achevée."

The good old man was moved even to tears. "My friend," said he to the farrier, "go to the neighbouring mountain, dig up the ground, and you will find veins of a black earth suitable for the forge." No sooner said than done. Houillos went to the spot pointed out, found the earth there as had been predicted, and having thrown it into the fire, he proceeded to forge a horse-shoe at one heating. Transported with joy, he would not keep the precious discovery to himself, but communicated it to his neighbours, and even to his brother farriers. A grateful posterity has bestowed his name (already stated to have been Houillos) to coal (in French, *Houille*), and in this respect he has been more fortunate than many other discoverers. His memory is still cherished by all the miners of Liége, who, of an evening, relate the story of the honest collier, or of the old coal-miner, as they delight in styling Houillos, the farrier of Plénevaux. The miners say it was an angel who revealed to him the spot where the coal was. On the other hand, archæologists pretend that he must have been an Englishman; because in a MS. of the period, it is written that "it was doubtless an Ang—;" the concluding letters having been destroyed by the worms.

The working of coal in Belgium was not confined to Liége. Fossil fuel was soon afterwards found also at Charleroy, Mons; and simultaneously with the use of coal, which is so well adapted in that country for the forge, the manufacture of arms developed itself, dating back with the Belgians from the earliest times. It is said that they even carried on this manufacture prior to the conquest by Julius Cæsar, and historians make use of this argument to prove that coal was known and used in Belgium from the remotest antiquity.

There is little to relate concerning the discovery and application of coal in France; but if we consult the ancient documents of some of its provinces, we shall find tha certain collieries have been worked since the fourteenth century. Thus in Forez, the lord of Roche-la-Molière in 1321 levied a fine on those of his vassals who worked the mines of earthy charcoal (*charbon terreuse*). Every landowner had the right of working the coal below the ground which belonged to him, it being always well understood that he paid a royalty to the seigneur or lord of the manor.

In Forez, as in Belgium, the use of coal seems to have been known prior to the earliest date mentioned by historians. At Saint-Etienne the manufacture of arms, and of steel implements in general, is the most ancient industry of the country; and the water of the Furens is said to give a particular temper to the metal. Although it is not proved that the people of *Gagäe*, in Cæsar's time, already forged swords like their brethren the Nervii of Belgium, yet it is doubtless from the names of their ancestors that the Stephanois of the present day, the *Gagas* as they are called, have derived their sobriquet or nickname.

Notwithstanding the accounts of tradition or history to the contrary, the working of the greater part of the coal-mines would seem to have commenced at much earlier periods than is generally supposed. Many of these localities must have been known at an early date, for the coal-seams make their appearance at the surface—they *crop out*, or *come to grass*, as miners say. But what must not be lost sight of is, that it is not the more or less distant date when such and such a working was commenced, but the importance of the workings themselves, and the quantities extracted, that more particularly interest us at the present day. The importance of coal-mines is not estimated by time, like that of metalliferous mines, which have been worked for ages; it is only measured by the quantities produced. From this point of view, coal-mines are affairs of yesterday. It may be said that the date of working on a large scale marks the date of its origin, and fixes, in the history of the human race, a glorious halting-place, where it is well to pause. The important discovery was not, in fact, that of the coal itself, but of its application on a large scale, like that of the present day, to the manufacture of iron and all the other metals, and to the heating of steam-boilers, whether fixed, locomotive, river, or marine. It is ever the story of Columbus and the egg—always the same. The glory of the immortal Genoese did not so much consist in the actual discovery of America, as in having pointed out the true route to the New World, and in having unveiled all the advantages which were to be derived from its virgin and fertile lands.

So it is as regards coal. Of what import is it for us to know that the ancients were acquainted with it, if they did not know how to use it? We will, therefore, say no more about the Greeks, Romans, and Chinese, of whom sufficient mention has been made. In the Middle Ages, and at the period of the Renaissance, have we not seen coal sought, got, and prohibited? Royal orders were at one time issued for the punishment, by fine and imprisonment, of the artisans who made use of it. Medical men were opposed to its use, and attributed imaginary ill effects to it. Acts of Parliament prohibited it, and the Sorbonne itself launched its thunders against it. All this showed a disregard of the *law of progress;* but we must bear with the ignorance of our ancestors, for that law has only been recognised in our day—in Europe, during the Middle Ages, the existence of such a law was not even suspected.

The true history of coal dates from the eighteenth century, and may be said to be connected with the history of modern civilization. And see how all is linked together! The steam-engine is invented amongst the coal-mines. In England deep pits were full of water in the Newcastle coal-field, which must be drained before the coal could be extracted; but the pump, which had remained the same ever since the days of Archimedes, was no longer sufficient for the purpose, and a more powerful engine became necessary. Savery, Newcomen, and Watt, succeeded each other. The *fire-pump*—the steam-engine—was discovered; Watt settled almost definitely its chief arrangement; and then it is no longer water only which the engine will raise to the surface of the earth, but coal as well, and that however large the quantity may be, or however great the depth.

The first application of steam to pumping machinery was not, however, made in a coal-mine. Captain Savery erected one of his "fire-engines" to lift the water from one of the Cornish mines in the Breage and Wendron district; but as the construction of this engine depended on the production of a partial vacuum in a metallic sphere, into which the water rushed, to be expelled again by the fire, its action was slow, and the work done exceedingly small.

Newcomen, of Dartmouth, was the real inventor of an atmospheric steam-engine. The piston was lifted by the steam; this being condensed, the piston was forced to the bottom of the cylinder by the pressure of the atmosphere. One of Newcomen's engines was, it appears, erected at Huel Vor, near Helston, and another at Creegbraws, in Gwennap. The mines of Cornwall must, however, have stopped working, had not the genius of Watt come to the aid of the miner, enabling him to overcome the difficulty of removing the vast accumulations of water in the deep shafts, about the middle of the last century.

Coal is a heavy, bulky article, selling at a low price. It is not enough to tear it from the bowels of the earth; it must be transported economically, often for considerable distances. What makes transport difficult and costly? The state of the roads. These were improved, without any doubt being felt with regard to the immense extent of the result which would soon be attained. After a time wooden tracks were devised, over which the wheels roll easily. These were applied in the underground ways in the first instance, but afterwards extended to those above ground. But wood soon wears and rots. The wooden tracks were at a subsequent and comparatively recent date replaced by others of cast iron, which were grooved in the first instance, and then made flat with a lateral flange. Wrought iron was soon substituted for cast; the ribbon of metal or *rail* was devised, and with it the *railway*, or iron-road. This happened in the collieries of Wales, just as those of Newcastle were the birthplace of the locomotive steam-engine. Still the invention was not yet complete. The Cornishman, Trevithick, constructed a locomotive with a simple cylindrical boiler, like those of stationary engines. But the heating surface, the production of steam, and the motive power were too small; and, besides, to gain a hold upon the rails, without which, it was then supposed, the wheels would merely turn without moving forwards, the driving wheel was toothed and worked with a rack. The speed was less than that of a carriage drawn by horses. Is then the invention on the point of being lost? Human genius does not halt in its discoveries. In England, the great engineer, George Stephenson, an old coal-miner, and Seguin, in France, completed the locomotive. Seguin, by the contrivance of those numerous tubes which traverse the boiler in a longitudinal direction, and which afford a passage to the heated flames given off from the fire-bars, increased the evaporating surface to an astonishing degree, and consequently the production of steam. Stephenson improved upon the ideas of his rival; and discharging into the chimney, by a direct jet, the steam which has just acted upon the piston, he augmented the draught of the furnace, which had been already improved by the invention of Seguin. From that time the locomotive was complete; as was the case with Watt's engine, it was only the details which received modifications.

Such is the true history of coal, and such are the means by which the combustible mineral has been produced. To effect such a result it has taken the whole of the eighteenth century, and the first thirty years of the present; but then what a conquest has been achieved! The steam-engine, which only ought to be used for drawing the water and the coal out of the mines, has been universally adopted; everywhere its work has been substituted for that of man; and the saying of Aristotle is confirmed, "that there will be no longer any slaves when the spindle and shuttle walk alone."

At the same time the working of the metals, with which the progress of civilization is so intimately connected, has undergone modifications at all points. The application of coal to the manufacture of cast and wrought iron and steel has remodelled the science of metallurgy.

The recent processes of elaboration of these various products have effected important innovations and modifications even in the art of war, which seemed to have been irrevocably fixed

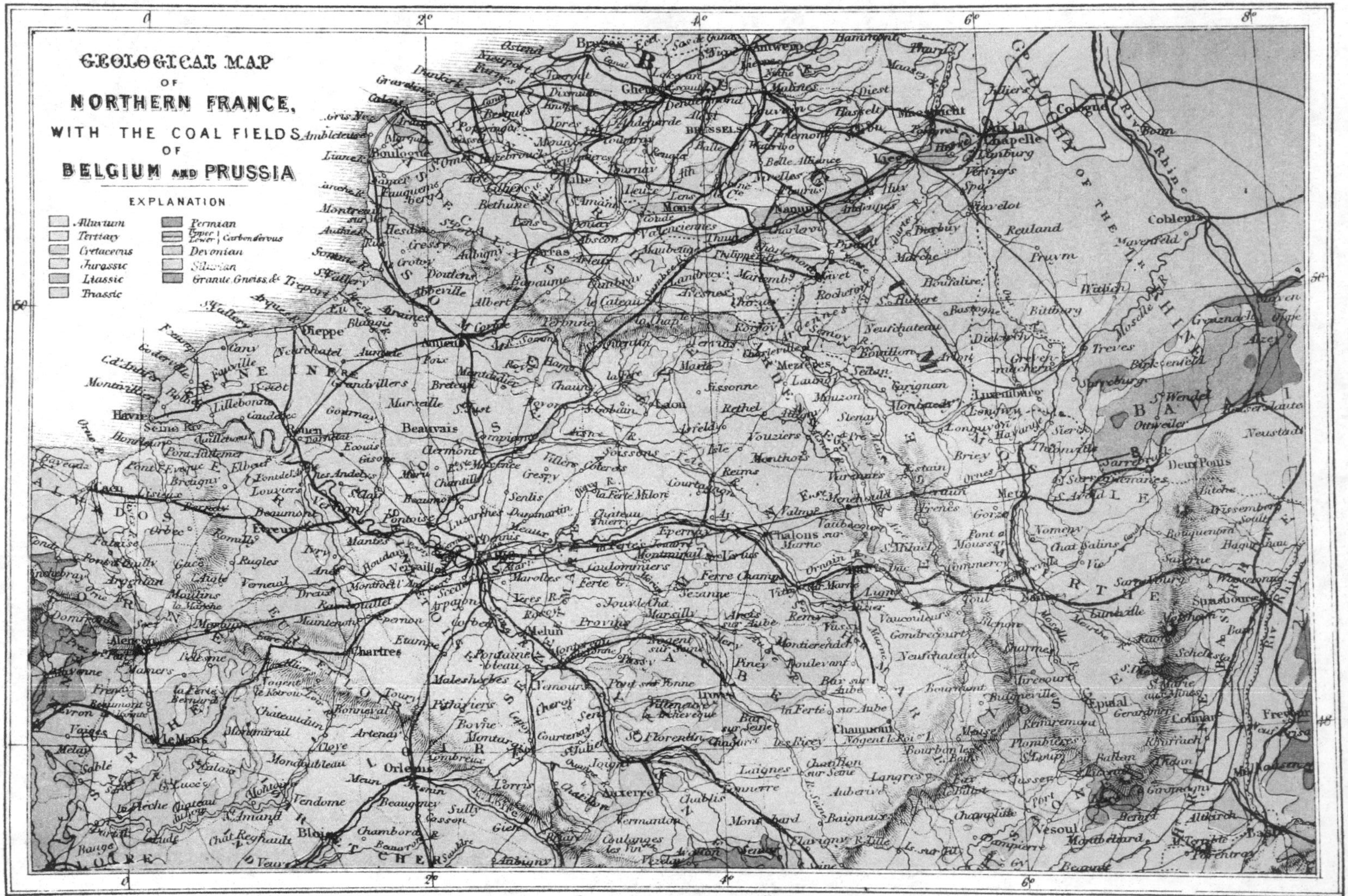

MAP II.

GEOLOGICAL MAP
of
NORTHERN FRANCE,
WITH THE COAL FIELDS
of
BELGIUM and PRUSSIA

EXPLANATION

Alluvium
Tertiary
Cretaceous
Jurassic
Liassic
Triassic
Permian
Lower Carboniferous
Devonian
Silurian
Granite Gneiss & Trap

English Miles.

CHAPMAN AND HALL, London.

Vincent Brooks Day & Son lith.

J.B. Jordan, del.

since the time of Napoleon. Perhaps it may even prove that the terribly destructive engines used nowadays by land and sea will have the effect, if not of putting an end to war altogether, at any rate of making it of less frequent occurrence; which would be the greatest benefit to mankind that industry ever devised.

Whilst the introduction of the steam-engine into factories, and the use of coal in metallurgy, have brought about these important results, railways, on the other hand, have proved themselves to be no less wondrous agents of progress. The locomotive, which at first was only intended to transport heavy and bulky articles, such as coal, was soon used for the conveyance of every kind of merchandise. It has even done more; for the very purpose for which it was originally designed has been changed, by its special application to the accelerated transport of man. By that means it has annihilated distances and frontiers; and if peoples lean towards a fraternal alliance now, it is to the locomotive, more especially, that they are indebted for the progress that has been made.

Coal-mines profited the earliest and to the greatest extent by all these magnificent applications. Already the locomotive carries the indispensable mineral afar, and at so small a cost as to seem almost incredible if the rates of transport were not published in the traffic-returns of the railway companies.° In France it is often less than one penny per ton of two thousand lbs. and per league of four kilomètres, which is nearly at as low a rate as by water conveyance.

The low cost of carriage has everywhere facilitated the distribution and the supply of coal, and producers have made every effort to satisfy all the demands as well as all the requirements of the consumers. The steam-engine, applied to the

---

* In the year 1866, 6,013,265 tons of coal were brought to London, of which 3,033,193 were carried by sea and 2,980,072 by railway. Of the latter quantity 1,188,966 tons were brought by the London and North-Western Railway, 1,006,277 tons by the Great Northern, and 367,908 tons by the Great Western Railway.

extraction of the combustible, draws it from the dark depths in enormous quantities at a time. By this means more than a thousand tons are drawn, at some collieries, from the mouth of a single pit in one day.* Everything has concurred to extend the application of coal, and to render its use common. Ships and railways carry it to countries which do not possess this mineral, and ordinary ways are placed in contribution when the distances are not too great. By all these means, the production has exceeded all bounds. For the last half century it appears to have had a tendency to double itself in England, France, and Belgium every fifteen years; every ten years in Prussia; and every five years in the United States. In 1866 England reached the formidable figure of 101,630,544 tons; in 1850 she only raised fifty millions. Sir Roderick Murchison, President of the Geological Section of the British Association, at the meeting of 1865, could not refrain from bestowing a glance on the future, and seemed to ask himself with anxiety, what England would become when its collieries were exhausted? More recently Mr. Hull and Mr. Jevons wrote on the subject, and Mr. Gladstone, struck by the facts brought forward, spoke of them in the House of Commons, in order that the country, warned in time, should not some day be taken by surprise. Since then, Parliament has ordered an inquiry by means of a Royal Commission, "to investigate the

* An idea of the enormous quantities of coal raised at some of the largest collieries of this country, in 1866 and 1867, may be formed from the following list :—

Collieries.	District.	Annual Quantity raised. Tons.		Daily Quantity raised. Tons.
Earl of Durham's Collieries, . . .	Durham, ...	1,185,697	...	3952
Earl Vane's Collieries, ...	Durham, ...	1,067,562	...	3558
Charlesworth & Co., ...	Yorkshire, ...	1,000,000	...	3195
Pease's Collieries, ...	Durham (1867), . .	870,080	...	2900
Roddy Moor, ...	Durham (1867), . .	853,416	...	2845
Hetton Coal Co., ...	Durham, ...	696,206	...	2128
Hetton Coal Co., ...	Durham (1867), . .	710,797	...	2369
Elscar and Park Gate Collieries, .	Rotherham, Yorkshire,	480,342	...	1534
Monkwearmouth Colliery, Wear-⎰ mouth Coal Co., ... ⎱	Sunderland, ... (1867),	388,245 390,576	... ...	1240 1302
Chester-le-Street, Durham, . . .	Tyne, ...	242,435	...	774
Shire-oak Colliery, ...	Derbyshire, ...	185,600	...	618

probable quantity of coal contained in the coal-fields of the United Kingdom, and to report on the quantity of such coal which may be reasonably expected to be available for use;" and Great Britain is now taking stock, as it were, of her subterranean domains.

As we have seen; the history of coal is full of instruction, and the enormous part which in our laborious age fossil fuel plays, which a century or two since it was attempted to banish from towns, is henceforth deserving of attention. We are not in an age of decadence, and those who bequeath to the future the astonishing inventions of the present century may depart with head erect. They will not say, with Titus, that they have lost their day; but, like Augustus, that they have executed their task with ability.

CHAPTER IV.

THE BLACK COUNTRY.

THE two most celebrated English coal-fields occupy a littoral position. One, facing the south, the South Wales coal-field, has borne to the four corners of the world the reputation of Cardiff, the coal preferred by foreign smelters, and which has taken its name from the place of exportation. The other, the Newcastle coal-field, saluting the rising sun, has spread over all the world the rival coal of Cardiff, *the Newcastle*, known by the name of the district from which it is derived. Wales produces annually above eight million tons; Newcastle more than twenty-four million: the last amount being double the production of the whole of France! In Wales, towns like Newport, Cardiff, and Swansea, places almost of yesterday, already possess populations of more than forty thousand persons, in consequence of the coal-trade. One of them has centered in enormous smelting works, owing to the low price of coal, the reduction of the copper-ores of the entire world. "Behold Swansea," exclaims the author of *La Géologie Appliquée*, M. Amédée Burat, "its poetical name has no influence on its prosperity. Formerly, under its ancient title (Swan-Sea), it was comparatively unknown. Now it is the great city

GEOLOGICAL MAP OF PART OF FRANCE after Heinrich Bach

J. B. Jordan, del.

Vincent Brooks Day & Son lith.

CHAPMAN AND HALL London

MAP IV.

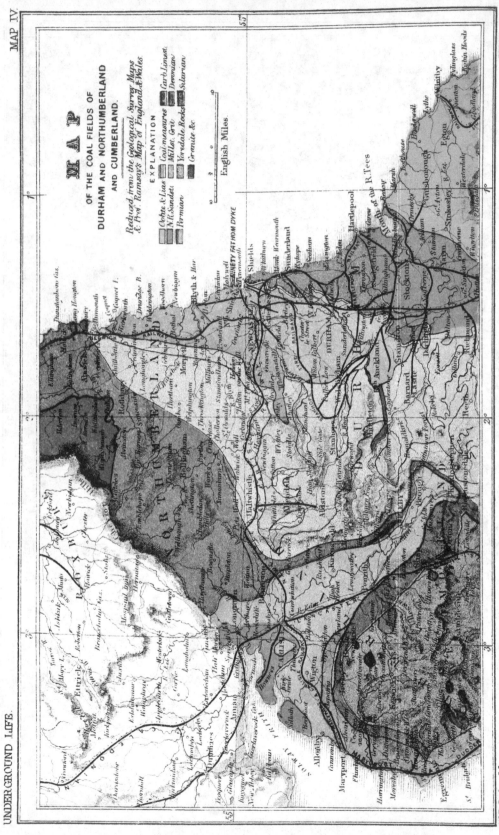

MAP

OF THE COAL FIELDS OF
DURHAM AND NORTHUMBERLAND
AND CUMBERLAND.

Reduced from the Geological Survey Maps
& Prof. Ramsay's Map of England & Wales.

EXPLANATION

Oolite & Lias	Coal measures	Carb. Limest.
N.R.Sandst.	Millst. Grit	Devonian
Permian	Yoredale Rocks	Silurian
	Granite &c.	

English Miles.

Jno. B. Jordan, del.

CHAPMAN & HALL, London.

W.Dickes imp.

of the smelters. She it is who sends ships to double Cape Horn to bring back the ores of Chile; it is for her, and to enrich her lords, that the negroes of Cuba, as well as the free populations of Coquimbo and La Paz, labour, and it is to coal alone that she is indebted for this power." *

Newcastle, also, is indebted to coal for its celebrity. Neither the Tyne which washes its quays, nor the wall which Septimus Severus built there to ward off the incursions of the Picts, nor the strong castle raised by Robert, the son of William the Conqueror, and which was taken and retaken so often by the Scotch, had been able to render it illustrious, or even to make it known: it is coal that has spread its fame beyond the seas. Coal has also founded Sunderland, which, with Newcastle, constitutes the principal ports where the coal trade of this part of the British coast is carried on. The English are proud of their collieries. They call them *the Black Indies*, to show the importance which they attach to this industry; and they would not exchange these Indies for those of Asia or America. A writer in the *British Quarterly Review* † says, "On one occasion about three hundred vessels, all coal-laden, were seen making sail together in a single tide, and distributing themselves over the ocean with their prows turned in almost every direction—all sinking deep into the waters, and weighed down with their mineral burden of far more worth to us than auriferous sands or Mexican mines."

Is it necessary to expatiate on the other coal-fields of the United Kingdom? those of the centre (Map V.), in the counties of Stafford, Derby, Lancaster, York, &c.; that of the north, or the Scotch coal-field, extending from Edinburgh to Glasgow Map VII.), and even from sea to sea, like that of Newcastle and Whitehaven (Map IV.)? There is coal even in Ireland, and the Carboniferous formation occupies a very extensive area in

---

* Much copper ore is now smelted in Chile and South Australia.

† *British Quarterly Review*, 1857; vol. xxv. p. 89.

that country, but is almost destitute of coal; the upper and
more valuable portions of the coal-bearing deposits having
been, probably, removed by denudation. The greater part of
the coal of the south of Ireland is anthracite (Kilkenny) coal,
but the north of Ireland furnishes a good "gas coal." Ireland
has only contributed 123,750 tons to the 101,630,544 tons
yielded in 1866 by the entire United Kingdom.

In the English collieries there is an unequalled animation
and life. Railways and canals traverse them in all directions;
frequently even two lines of railway, at different levels, follow
the same direction, so great is the locomotion to which the
combustible gives rise. Factories and manufactures abound
on all sides. Here is Sheffield, the country of steel; there
Birmingham, where metallurgical and mechanical workshops,
buildings of all descriptions, are piled up and crowded together,
producing machinery, tools, brass, bronze, Britannia metal,
steel pens, pins, needles, iron bedsteads, together with coins,
medals, and ornamental ware, &c.; more distant are the Staffordshire towns, where pottery and china are made, occupying
in that county the district known as the Potteries, as that of
the coal-mines is called the Black Country. Then there is
Manchester, the first industrial town in the world, the city of
the weavers, where King Cotton employs nearly all heads and
arms, as is also the case at Liverpool, where the cotton is
imported. Lastly there are South Staffordshire and Wales,
the countries of great forges, where the iron-ore is found
interstratified with the fuel, and occasionally with the limestone itself, which serves as the flux;* thrown into the furnace
mixed with the ore, it assists in its reduction. The three substances which are indispensable in every metallurgical operation, the fuel, the ore, and the flux, by a singular prodigality

* The clay iron-ore is found in the Coal Measures of South Staffordshire, as well as in those of the
South Wales coal-field; but as it does not occur in sufficient quantity to supply the enormous demands
of the iron-works of either of those districts, other ores of iron are introduced from various localities to
supply the deficiency.

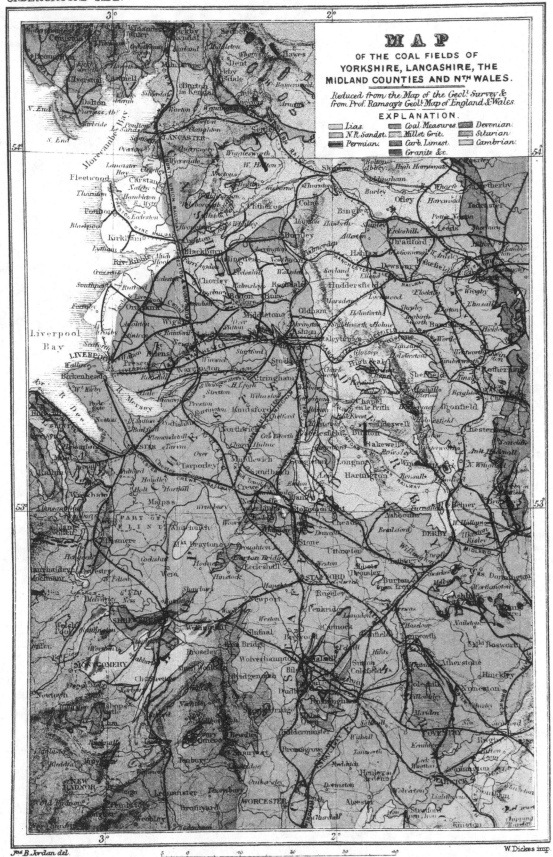

MAP
OF THE COAL FIELDS OF
YORKSHIRE, LANCASHIRE, THE
MIDLAND COUNTIES AND Nᵗʰ WALES.

*Reduced from the Map of the Geolᵗ Survey &
from Prof. Ramsay's Geolᵗ Map of England & Wales.*

EXPLANATION.

Lias.	Coal Measures.	Devonian.
N.R. Sandst.	Millst. Grit.	Silurian.
Permian.	Carb. Limest.	Cambrian.
	Granite &c.	

English Miles.

Joˢ B. Jordan del.   W. Dickes imp.

CHAPMAN & HALL. London.

of nature, are thus found associated together in the same geological series.

The little kingdom of Belgium, which was added to the map of Europe in 1830, owes its importance almost exclusively to coal and iron. If she holds a distinguished rank at the present day amongst the European nations, it is not because she has much weight in politics or in the decisions of the statesmen who regulate the balance of power in Europe (her geographical position and small extent of territory cause her to rank amongst second-rate powers) ; it is because she comes next after England, Prussia, and France, in the production of those two great agents of the power of modern states—Coal and Iron. Her liberally-developed mineral industry has created for her the most widely-spread relations on the globe, and it is especially to coal that Belgium is indebted for so flourishing a condition. Nature, in providing her with coal, has enriched her on all hands.

The Belgian coal-basin, developed between Liége and Mons, and passing by Namur and Charleroy, extends from east to west over a distance of forty leagues, with an average width of three, and contains three hundred thousand acres (Map II.). Everywhere there is nothing but collieries, metallurgical factories, and workshops for the manufacture of machinery; while the roads at each step are intersected and crossed by canals and railways.

The small extent of the Belgian coal-field is compensated for by the great number of the coal-seams, and by their numerous and singular zig-zag contortions, which have the effect of increasing the workable surface. These foldings of the seams are caused by the elevation and compression of the Carboniferous strata by eruptive rocks, after the formation of the former, which bent readily under the pressure to which they were subjected. All the seams, even to the very thinnest bands, have yielded to this force without being fractured

E 2

except at certain points; the sharp bends in the angles having been produced without undergoing fracture, while the parallelism of the beds is preserved—thus furnishing one of the most curious examples of disturbances resulting from the elevation and compression of sedimentary deposits (Map II.).

Many qualities of coal are met with in the Belgian coal-field; one peculiar kind, forming the upper series of seams in the Mons basin to which it is confined, called *Flenu*, unlike any coal found in Great Britain, except at Swansea, is good for making gas and for furnaces, and is even in request for use in the Parisian factories. It burns rapidly with much flame and smoke, not giving out an intense heat, and having a somewhat disagreeable smell. Commercially Belgian coals are distinguished as *gaillettes* and *gailletteries*, or large and middlings, *fines* or *menus*, or smalls and slack, and into *tout-venant*; the latter term being applied to the coal as it is taken out of the mine, without being screened or picked. Classed according to the industrial uses for which they are applicable, coals are *dures* (or hard), *grasses* (fat or coking), *demi-grasses*, or *maigres* (dry and burning with small flame), according as they contain more or less volatile matter, or are more or less bituminous, burn with a longer or shorter flame, and cake or swell up in the fire, or the reverse. These four great divisions are recognized in nearly all the mines.

In France, the industrial centre which most resembles Belgium are the basins of Rive-de-Gier and Saint-Etienne, comprising about fifty thousand acres, and stretching in a direction from north-east to south-west, between the Rhone and the Loire. In the south the Coal Measures abut on the flanks of Mont Pilat, which separates the tributary waters of the ocean from those which flow into the Mediterranean; on the north they abut on the terminal spurs of the mountains of the Lyonnais and Forez. Between these two limits they occupy a narrow strip

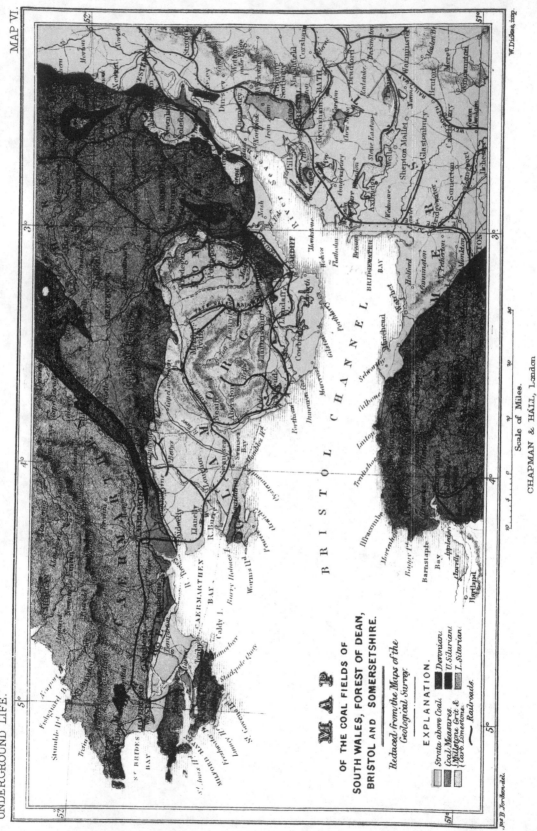

M A P

OF THE COAL FIELDS OF
SOUTH WALES, FOREST OF DEAN,
BRISTOL AND SOMERSETSHIRE.

Reduced from the Maps of the
Geological Survey.

EXPLANATION.

Strata above Coal.   Devonian.
Coal Measures.   U. Silurian.
Millstone Grit &   L. Silurian.
Carb. Limestone.
——— Railroads.

Scale of Miles.

CHAPMAN & HALL, London.

Jos B. Jordan del.   W. Dickes, imp.

which widens out sensibly from the Rhone in the direction of
the Loire (Map III.).

In travelling from Lyons by the railway which skirts the
Rhone as far as Givors, and then ascends the beautiful valley
of the Gier with its wooded and verdant hills, the coal-mining
district commencing at Rive-de-Gier is soon reached. From
this latter point, onwards, the country is full of pits and
galleries, opened for getting out the coal, or for draining off
the water. In some parts of the basin this is particularly
abundant, and forms actual subterranean streams. Here and
there are long lines of coking-ovens where the coal is charred,
as wood is in the forests. The fires by night light up different
points in the horizon, and present the appearance of a volcanic
country full of *fumerolles*.

The towns on the route, Rive-de-Gier, Saint-Chamond, &c.,
are far from presenting an agreeable prospect to the eye. The
mere tourist finds little to attract his attention, everything is
given up to the calling of the coal-miner. The streets are full
of thick black mud; the fronts of the houses are blackened by
smoke and coal-dust.

This penetrating dust respects nothing; it soils and blackens
the leaves of the trees, linen, the face of man, and everything
else it comes in contact with, and the town of *Terre-noir*,
passed on the road, is worthy of its name. Carts or wagons,
heavily laden and lumbering vehicles, are crowded together at
the approaches to the railway stations and the centres of population. Frequently the railway itself crosses the street; or
the rails, by right of conquest, are laid on the highway. The
factory chimneys discharge into the air their canopy of flame
and smoke; the atmosphere is impregnated with the bituminous odour peculiar to burning coal; the clanging noise of
hammers and rollers resounds on all sides; the fictions of antiquity have almost assumed a definite form, and one might call
it the country of the Cyclops. It is the land of the brave

colliers ; the basin of the Loire or of Saint-Etienne, so called after the town that has been especially transformed and enriched by the coal-trade.

At the beginning of the seventeenth century, Saint-Etienne was merely a village inhabited by some few hundreds of workmen skilled in the art of forging arms and tools. Two centuries afterwards it scarcely held twenty thousand inhabitants, although to the manufacture of arms had been added that of cutlery or of cutting tools, together with that of heavy ironmongery and the weaving of ribbons. But no sooner had the collieries of this interesting district become developed by making iron after the English method, than the population of the place increased more and more. The present total exceeds one hundred thousand souls; at which point the state, doing justice at last to its repeated demands, caused the head-office of the department of the Loire to be transferred from Montbrison to Saint-Etienne.

It is only two centuries since Saint-Etienne was but a modest village, and Rive-de-Gier and Givors were not even in existence ; and now they are important cities, owing their origin to coal and iron. Saint Chamond was only celebrated for its immense castle, held by the counts of Forez ; the castle is now in ruins, but at its feet a city rises, which has been rendered populous and prosperous, much more by metallurgical industry and the coal-trade than by the manufacture of silk laces, which is also carried on there.

It is also to the collieries of the Saint-Etienne basin that the country is indebted for the first two railways which were constructed in France : that from Saint-Etienne to the port of Andrézieux on the Loire, authorized in 1823, then marked out like a common road with very steep inclines, and worked by horses ; and that from Saint-Etienne to Lyons, dating from 1826, which was the first French railway worked by locomotives. At that period, so little thought was bestowed on the quicker conveyance of passengers, that these two railways were only established to facilitate the transport of mineral fuel ; and when, in 1834, mention was made in the Chambers of opening railways radiating from Paris to the provinces, a minister went so far as to pretend that from four to five leagues a year might easily be constructed, and that these new ways were only fit for amusing the cockneys who run up to see the locomotive go by. A great savant added, that people would be suffocated by the steam in the long tunnels; a celebrated economist, that France never could produce all the metal wanted for the iron roads; while, lastly, a deputy from the High Alps exclaimed that the trains and the road would be precipitated into the valley. The miners let them have their say, and gave the country its earliest railways.

The animated spectacle presented by the coal-basin of the Loire, is repeated in France at many other points. It is a remarkable fact that the coal-trade, in every country where it is established, assumes a very strikingly uniform aspect. Thus the northern basin, a prolongation of the Belgian coal-field (Map II.), presents about Dinain, Anzin, Valenciennes, and in the Pas-de-Calais between Lens and Béthune, the same scene as the basin of the Loire at Saint-Etienne, Rive-de-Gier, and Saint Chamond.

In the department of Saône-et-Loire, Epinac and Blanzy rank amongst the most important of the French collieries; and Creuzot, a dreary valley almost uninhabited a century ago, when it was little known by the name of *Charbonnières*,* which it then bore, is now one of the most busy industrial centres (Map III.). The extraction and the transport of coal and iron-ore, the manufacture of cast-iron, and the construction of machinery, give employment there to upwards of ten thousand workmen. This establishment is the fortunate rival of the most famous of the kind in the world. England, Belgium, the

* It was also called *Creux*, which has been corrupted into *Creuzot*.

United States, can boast of none surpassing it. This astonishing development, which may be said to have originated from a piece of coal, has only reached its present importance since 1837, under the able direction of the MM. Schneider.

In the Gard, the prosperity of the coal-fields of Alais and Grand'Combe is also of recent date; having only in fact began from the day when the railway conveyed these deposits to the Rhone in 1840. Alais, which up to that time had only been remarkable in the history of the southern provinces of France for its religious wars and its trade in silk, has since become an essentially industrial city, in which the old quarrels between Catholics and Protestants have been assuaged, where the importance of silk-spinning and of glass manufactories have little by little disappeared before that of metallurgical works. Bességes and Portes have not been slow in following the footsteps of Alais and Grand'Combe, and now the Department of the Gard, which only lately was all but forgotten, is classed amongst the most interesting in France, in consequence of the great progress which mineral industry has made there. Thus the Department of the Gard holds the third place in the list of French coal-producing districts (the yearly production of that coal-field being now above twelve hundred thousand tons), and is only surpassed by those of the Loire and the Nord, which stand at the head, each contributing more than a fourth part of the total annual production.

The basin of Aubin, in Aveyron, only dates back for a little more than thirty years. Decazeville owes its origin to a minister of the Restoration; and it is only since 1826, when the English foundries and forges were erected, that the coal-mines of Aveyron received their first development. These mines have also had the advantage of yielding in greater abundance than those of Saint-Etienne and Creuzot, the clay-ironstone which is so abundant in the British Coal Measures.

In order to promote the extraction of the coal, it has been

MAP VII

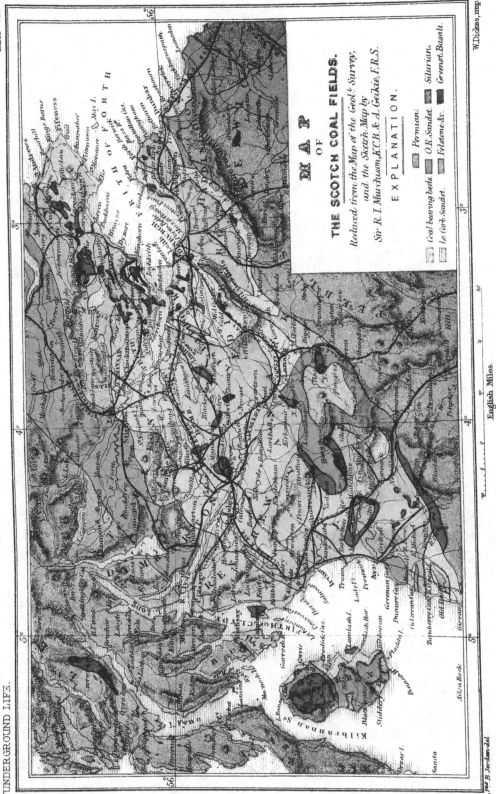

MAP
OF
THE SCOTCH COAL FIELDS.

Reduced from the Map of the Geol.ˡ Survey,
and the Sketch Map by
Sir R. I. Murchison, K.C.B. & A. Geikie, F.R.S.

EXPLANATION.

Permian.

Coal bearing beds.

O.R. Sandst.

Lo. (or) Sandst.

Felstone &c.

Silurian.

Greenst. Basalt.

Jas. B. Jordan del.

English Miles.

CHAPMAN & HALL, London.

W. Dickes, imp.

everywhere necessary at the outset to make a point of consuming the greatest quantity on the spot. The ironworks which for a given weight of metal consume as much as five times the same weight of fuel in producing it, are those which have been erected in the neighbourhood of the coal-mines. In this way the great establishments of Terre-Noire, Saint Chamond, Givors, Creuzot, Alais, Decazeville, Commentry, Denain, Anzin, and many others have been erected in France. Glass-houses, the manufacture of mirrors and porcelain, furnaces for burning bricks and tiles, have also sprung up around the most important collieries. Lastly, in certain localities, as at the mines of Basse-Loire, and of Sarthe and Mayenne (basin of Le Maine), they have employed a badly-burning coal for making lime, as a manure for the clay-lands of those districts. The whole secret of the successful working of collieries, when the quality of the coal will not admit of a long carriage, or when the roads themselves are bad, consists in the conversion of the coal into some other material—iron, glass, pottery, lime, etc.—of ready sale, or at any rate more easily and certainly disposed of than the coal itself.

In the mining district of Le Maine, in the course of a few years, the country has been, in this way, regenerated by the application of coal to lime-burning. Previously they only cultivated rye and buckwheat, and the land presented a desolate appearance. Now, thanks to the adoption of calcareous correctives, wheat of superior quality is raised in such quantity that one-half of the crop is exported. Landed property has become trebled in value, and persons who knew the country thirty years ago say that it is now scarcely recognizable. If the working of the collieries has sometimes been injurious to agriculture, in this instance the ill has been largely atoned for.

In central Europe, in certain favoured countries, as in Prussia for example, coal-fields occur comparable, if not to those of England, at any rate to those of France and Belgium

(Maps II., III., and X.). On the left bank of the Rhine Prussia possesses, near Saarbrück, a greater thickness of Coal Measures and of coal than exists anywhere else in Europe.* She also possesses the coal-basins of the Inde, that of the Wurm, near Aix-la-Chapelle, as well as important deposits in Westphalia and Silesia, the total production of which, a third greater than that of France and Belgium, reached seventeen million tons in 1865, and even exceeded twenty-three million tons in 1866, if all kinds of fossil fuel (such as brown coal) be taken into account. The rich coal-fields of Westphalia, in the valley of the Ruhr, extend over a surface of 115 square miles, and are calculated to contain about four hundred thousand million tons of coal. In 1866 the production amounted to 8,583,362 tons, and the number of hands employed was 37,686 (2185 less than in the previous year), the war having withdrawn many hands from productive labour. The price of coal at the pit's mouth was, on an average, for the best, 5s. 6d. per ton; and the wages of the miners varied from about 1s. 1d. for boys to 2s. and 2s. 6d. for men, per day of ten hours. The carriage of coals on the railways was, on an average, four-fifths of a halfpenny per ton per mile. None of the mines exceed 150 fathoms in depth, and the seams, which are no less than 117 in number, vary in thickness from $2\frac{1}{2}$ feet to $4\frac{1}{2}$ feet, few exceeding that amount. The quality of the coal is various, but as a general rule the Prussian coals are soft and tender.† Bavaria, Saxony, Austria with its mines of Bohemia, come next to those of Prussia. Russia possesses immense carboniferous deposits, stretching from the White Sea to the Sea of Azof, and along the Ural. Some unfortunately contain but little coal, while others have not yet received all the development of which they are susceptible. A saying of Peter the Great is

* "Coal and Coal Mining," by W. W. Smyth, p. 79.

† A concise account of the Prussian coals and coal-fields will be found in the admirable catalogue, drawn up by Dr. Wedding, descriptive of the collections contributed by the Zollverein to the International Exhibition of 1862.

related in reference to the coal-fields of his vast empire: "These mines will make the fortune of our children." Is it really true that he made this remark? In his time the numerous uses of coal were not known; and is it not more likely to be one of those sayings afterwards invented by history? An account of the coal of Russia has lately (1866) been published* by Lieutenant-general G. de Helmersen. In this memoir the greatest development of coal of Carboniferous age in Russia in Europe, is stated to be as follows:—1. On the eastern and western slopes of the Ural, associated with limestones, sandstones, and argillaceous shale. 2. In the governments of Novgorod, Tver, Moscow, Kalouga, Toula, and Riazan, the coal occupies a broad (plat) elliptical basin 600 versts in length by 400 in width, in the centre of which the city of Moscow is situated. 3. In the government of Simbirsk, at a place between Stavropol and Sysranë, in the small peninsula of Samara, which is formed by the river Volga. 4. In the government of Ekaterinoslav, and in the military settlement of the Don, where the Coal Measures form a low chain of mountains, called the Donetz, and consist chiefly of sandstones, accompanied by argillaceous shales, limestones, &c., together with abundant deposits of iron-ore, which up to the present time have not been worked for economic purposes, though they would well repay the cost. In the chain of the Donetz the coal forms numerous seams of very variable quality; varying from the best kinds of anthracite and caking coal to the worst. The coal-seams rarely attain a thickness of seven feet, but are usually from two to three feet thick. M. Le Play, in a work † published by him in 1842, states that he had counted no less than 225 beds of workable coal, having an average thickness of one foot nine inches.

The essential difference between the coal-deposits of Russia

---

* "Des Gisements de Charbon de Terre du Russie," par G. de Helmersen.

† "Voyage dans la Russie Méridionale et la Crimée," tome 4. "Exploration des Terrains Carbonifères du Donetz," par E. Le Play, Paris, 1842.

in Europe and those of Western Europe consists in the entire
absence in the former of true or upper Coal Measures, no
traces, even, of which have as yet been discovered. Up to the
present time only the lower Coal Measures, or those associated
with the Carboniferous Limestone series, have been met with
in the former area.

Southern Europe is poor in coal-mines, if we except Spain,
where we find in the north the coal-fields of the Asturias, of
Old Castile, and of the province of Leon, which on both sides
of the chain of the Pyrenees display beds of coal by the hundred, and turned up nearly on end. The quality of the coal is
excellent, but unfortunately an outlet is wanting, such roads
as there are being very bad. Other coal-fields, as that of San
Juan (de las Abadessas) in Catalonia, or that of Espiel and
Belmez in Andalusia, are likewise worked to a small extent,
affording reserves for the future.

Italy derives almost all its mineral fuel from Britain. Some
small islands are worked for coal along the Gulf of Spezzia, in
the Tuscan Maremma, and in the Calabrias; but the true Coal
Measures are wanting in Italy, except in the island of Sardinia,
and at the foot of the Alps, and even there they only occur in
very insignificant quantities. Tertiary lignite occurs in numerous places in Italy, and is often of the very best quality, but
the basins are small and might soon be exhausted if actively
worked. Monte-Bamboli, in the Maremma, furnishes excellent
black lignite of so bituminous a nature as to be distinguishable
with difficulty from Newcastle coal, to which it is considered
equal for metallurgical purposes and gas-works, and for steam-
vessels—the French government allowing it to be used by the
imperial navy.

The rest of southern Europe is still less supplied. In
Greece, in some isles of the Archipelago, on the coasts of the
Black Sea, we only find remains of Coal Measures; but at
Heraclia, in Asia Minor, coal exists to some extent.

In Africa many carboniferous centres have been discovered. In the French province of Algeria merely insignificant outcroppings of lignite have been met with; but in Choa and Abyssinia there is probably coal. Unfortunately the Grand Negous Theodore is not a prince favourable to industrial pursuits; and although very willing to have the mines worked, it is for the benefit of the crown, and not for that of his subjects.

On the other side of Africa coal-fields have also been noticed, as for instance at the Cape of Good Hope, at Natal, and on the coast of Mozambique, as well as along the banks of the Zambesi. Coal extends over a very large area in the upper parts of Natal, and has been worked at Biggarsberg and Newcastle. The coal from the former locality is reported to be semi-bituminous, free-burning, and to produce a very hot fire, for which reason it is preferred by some of the blacksmiths in Pietermaritzburg to ordinary English coal. The mineral discoveries lately made in this part of Africa promise most important results.

In the great island of Madagascar there are likewise coal-mines. On the western coast, towards the bay of Passandava and the bay of Bavatou-bé, the beds have been traced for a very great distance; but they pass under the sea, probably uniting with the coal-field of the African coast, and thus forming an almost entirely submarine basin. The coal of Bourbon (Isle of Réunion), a kind of lignite of tolerable quality, was tried in 1861 by M. Simonin, who deposited specimens of it in the Natural History Museum of Saint-Denis.

A Frenchman, Mons. J. Lambert, when established as a merchant in the Mauritius, formed a plan for working the mines of Madagascar, and Mons. d'Arvoy, formerly French consul at Port-Louis, was intrusted with the direction of the works. Queen Ranavalo regarded with an unfavourable eye this intrusion into her empire of the Whites or *vazas*, as strangers are called in that country, and excited the Hovas and Saklavas against them. One night the miners were suddenly surprised and massacred; Mons. d'Arvoy himself being killed with spears, after undergoing shameful tortures.

In India, Birmah, Cochin China, China, Japan, and even in Central Asia (Siberia, Persia, &c.), nature has shown herself to be more bountiful than in Africa. There are very important deposits of coal in those countries, some of which are actively worked. Those of India, owing to the energetic enterprise of the English, are better worked than the Chinese; and probably, owing to the interference of the French, similar results will follow in Cochin China.

Dr. Oldham, in his Report on the "Coal Resources and Production of India,"* states that the British territories in India cannot be considered as either largely or widely supplied with this essential source of motive power. Extensive fields do occur, but these are not distributed generally over the districts of the Eastern empire, but are almost entirely concentrated in one (a double) band of coal-yielding deposits which, with large interruptions, extends more than half across India, from near Calcutta towards Bombay, extending through about five degrees of latitude.

Up to the present time, it may be said that little more than surface-workings have been carried on in India. The deepest pit in that country scarcely exceeds seventy-five yards; while certainly one-half of the Indian coal which has been used up to the present time, has been produced from open workings or quarries, in which the coal has been worked like any ordinary stone. In parts of the Raneegunge coal-field, the only Indian field which has as yet been worked to any extent, these open workings are of marvellous extent and size, covering hundreds of acres. The area included by the coal-bearing rocks of this well-known coal-field, at a distance of 120 to 160 miles north-west from Calcutta, is about 500 square miles; and in this

* Return called for by the Right Honourable the Secretary of State for India; by Dr. J. Oldham, Superintendent of the Geological Survey of India: Calcutta, January, 1867.

series there is a thickness of workable seams of coal of from 100 to 120 feet.

Patches of coal or lignite of Tertiary age have been found in several places along the outer range of the Himalaya Mountains. In many cases these lignites have consisted of a bright, clean, jetty substance, and burn well; thus giving rise to hopes of continuous deposits of coal, which have not been justified by the facts.

The *average* composition of Indian coals, as determined from seventy-four localities, would appear to be—fixed carbon, 52·2; volatile matter, 31·9; ash, 15·5 : while the average composition of English coals sold in the Calcutta market during the last three years, was found to be—fixed carbon, 68·10; volatile matter, 29·20; ash, 2·70.

These figures show the great inferiority of the Indian to ordinary English coals. The very best coals of Indian fields only touch the average of English coals, and are not capable of more than two-thirds, in most cases not more than one-half, the duty of the latter; and this inferiority of the Indian coals has prevented their being used in any case where the fuel is not to a large extent locally attainable.

Mr. Raphael Pumpelly[*] gives a hypothetical map of the structure of China, from which it appears that more than two-thirds of the whole area of that country is occupied by coal-bearing strata, which, yielding both anthracite and bituminous coal, have been extensively worked for ages, and appear, from the plant-remains, to be of Mesozoic date.

The characteristic plants resemble those of the European Oolitic flora; and differ from the Indian and Australian in the absence of the genera Phyllotheca and Glossopteris.

Coal-mines are worked in the province of Chihli, in the immediate neighbourhood of Nanking, and in the immense

[*] "Geological Researches in China, Mongolia, and Japan during 1862-65:" published at the Smithsonian Institution, Washington.

coal-basin of Sz'chuen. The Futau mine, which lies less than two miles S.S.E. of Chaitang, produces a "steam coal" that is equal, if not superior, to the best Welsh variety. All the coal worked in the district of Fangshan and in the eastern portion of the Wangping field is anthracite, as is also that of the Kwei coal-field.

Coal of comparatively recent origin is found in depressions of red sandstone at Kelung in the island of Formosa.

In Japan coal is worked in the neighbourhood of Nagasaki, on the west coast of the island of Kiusiu; and there is a coal-bearing series of more or less metamorphosed rocks containing fossil Equiseta, beyond Iwanai, near Ousubetz (north), in the island of Niphon; the coal of which is highly bituminous, and probably of value as a gas-producing material.

Mr. Cuthbert Collingwood, in a communication lately made to the Geological Society, describes coal from several localities in Japan as bright, clean, and like Sydney coal.

Excellent coal is found at Dui in the island of Saghalien.

Coal is also found, often over large areas, in the Sunda Isles and Sumatra, in Australia, New Zealand, and New Caledonia: but its quality is not always first-rate, or else the means of transport are wanting. Nevertheless the coal-fields of Australia are worked by the English, and the French station of New Caledonia has, at one time, supplied coal to Bourbon.

The discovery of coal has lately been announced on the Irwin River in the Victoria district of Western Australia; and, also, on the southern coast near the Fitzgerald River. It is of the character of Welsh coal, and well suited for engine purposes.

There is a great quantity of coal at the northern end of the island of Labuan, for a distance of four and a half miles in a W.S.W. direction; and also on the island of Moarro or Moaro, sixteen miles south of Labuan, at the entrance of the Brunai

River. In both instances the seam of coal is generally nine feet thick, and appears to be a sort of cannel or lignite of very recent date, containing much resinite (*Dammara resin*) in small lumps. In Borneo coal is worked in several places, especially in Borneo Proper and Bandjermassin.

On the other side of the Pacific, in South America (Map XI.), the coal-mines of Concepcion, Lota, Valdivia, &c., are situated on the coast of Chile. They traverse Araucania, the ephemeral kingdom of the good M. de Tonneins, the lawyer of Périgueux, and extend as far as Patagonia. Coal is also obtained at the base of the Andes in Peru; in the different provinces of Brazil; in Venezuela; in the island of la Trinité, &c.: but the quality is frequently indifferent, at the same time that English coal opposes a fierce competition to the native fuel. In Brazil Viscount Barbacena has ascertained the existence of a series of coal-beds in St. Catherines, at nine different levels, underlying a sandstone formation, horizontally disposed, and varying in thickness from 1½ to 10 feet. Analyses of specimens of the coal prove it to be of good quality, its profitable working depending solely upon the facilities for transport. The difficulties of working and selling are almost insuperable in this country, where roads are wanting, and where there is scarcely any industry beyond that connected with the precious metals and copper. The occurrence of beds of coal has been lately announced by Mr. W. Wheelwright on the eastern slope of the Andes, between the cities of Cordova and San Juan, about twenty-five leagues from the latter city.

Let us end our course by way of North America. There are coal-fields there extending to the Pole, in Greenland, in Baffin's Bay; but it has been ascertained by a recent examination of the country, conducted by Mr. Alexander Murray, of the Geological Survey of Canada, that the coal-field of Greenland is of much less extent than was previously supposed; and it is, besides, covered by ice. A large tract of country is

F

reported by Mr. Murray to be spread over by rocks of Carboniferous age, but it is doubtful whether they contain seams of
coal sufficiently thick to be of commercial value. Mining for
coal has been seriously attempted at three places only, viz., at
the Terra-Nova Mine, in Little Bay or Bay Vert; the Union
Mine, at Tilt Cove in Notre-Dame Bay; and at the La Manche
Mine, at the head of Placentia Bay.

Some parallels of latitude lower down, on the Pacific coast,
there are the carboniferous districts of California and Oregon;
the one being still in its infancy, while the others occupy an
exceptional position on the very shores of the bay of San Francisco. The coal is discharged from the trucks in which it is
brought out of the galleries of the mines, directly into the
holds of the ships. When the railway is finished between the
Atlantic and Pacific, San Francisco will, together with New
York, become the great port between London and Pekin.
After the Californian deposits come those of Utah, worked by
the Mormons in the neighbourhood of the Great Salt Lake.

On the coast of the Atlantic we notice, first, the extensive
and productive coal-fields of British America, at Cape Breton,
at the eastern extremity of Nova Scotia, and in the southern
provinces on the Gulf of St. Lawrence. The coal-fields of the
United States are the largest in the world, extending from the
foot of the Alleghany mountains on both sides of the chain, but
especially on the western side, as far as Missouri and Arkansas,
reappearing at the base of the Rocky Mountains. (Map
VIII.) The states of Pennsylvania, Virginia, and Tennessee
are indebted to them for a portion of their prosperity. The
hard anthracite coals of Pennsylvania have been applied with
the greatest success to the manufacture of iron, and furnish
half the entire quantity of coal raised in the United States.
Further eastward there are the basins of Illinois, Indiana, and
Missouri; and lastly, the Michigan basin south of the great
lakes.

MAP VIII.

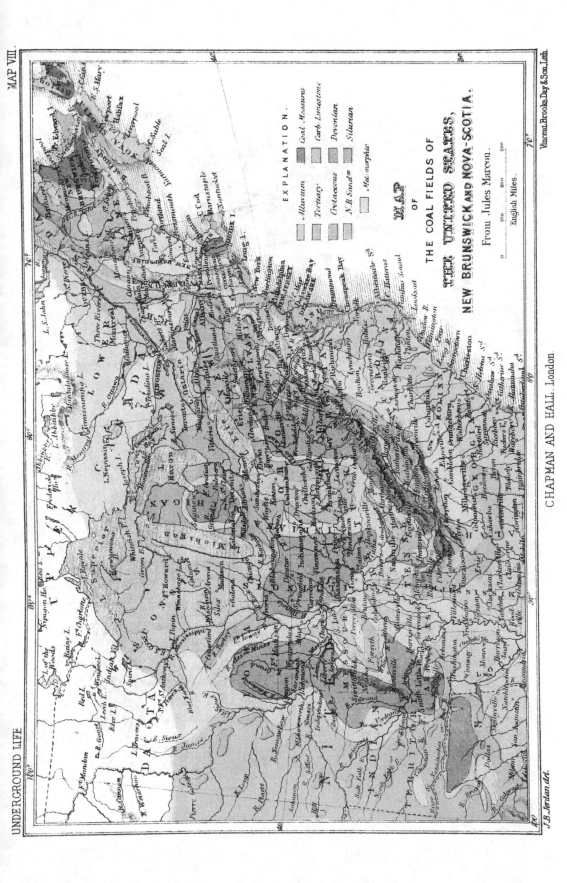

EXPLANATION.

Alluvium
Tertiary.
Cretaceous
N.R Sands
Coal Measures
Carb. Limestone
Devonian
Silurian
Met. morphic

M A P
OF
THE COAL FIELDS OF
THE UNITED STATES,
NEW BRUNSWICK AND NOVA-SCOTIA.
From Jules Marcou.

0   100   200   300
English Miles.

J.B. Jordan.del.

CHAPMAN AND HALL. London.

Vincent.Brooks.Day & Son.Lith.

Fig. 21.—Oil-wells at Tar Farm, Pennsylvania.

Covering an area eight times larger than all those which we have just noticed, and worked for barely forty years, the produce of these mines already equals the output of France and Belgium together. The coal-formation of the United States is of immense extent, occupying a fourth part (often a third) of the area of the states. It is the great reserve for future ages.*

It is a remarkable fact that mineral fuel is most abundantly diffused over the temperate regions of the northern hemisphere, as if the peoples who are now the most civilized had naturally settled around the coal—that most marvellous agent of civilization and progress.

In North America, to geological conditions already so favourable others are added of an exceptional kind. It is in the coal-producing states that most of the petroleum springs have been discovered, which have been worked since 1858 (fig. 21). The Americans have glutted the markets of the world with this *rock oil*, and Petrolia has proved to Yankee adventurers a second California. In the year 1866 the production of Western Pennsylvania alone sometimes amounted to not less than 15,000 barrels of crude oil per day, and that of the United States to not less than 18,000 barrels. There has been an oil-fever, as there was a gold-fever. The lovers of grand sights have found in them materials for dramatic sensations. The petroleum has caught fire in some of the wells, and an immense conflagration has ravaged the works. In Pennsylvania, Oil Creek has carried the flames over an extent of several miles.

M. Simonin, in passing through the country in 1859, was everywhere struck with admiration of the vigorous republic, and compared the flourishing condition of these young countries with the neglect and misery of most of the republics of Spanish

---

* It is not within the scope of this work to notice all the localities where coal occurs; a list of them may be found in the lectures by Professor J. Morris, on "Coal, its Geological and Geographical Position."

America. "Do the descendants of Pizarro and Ferdinand
Cortez bear any resemblance to the children of the brethren of
Penn, or to those of the Independents?"* Lofty black furnaces, before which are heaped up the produce of the coal and
iron mines, rise on both banks of the Hudson River, down
which scarce half a century ago floated the first steamboat.

Fulton, whose invention had not been appreciated in France,
met with the coldest indifference in his own country. On the
day when he made a trial of his boat one passenger alone presented himself with sufficient hardihood to accompany him;
and when thanked by the inventor, the tourist replied, "It is
of no consequence, Sir; I have made up my mind to kill myself,
and I hope we shall blow up on the voyage."†

* Chateaubriand, "Voyage en Amérique."

† This traveller was a Frenchman, Michaux, the celebrated botanist, author of the still classical
work on the forests of America ("North American Sylva"). Michaux was on his return from an
excursion to the borders of Lakes Erie and Ontario, when he heard that the trial of Fulton's steamboat was about to take place. On his arrival he found nobody on board except the captain and his
crew, with whom he made friends. Probably the saying above given was only intended for a joke.

CHAPTER V.

HOW COAL IS DISCOVERED.

Coal-basins.—Discovery of the Mines of La Sarthe and Anzin.—Discovery of Coal by Viscount Désandrouin in French Hainaut.—Tubbing.—Steam-engine first used in France in Draining the Mines of Hainaut.—Discovery of the Coal-field of the Pas-de-Calais.—Watery ground or *Torrents*.—Use of the borer in Coal-mining.—Herr Kind.—Coal-basin of the Moselle; number and cost of borings.—Cost of some English borings.—Borer used for discovering Coal-fields, and proving their extension beneath rocks of newer age than the Coal Measures; as in Deps. Gard and Haute Saône.—Boring-rods and tools.—Free-fall cutter.—Simplest mode of boring.—Modern improvements in boring.—Boring at Mouille-longe stopped by accidental breaking of borer.—Experiments of Walferdin at Mouille-longe on subterranean heat.—Rate of increase with the depth in the mines of this country.—Principles by which the search for coal should be guided.—Probabilities as to finding coal under London and Paris.—Well-borings at Passy and Grenelle.—Opinion of Mr. Godwin-Austen as to the occurrence of coal in the south-eastern counties of England ; Opinion of Sir R. I. Murchison.—Well-borings at Harwich and Calais. —Conclusions to be drawn.

IN common with miners and geologists, we have designated under the term coal-basins the areas, often of considerable extent, which are occupied by the measures, as the strata are called, which are associated with and contain the mineral fuel. This nomenclature is justified by the facts ; for coal and the rocks with which it is found associated usually occupy the beds of ancient lakes or the bottoms of valleys and seas which are no longer in existence. The strata descending beneath the surface rise again to make their re-appearance on the flanks of these valleys, on the borders of these old lakes and estuaries, or on the shores of these pre-existing seas. They thus affect the disposition termed in mining language a basin (*fond de bateau*), fig. 22.

At the same time it is by no means meant to say that the present boundaries of the coal-basins are the original limits of

the deposit, it being well known that the Carboniferous rocks
must have extended over larger areas than they now occupy,
and from which they have been partly removed by denudation.

At the bottom of the formation, and forming the base of
the Carboniferous strata, are breccias, conglomerates, grits,
and sandstones of the Old Red Sandstone and the older
Palæozoic and crystalline rocks; composed of fragments of
schists, granites, porphyries, large pebbles of quartz, mingled
and cemented together (Plate XI., fig. 3). These fragmentary
deposits, detached from the older rocks, constitute a valuable
geological horizon. They select a suitable geological horizon
for guidance in the search for coal, which very often crops out,
or comes "to day" or "to grass," as miners say—meaning that

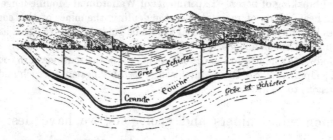

Fig. 22.—Cross-section of the Coal-basin of Rive-de-Gier, after Elie de Beaumont.

the extremity of the beds makes its appearance at the surface
—in which case the coal is easily discovered. Generally the
beds thus exposed are weathered and earthy; but at a slight
depth they are unchanged, having been protected by a covering of solid rock from the influences of water, air, and light.
In that case they present that colour, brilliancy, and compact-
ness peculiar to coal (Plate I.). The preliminary examination
is conducted by cutting trenches, galleries, and shallow pits;
then if the conditions are favourable, the actual working may
be commenced. At certain suitable points this may even be
done by means of open workings, which are carried on after
the manner of an ordinary quarry.

Most coal-basins have been discovered and proved in this

way; but when the coal is concealed beneath the soil, or when nothing is discoverable to the sight, it is chance or geological reasoning which reveals it ; and then begin operations of a far more complicated character, requiring much engineering skill and mining experience.

The following are instances of this fact :—In 1813 a well was sunk in La Sarthe, in the environs of Sablé, for the supply of an estate. Amongst the rubbish a blackish earth was noticed, a specimen of which was sent to the Society of Arts of Le Mans, the members of which held an extraordinary meeting, when some of those present expressed an opinion that the substance in question might, possibly, be coal. It was immediately tried in the stove of the room in which the meeting was assembled, and the earth burned amid the acclamations of the spectators. The Society in its reports gave great notoriety to this result, which in this instance, at least, proves that provincial societies, which it is so often the humour of the public to ridicule, are occasionally good for something.

The person who ought to have been enriched by this discovery intrusted the works to inexperienced men, who passed through the seam of coal, which was probably disturbed or altered, without noticing it. It was only in 1816, three years afterwards, in making a cutting for a road, that the seam was struck upon afresh, and this time without affording room for doubt. The neighbouring proprietors immediately set to work, and the state soon afterwards granted them concessions which have continued to prosper ever since.

Chance does not always aid the miner as it assisted him in the coal-basin of Le Maine. The ancient adage, " Help yourself and Heaven will help you," is more applicable to mining operations perhaps than to any other kind of enterprise on earth. It is only after invoking the aid of geology, and making careful researches under the guidance of that science, that in general the existence of a new coal-field is proved. An instance of this is afforded by the French northern coal-field, where a vast amount of patience, courage, and money were required, combined with great intelligence, to endow the country with that valuable deposit. The basin being entirely subterranean, or covered by strata of newer age than the Coal Measures, nothing is revealed to the eye; and it required the penetration of the geologist, while geology was yet in its infancy, to arrive at the discovery of its existence. In the year 1716, Viscount Désandrouin, a skilful coal-miner of the province of Charleroy, in Belgium, noticed that the measures of that country had a constant direction from east to west, and that in French Hainaut they were overlaid by the Cretaceous strata (Map II.). He then formed the idea of traversing these strata by means of shafts or pits, and of working the underlying coal. In the course of less than four years his researches were crowned with success; but the works were inundated by extraordinary quantities of water. In the country in question the accumulation of water in the strata is very great, and it became necessary, in order to continue the mining operations, to introduce into the shaft those wondrous wooden shields called tubbing, the name of which is borrowed from the form. The different parts composing the tubbing— formerly of wood, now frequently made of iron—press against the walls of the shafts like the staves of an immense cask; except that in this instance the water has to be kept outside the barrel, and must never be allowed to obtain admittance into its interior.

It was also at these mines of Hainaut that the steam-engine, which had lately been employed in England—first by Savery, and subsequently by Newcomen in some Cornish mines—was applied for the first time in France. For the rest, dry coals only, and those of inferior quality, were met with; and it was only in 1734, after eighteen years of uninterrupted labour, that operations were altogether abandoned. It was time that they were, for the mines of Anzin were discovered in the meanwhile, and the Viscount Désandrouin and his enterprising partners had spent nearly all their fortunes.

It is not necessary to dwell upon the various phases through which the operations in this coal-field had to pass before it was fully and successfully developed. It appears to be a peculiarity of most mining operations to excite competition, and to reward only the second, not unfrequently the third, generation of resolute miners who devote themselves to the pursuit.

In later years, researches similar to those which Viscount Désandrouin had first commenced around Valenciennes, have been renewed in the neighbourhood of Douai, and continued towards Lens and Béthune, where again long-continued and persevering efforts have been rewarded with that great success which has been the cause in the department of the Pas-de-Calais of sudden and unexpected fortunes being made, as well as of an almost boundless industrial prosperity. Chance and geology have contributed to promote this happy change; chance even has effected nearly all—for geology, having got on a wrong scent, had been searching to no purpose around Arras for the continuation of the Valenciennes basin (Map II.). The normal prolongation of the strata was, in fact, in that direction; but the geologists failed to perceive that there was, as it were, an underground ridge of the older rocks taking a sharp bend backwards from its former direction at Valenciennes in the direction of Douai. The sea of the Coal Measure period must have extended along this shore, and the strata deposited upon it must have followed this bend on leaving Valenciennes, so that the Coal Measures must make an elbow, or be sharply deflected in a north-westerly direction, instead of being prolonged in a straight line. This was an unexpected deviation, which the explorers could not have foreseen. After years of fruitless efforts, they had everywhere despaired of finding the coal, when a lucky accident opportunely occurred to revive their courage, and to give a vigorous impulse to the works.

In the year 1847, during a search for artesian springs at Oignies, near Carvin-Epinoy, in the Pas-de-Calais, the borer all at once and quite unexpectedly revealed the presence of Coal Measures beneath the Cretaceous strata.

Scarcely was the news made known than everybody set to work, and the tool was no longer used to search for water, but for coal. So many borings were immediately made, over a length of about twenty leagues, and an average breadth of four, that the ground was pierced like a colander by a series of borings, all accurately laid down on a plan drawn to a tolerably large scale, reminding one of the constellations of stars as they are figured on celestial charts (Map XI.). Success exceeded all hopes. The subterranean beds of watery ground, (called *torrents*), which are so abundant in these districts, caused the most serious obstacles to the miners; but they ultimately succeeded in overcoming all impediments. The mineral wealth of France became augmented by *a hundred thousand acres* of coal-fields. Twenty-seven companies were formed to work the new concessions. Some forty pits have been sunk through, on an average, from 109 to 164 yards (100 to 150 mètres) of overlying ground, to depths varying from 197 to 328 yards (180 to 300 mètres); one pit, that of Ferfay, having its workings at the depth of 503 yards (460 mètres). In fifteen years the produce of this basin has steadily increased from less than 5000 tons in 1851 to 80,000 tons in 1854, while it attained upwards of 1,600,000 tons in 1866; a tenth of the entire produce of France. And all this originated in a search for water. The borer, which from finding the subterranean springs of Artois had passed into use in the coal-fields, discovered in its turn Coal Measures in Artois.

It is necessary to describe the search for coal by boring, the surest practical aid which comes at the present day to assist the speculations of geologists in the discovery of Coal Measures. This ingenious method proves, at the same time, to what astonishing results the patience and skill of man can reach.

The distance, sometimes enormous, which separates the Carboniferous deposits from the surface is passed through slowly and with difficulty. The ground is perforated and pierced with steel tools even to its remotest depths. When nodules of flint are encountered, the hardest and best-tempered steel becomes blunted, breaks, and more metallic powder is drawn up than fragments of rock. Sometimes a bar is broken, or the tool is damaged, when it is grasped by means of grapnels; at others, the passage through running ground like quicksands, or tumbled and fallen ground, and springs of water, oppose the greatest difficulties to the progress of the work. The bore-hole must then be protected by a column of metal tubing, the parts of which—metal pipes—are sent down one by one.

The boring-rods are of wood or iron, and screwed together as the work proceeds. The steel chisel or bit (fig. 23) which strikes the rock and wears it away, is called by the workmen a trepan. Several modifications of these boring-instruments have lately been patented

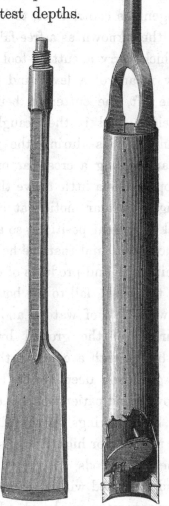

Fig. 23.—Trepan or Chisel.　　Fig. 24.—Shell-pump with clack.

Degousée and Laurent's Methods. Scale ¹⁄₁₀.

by the Messrs. Kind, which, they believe, will suit any
future operations in the Vosgian strata of the Moselle district,
some of the sandstones and conglomerates of which are
extremely hard.

In very deep borings, where the weight of the rods is so
great that the cutting tool would be likely to get bent or
broken if the whole mass of the apparatus were allowed to fall
together, the tool is often mounted in such a manner as, by an
ingenious contrivance, to become detached from the rods, and
is then known as a free-fall cutter. At the surface the rods,
which carry a cutting-tool of considerable weight, are raised
by means of a lever and chain, and on reaching the top of
the lift, the cutter on being released falls alone by its own
weight, and is then caught and lifted again by a pair of
hinged clips during the upstroke of the rods. A workman holding a cross-bar or an iron key, which clutches the
upper rods a little above the surface, gives to the apparatus a
slight circular motion at every stroke, in order that it may
take different positions, so as to strike all points of the rock in
succession, and that the hole may always be round. The dust,
chippings, and products of excavation caused by the percussion
of the chisel, fall to the bottom of the bore-hole, which is kept
always full of water, and the soft matter is raised to the
surface of the ground by a lifting pump, consisting of a
cylinder with a clack at the bottom (fig. 24). The bore-hole
is gradually deepened in this way; but the work progresses
slowly, and patience is one of the essential qualities for a borer.

Some borings are made in an altogether primitive manner.
A triangle, or high pair of shear-legs, for lowering and raising
the boring-rods, a lever or a windlass worked by hand-power,
and furnished with a strong stage or flooring, are the preliminary apparatus employed. In deep borings of large diameters,
the work is carried on in a more costly manner, and the aid of
a small steam-engine is called into requisition (figs. 25, 26).

Fig. 25.—Boring Operations (System of Mulot and Saint-Just-Dru): Boring the Hole.

Fig. 26.—Boring Operations (System of Mulot and Saint-Just-Dru): Changing the Rods.

When borings are made for the purpose of finding coal, operations are begun, at least in many instances, in a formation of more recent date than the Coal Measures. After the overlying strata are sunk through, generally consisting of limestones, sandstones, marls, and clays, the true Coal Measures are reached, the latter being recognizable by black shales and silicious sandstones, laminated with brilliant scales of mica. Thin seams of coal are, perhaps, soon met with, the precursors of the seam which is sought for. At this stage of the proceedings the anxiety of the sinkers is doubled, lest, owing to any disturbance of the beds, the usual horizon occupied by the coal should be passed through without encountering it. In making his calculations on similar borings, should any have been already made in the same locality, this depth is known approximately. It is a day of victory when the borer brings up coal—black, shining, compact coal—which leaves no room for doubt as to the ultimate success of the undertaking.

The mode of boring which has just been described, is such as has been practised until very recently; but it has been somewhat modified of late, and is now carried to such exactitude as to leave little room for further improvement. The tools have been rendered not only very simple but fewer in number. One of the inconveniences attendant on the old methods of boring was, that only shapeless fragments, dust, and mud were brought to the surface, from which little or no information could be derived; whether (for instance) the sandstones sunk through were for certain those of the Carboniferous age, or whether the black matters were derived from true coal or merely from black shales; what were the direction and thickness of the beds, or whether the rocks presented any impressions of organisms characteristic of the Coal Measure formation. All these requirements have been met by the improved means by which the borings are now carried out. A gouge in the form of a hollow cylinder is employed, furnished at the base with a row of teeth, or only with four or six cutting blades made of cast steel (fig. 27). It is worked like an ordinary boring-auger, and it cuts a solid column or cylinder (*carotte*) out of the rock, as regular in shape as though it had been turned in a lathe. When this cylinder has been detached laterally from the main mass of the strata sunk through, it is broken off at the base by means of the gouge-bit, a regular spring-tool or grapnel which holds it and brings it to the surface (fig. 28). As the hole may now be cut 8 inches in diameter, a broken pillar of rock 4 inches in diameter and as long as the hole sunk can be brought up piecemeal, an unexceptionable witness, by means of which the fossils included in the formation may be studied, as well as the structure of the strata, their thickness and inclination to the hori-

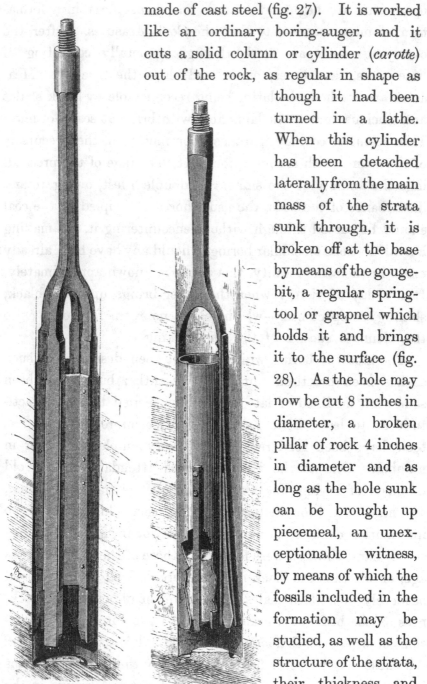

Fig. 27.—Cutting Gouge.   Fig. 28.—Grapnel
Systems of Degousée and Laurent.   Scale $\frac{1}{10}$

zon, fig. 29. When the cylinder is raised without moving it from its natural position, the direction, as well as the amount of dip of the beds, may be also ascertained.

In the hands of the Messrs. Kind & Son, skilful Saxon borers (the father has risen from a common miner to be one of the most distinguished engineers of the present day), the borer has performed wonders. Germany styles Herr Kind *the Napoleon of well-borers;* and we must not fail to remember

Fig. 29.—Specimen of argillaceous shale with plant-impressions, brought up from a boring made by Herr Kind at Stiring (Moselle). Scale ½.

that he is the venerable father of his art. It is rather in the search for coal than for artesian springs that he has distinguished himself; having made trials, by borings, to a greater depth than most men. In the exploration of the Moselle coal-field Herr Kind is in France without an equal. He has bored holes of dimensions before unknown; he has bored mechanically even shafts of mines in the midst of rising springs of water. Mulot and Saint-Just Dru, Degousée and Laurent, who are acknowledged in France to be masters of the art of boring, have followed Herr Kind in this respect, and in the Moselle as in the Pas-de-Calais, their names are associated with the most successful workings. Messrs. Easton & Amos, and Mather & Platt, amongst others, have distinguished themselves in this country in this branch of engineering. By the system invented by Mr. Colin Mather, many deep borings have been made; one at Middlesborough, near Newcastle, being carried to a depth of 1312 feet, with a diameter of 22 and 18 inches.

The coal-basin of the Moselle deserves to occupy especial attention in the history of the discovery of coal in France. In this field, thrown open by their patient investigations, the French miners have made themselves celebrated, and deserved well of their country. By their success they have augmented in a sensible degree the subterranean wealth of the nation; a reward for perseverance which had not been deterred by expense. More than fifty borings have been undertaken, some of which exceed 547 yards (500 mètres) in depth (about half of which have been attended with success), under the most difficult circumstances and continual irruptions of water, at a cost of about £400,000.

When the allies revised the frontiers of France in 1815 they endeavoured to define them in such a manner, on the side of Rhenish Prussia, that all the rich coal-basin of Saarbrück, which has been worked for twenty years, should lie outside the new boundary (Map X.). It seemed to the Prussian engineer of mines who inspired the diplomatists with the bright idea of adopting such a boundary, that the strata, if they did not unpolitely turn their back upon France, were nevertheless situated at such a depth beneath the surface that no more coal could be expected to be got on the French side. The enemy had not reckoned upon the bold and venturesome initiative taken by the inhabitants of the Moselle. The successful attempts made in the Department of the Nord had attracted the attention of these intelligent people, and were remem-

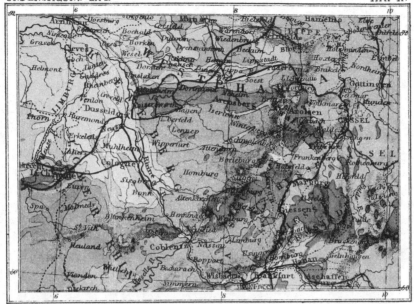

THE COAL FIELDS OF WESTPHALIA.

0   10   20   30   40   50   100

English Miles.

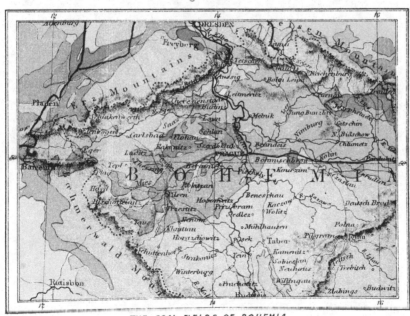

THE COAL FIELDS OF BOHEMIA.

EXPLANATION.

	Alluvium		Jurassic		Permian
	Tertiary		Liassic		Coal-bearing strata
	Cretaceous		Triassic		Lower Carboniferous
	Silurian		Igneous & Metamorphic		Devonian

J.B.Jordan del.   CHAPMAN AND HALL London.   Vincent Brooks Day & Son Lith

bered by them. They set resolutely to work at once, in the
first instance in the environs of Forbach; the ground was
bored, pits were sunk, and in spite of the length of time and
the patience which are always necessarily required in the
prosecution of such undertakings, nothing discouraged the
explorers. Capital when spent was followed by the subscrip-
tion of fresh funds; and if bore-holes and pits afforded no
results, others were made. It was also necessary to contend
against the water which filled the borings, causing falls in
them, or rising in artesian jets. In short, after many years of
continuous efforts, the moment of triumph arrived, and man
rested victorious in this contest with the ground. At the opening of the Chambers in 1858, the Emperor Napoleon III.
announced the discovery of the Moselle coal-basin, an extension
of the vast and productive basin of Saarbrück. The fact was
thenceforth placed beyond doubt. It is always to be regretted,
after so patient and meritorious a struggle, in which amongst
others De Wendel and Pougnet rendered themselves famous,
that the success achieved has not been so brilliant as in the
Department of the Pas-de-Calais, where the attempts were
less prolonged, if not of a less costly nature.

It is not only for the discovery of new coal-basins that the
borer is used, but it is also employed in a less ambitious manner,
to prove the more or less probable extension of a known coal-
field. The problem to be solved is the same in either instance.
In the latter case the question always is to find coal beneath
strata of a later age, by which the true coal-bearing beds are
covered up and concealed. In the Department of the Gard at
Grand'Combe and Alais, and also at Ronchamp in the Haute-
Saône, the prolongation of coal-basins which were previously
believed to be of more limited extent, has been proved by
means of borings. Sometimes eruptive rocks were encountered
before the coal was found, and threw a doubt upon the continuity of the beds; but beneath the porphyry extended more

G

modern rocks, New Red Sandstone or Oolitic limestones, through which the adventurous borers have sunk and found the coal which these strata only covered.

One of the numerous incidents connected with the greater number of these researches, in which the highest applications of underground geology and the physics of the globe are mingled, has an adventure of a dramatic nature connected with it, and may therefore be related :—The coal-seam of Creuzot crops out at the foot of the valley, where the iron-works are built, and disappears beneath the surface at a very high angle, nearly perpendicularly. At the depth of $262\frac{1}{2}$ yards (240 mètres) it reposes conformably on the underlying rock, over which it forms an undulating covering, and is itself again overlaid by the sandstones and shales of the Carbon-iferous formation. The latter in turn constitute the base of the *variegated sandstones* of the New Red Sandstone formation, which are so called by geologists because their colours are often mottled, passing from red to green or yellow (Plate XI., fig. 2).

The coal raised amounts to two hundred thousand tons per annum ; but the consumption in the iron-works is so great as to render necessary the purchase of as much again. Not only is the local supply deficient, but in addition, however carefully the coal may be worked, its exhaustion may be foreseen, inas-much as the seam soon abuts against what is termed in geology *a fault* (a dislocation of the rocks), the seam being cut off by a ridge of barren rock which passes sharply through the Coal Measures. The fractured beds have been disturbed in such a way as to be thrown down to a lower level probably, on the other side of the fault, than that they would otherwise have occupied. But the fault not having been driven through, it has to be proved whether the coal is actually to be found on the other side; and, should the Coal Measures be there, at what depth it is likely to occur.

The former question receives a solution, beyond a doubt,

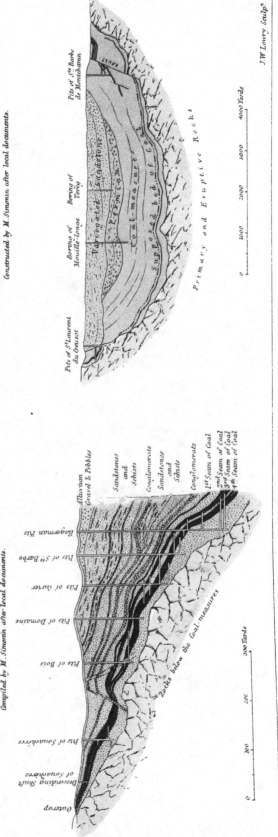

Geological Sections in France

Plate XI.

Coal Mines

Hypothetical Section across the Paris Basin from West to East, along the line A B on map.

West

Alençon   Mortagne   Dreux   Versailles   Grenelle Well   Epernay   St Menehould   Verdun   Metz   Saarbrück

PARIS

East

Transverse Section of the Strata between the Pits of St Barbe de Montchanin & St Laurent du Creusot.

Constructed by M. Simonin after local documents.

Pits of St Laurent du Creusot   Boring of Mouille-longe   Boring of Torey   Fits of Ste Barbe de Montchanin

Variegated sandstone   Permian   Coal measure   Supposed bed of Coal

FAULT

Primary and Eruptive Rocks

0   1000   2000   3000   4000 Yards

Section of the Coal-measures of Epinac (Saône-et-Loire) between the Souachères & Hagerman Pits.

Compiled by M. Simonin after local documents.

Descending Shaft of Souachères   Pit of Souachères   Pit of Bois   Pits of Domaine   Pits of Curier   Pit of St Barbe   Hagerman Pit

Alluvium   Gravel & Pebbles   Sandstones and Schists   Conglomerate   Sandstones and Schists   Conglomerate   1st Seam of Coal   2nd Seam of Coal   3rd Seam of Coal   4th Seam of Coal

Outcrop

Rocks below the Coal measures

0   100   200   300 Yards

J.W. Lowry Sculpt.

Chapman & Hall, London.

Vincent Brooks, Day & Son, Lith.

by the reappearance of the Coal Measures at Montchanin, a distance of $3\frac{3}{4}$ miles (6 kilomètres) from Creuzot, along the Canal du Centre. It is, therefore, probable that the coal is continuous over the intervening distance, except where it may be affected by disturbances in the strata; so that the second problem only remains to be solved—to ascertain the depth of the Coal Measures beneath the surface, between Creuzot and Montchanin.

M. E. Schneider, the able manager at Creuzot, courageously undertook to work out this difficulty, and in 1853 a boring was consequently determined upon. M. Fournet, professor of geology at the Faculty of Sciences of Lyons, having been invited to select the most promising spot, made a careful preliminary examination of the ground, and at the end of some months he indicated the place called *La Mouille-longe*, between Creuzot and the Canal du Centre, as that which appeared to him to be best suited for the site of the intended boring. Soon afterwards Herr Kind was entrusted with the execution of the work. The most improved tools were employed, and some were even contrived for this special case. Solid cylinders of stone $1\frac{2}{10}$ inch (30 centimètres) in diameter, and about a yard long, were raised one after the other, and the products of the boring when brought to the surface, after being examined with the utmost care, were labelled and arranged; the whole suite forming a most accurate and interesting geological section.

The boring at Mouille-longe extended over four years, and was only stopped in 1857 at the enormous depth of $1006\frac{1}{2}$ yards (920 mètres); while the bore-hole, which at the outset had a diameter of $1\frac{2}{10}$ inch (30 centimètres), retained one of $\frac{5}{8}$ of an inch (16 centimètres). The Coal Measures had not been passed through; and certain impressions of plants brought up by the boring-tool having been submitted to the examination of M. Adolphe Brongniart, were recognized by him as

G 2

those of Annularia longifolia, one of the plants characteristic of the Coal Measures (fig. 12).

An accident, which it was impossible to foresee, unfortunately happened to arrest this boring, which is perhaps as deep as any which has ever been executed, and which has almost acquired a legendary interest amongst persons concerned in the sort of work in question. The boring-tool got broken at the bottom of the bore-hole. This is a rare occurrence; no serious accident had happened before in that narrow opening, passing nearly vertically downwards, and up and down which the steel chisels had to be raised and lowered in screwing and unscrewing the wooden rods one after another. Herr Kind, whose long and arduous career as a borer had been chequered by so many sorts of events, found his experience at fault this time. None of his grappling irons could be made to take hold of the broken tool; the chisel firmly lodged at the bottom of the bore-hole resisted every attempt to withdraw it, and it became necessary, at the end of six months of fruitless efforts, to abandon the boring without a hope of ever renewing it. A million of francs would willingly have been subscribed by the people of Creuzot, rather than the work should have been stopped.

Some days after its final abandonment, Gentet, the foreman of the works, mounted the staging and made another attempt, as a last effort, to raise the broken tool; wishing, in one of those moments of prescience which are not uncommon amongst miners, to overcome the obstacle which seemed to him to exist at the bottom of the bore-hole. The whole power of the steam-engine, by means of which the boring had been carried on, was exerted in pulling the end of the rods, which Gentet was shaking violently, when all on a sudden an ominous cracking noise was heard, arising from the breaking of the rope. Gentet's hand being placed on the end rod, very near a staging plank, through a passage in which it worked, remained fixed in a vice as it were, crushed by the enormous weight of many thousands of pounds. As his comrades, losing their presence of mind, did not know for the moment how to extricate him, the sufferer, who alone retained his self-possession, told them to saw the rod below, as the only means of putting an end to his horrible torture. Then holding the remains of his crushed hand (the right) in the uninjured one, he walked on foot the distance of a league to Creuzot; and there, without uttering a complaint, he underwent amputation at the wrist. Instances of such brave courage are not rare amongst miners.

Since this accident happened the boring of Mouille-longe has been altogether abandoned. The framework which, besides covering the bore-hole, sheltered the steam-engine and its tackle; the workmen's village built in the neighbourhood in anticipation of the operations which were about to commence —all has remained deserted. On one occasion only an indefatigable observer, M. Walferdin, so well known for his differential thermometers,[*] visited the boring to verify once more the law of the increase of subterranean heat with the depth. In experimenting throughout the whole depth of the hole, he observed that the mean rise of the thermometer was one degree Centigrade for every 88½ feet (27 mètres) of depth beneath the surface, a result corresponding to the law generally admitted by physicists, who calculate a rise of one degree (Cent.) of the thermometer for 89 to 98½ feet (25 to 35 mètres) of vertical depth.[†] If this law proves true to the greatest

---

[*] M. Walferdin has made several experiments upon the borings of Mouille-longe, and of Torcy, near the former. He found a mean rise of one degree Cent. for a depth of 34 yards (31 mètres) in these borings, to a depth of 601·7 yards (550 mètres); a result agreeing with that he had arrived at with Arago at the well of Grenelle. Beyond 550 mètres (601·7 yards) the boring of Mouille-longe gives slightly different figures. The later experiments of M. Walferdin have not yet been published, and he has only given M. Simonin, as an approximation, the above-mentioned mean of 89 feet (27 mètres) for all the pits of Mouille-longe.

[†] The increase noted in the collieries of this country is at the rate of one degree Fahr. for every 58·3 feet of descent, below a certain point of invariable temperature. A depth of 50 feet and 50° Fahr. may be considered the depth and temperature of this invariable stratum over the greater part of central England. A series of observations made in the Cornish mines by Mr. Robert Were Fox appears to show that the increase of temperature is in a diminishing ratio with the depth. (See Reports of British Association for 1837, 1840, and 1857; also Edward Hull, "On Experiments for ascertaining the Temperature of the Earth's Crust"—*Quarterly Journal of Science*, vol. v. p. 14; also footnote, p. 111.)

known depths, there is no reason why it should not prove true beyond that limit, and it would then follow that at the depth of a league under ground the temperature would be that of boiling water, or 100° Cent. (212° Fahr.) ; and at twenty leagues all rocks and metals would be in a state of fusion. This may afford an explanation of volcanic eruptions, earthquakes, and in the past ages of our planet of the elevation of mountain chains, the formation of metallic veins, the origin of thermal or hot springs, etc.

But the physics of the globe have not alone derived advantage from the boring at Mouille-longe ; the geology of the Coal Measures has also derived from it knowledge of the greatest importance. The speculations of science have now become a fact which has received a practical verification. The Coal Measures and coal itself have been demonstrated to occur between Creuzot and Montchanin, and the miner will some day carry his pick to the bottom of this ancient Carboniferous sea, the riches of which are reserved for the future.

The examples of the different borings which have been given, show the principles by which investigators ought to be guided in their search for coal by means of that ingenious process. Except in the ancient Carboniferous formation, there is very slight hope of meeting with mineral fuel in very large quantities or of the best. quality. It is well known that the carboniferous deposits contained in the red sandstones of the Permian* and Triassic† formations, as well as those of the Jurassic, Cretaceous, and Tertiary ages, are of very limited areas, and that, with very few exceptions, they all display the varieties of coal termed lignites, and those often of but indifferent quality. In the formations in question, as well as in those of Primary age, underlying the Coal Measures, and which occasionally contain anthracite, it is not advisable to search for

---

* Named by Sir Roderick Murchison after the province of Perm in Russia, where the formation is especially developed.

† From the Greek τρίας, 'triple,' because this formation generally comprises three subdivisions.

coal in the absence of reliable and direct indications. But the
Coal Measures themselves may be sought for by means of
borings beneath the more recent strata which may perchance
conceal them. Nevertheless the extension of the ancient Carboniferous deposits not being by any means indefinite, it will
not be wise to act at random, but rather to be guided by analogy
with the adjoining coal-fields. To act otherwise, or to pay
attention to the stories of pretended discoveries of coal in
unlikely places, would be to deceive one's self or to become the
dupe of others. Pompous announcements are often made of
the existence of imaginary coal, and on the strength of them
the unwary are frequently induced to lend their money for the
promotion of foolish speculations, which ought in some cases to
be exposed in a court of justice. Although the neighbourhoods
of London and Paris are most unfavourable for such enterprises, so far as geological conditions are concerned, it must be
remembered that those present all the attractions likely to
excite the imagination of the multitude. What a sensation
would be created by the startling announcement of the discovery of coal under those cities!

When the probable discovery of coal under Paris and
London is spoken of, all consideration of Tertiary coal
(which does not occur in the basins in question) must be
omitted, with the exception of a peaty or ligneous mass of
a chocolate colour, which is met with in the Paris Basin.
The fuels of the older geological formations alone have to
be considered. The Tertiary deposits of the London and
Paris Basins both rest on the Chalk, in which no combustible
mineral has been met with. If it be supposed that the
geological series is complete, and that the Jurassic beds under-
lie the Cretaceous, while under the former again the New Red
Sandstone occurs, then it may appear that we should have the
true Coal Measures; but as the fossil fuel of the Jurassic and
New Red formations are only accidents, as it were, it is only

on the presence of the true Coal Measures that any hopes can be founded. But before reaching the last, supposing them to exist, all the overlying deposits must be sunk through. The Chalk will be met with beyond doubt, for the borings have always reached it in both the London and Paris Basins. The presence of the Jurassic series underground in the Paris Basin is almost a certainty, because limestones of that age are shown on the geological maps of France, as a continuous band around the Cretaceous rocks which bound the district in question (Map II.).

Now, on what do the Jurassic formations rest beneath Paris? No direct evidence can be brought to bear on this question, and the underlying rock may be either granite, Primary deposits, or the Carboniferous formation itself, and this last may be covered by a thick mass of the New Red. If, then, according to this view, the geological scale be complete, the order of succession would be as follows :—Cretaceous and Jurassic series, New Red Sandstone, and Coal Measures. Supposing no irregularity either of deposition or disturbance to have taken place, at what depth then would coal probably be met with? At 1640 yards (1500 mètres) at the very least; that is to say, at a depth at which at present it is regarded by some as unworkable.

That the above is the approximate depth may be shown in this way. The well-borings at Passy and Grenelle (both ended in the Lower Greensand) prove the Tertiary and Cretaceous series together to be more than 1924 and 1800 feet thick at those places. Now it will be granted that the thickness of the Jurassic beds, a series so well developed in France, would prove to be at least as great as that of the Cretaceous series, and the same may be said of the New Red Sandstone. There is then a total depth of from 5772 to 5400 feet to be passed through before reaching the Coal Measures, though not necessarily the coal itself. Then again, will the Coal Measures be found at all ? There is no reason to expect that such will be the case, because there is nothing on the geological map of France to lead us to infer that coal exists under Paris. Neither the outcrops in Belgium or Rhenish Prussia, nor those in the French coal-fields, authorize us to form any such opinion. Hence if, for the purpose of estimating the probability of success, the area of the French coal-basins be compared with that of the whole country, it will be found nearly in the proportion of 1 to 200 ; it is clear, therefore, that there are 199 unfavourable chances against success, or to speak familiarly, the odds are 199 to 1 against finding deep-seated coal in the horizon of Paris. It is true that it may there be represented by a purely marine deposit, which has never risen from beneath the sea. Delightful hope ! Let us then leave to our grandchildren the task of exploring this store of coal when our present coal-fields are exhausted, when a piece of coal will be worth its weight in gold, and there will be no other mode of obtaining heat cheaply, except by bottling sunbeams.

Mr. Godwin-Austen, in a very original and philosophical memoir, has suggested that Coal Measures might possibly be found under London and the south-eastern part of England, arguing that as coal is worked under the Chalk at Valenciennes in France, and had been found to a small extent in recent sinkings under the Cretaceous deposits ranging westwards towards Calais, it might extend further across the Channel, and occur under similar Cretaceous rocks in the south of England ; he considers that "it will not be too much to say that we have strong *à priori* reasons for supposing that the course of a band of Coal Measures coincides with, and may some day be reached along, the valley of the Thames, whilst some of the deeper-seated coal, as well as certain overlying and limited basins, may occur along and beneath some of the longitudinal folds of the Wealden denudation."[*]

* *Quart. Jour. Geol. Soc.*, vol. xii. p. 73.

Sir Roderick Murchison appears to take an opposite view of the question, and is of opinion that the strata underlying the Cretaceous and Wealden series of the south-east of England, if not in part of Jurassic age, will probably prove to be a thin band of Carboniferous Limestone without any productive coal, or more probably Devonian rocks only. * This inference is strengthened by the fact, that in the south-west of England the Chalk and Upper Greensand cut off all the lower rocks in succession, until at last they rest on the Devonian beds.

Since Mr. Godwin-Austen's memoir was published, borings have been made at Harwich and Calais which, while confirming in a remarkable manner the correctness of his theoretical deductions as to the occurrence of a tract of the older rocks beneath the Weald and the valley of the Thames, tend also to render it very doubtful whether Coal Measures or actually productive coal-bearing strata are present beneath those areas. At Harwich the boring was abandoned at the depth of 1070 feet in a hard and slaty Palæozoic rock, immediately underlying the Gault; showing distinct cleavage, and said to contain Posidonomya : the lower members of the Cretaceous series, together with the whole of the Jurassic and Triassic strata, being absent. A general notion of the way in which this underground ridge of old rocks may occur under London may be got from the annexed section (fig. 30), slightly reduced from one by Mr. Whitaker,† of the Geological Survey.‡

It will be apparent, from the foregoing remarks, that the question as to the occurrence of coal under London is fraught with the greatest difficulties. In cases like this, where the evidence is so small, and the data for reasoning upon are so few, and those few for the most part of a purely hypothetical nature, the results deduced from them must necessarily be

* Brit. Assoc. Report for 1866. " Trans. of Sections," p. 57.
† " Geological Survey Memoir on Map 7," p. 108, 1864.
‡ This woodcut is used by the kind permission of the Messrs. Longman.

to a great extent merely conjectural, and cannot safely be extended beyond probabilities or possibilities. The boring at Harwich, and the strike of Palæozoic rocks older than the Coal Measures in a westerly direction towards Calais, prove the existence of the ridge of ancient strata, foretold by Mr. Godwin-Austen to extend across the Channel, and that the geological scale is not complete in those districts; many of the formations between the upper Cretaceous and Palæozoic strata being altogether absent. Beyond and in an easterly direction direct evidence fails us, unless the boring at Kentish Town may be considered to throw any light upon the subject.

The first reappearance of Palæozoic rocks westward from Harwich is in Gloucestershire and Somersetshire, where the Carboniferous Limestone is seen at the surface, forming a nearly continuous belt around the coal-fields of those counties on their eastern side, with a dip to the west; while the Secondary rocks, which are based unconformably upon them, dip in an opposite or easterly direction, concealing them altogether from further observation.

The existence of coal, then, between the above-mentioned extreme limits depends upon the mode of occurrence of these

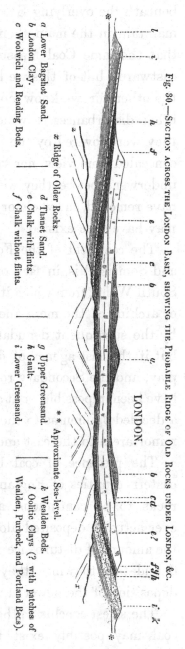

Fig. 30.—Section across the London Basin, showing the Probable Ridge of Old Rocks under London, &c.

a Lower Bagshot Sand.
b London Clay.
c Woolwich and Reading Beds.

x Ridge of Older Rocks.

d Thanet Sand.
e Chalk with flints.
f Chalk without flints.

g Upper Greensand.
h Gault.
i Lower Greensand.

* * Approximate Sea-level.

k Wealden Beds.
l Oolitic Clays (? with patches of Wealden, Purbeck, and Portland Beds).

older rocks, and of their undulations in an easterly direction
beneath the overlying Secondary rocks. If the Primary rocks
maintain, in the main, their dip to the west, it is most probable
that the same Coal Measures would have no existence to the
eastward ; but of this we have no direct evidence one way or
the other, for we know nothing of the upheavals, foldings, or
other disturbances to which they have been subjected before
they were covered by the newer strata by which they are now
concealed from view, nor yet of the amount of denudation they
underwent before they were so covered up, and which may
have removed, in whole or in part, any Coal Measures which
may have once existed.

The coal-field of the Forest of Dean (like those of Bristol
and Somerset) is, in fact, only a part of the great coal-field of
South Wales, from which it has been separated by the upheaval
or arching of the more ancient underlying rocks, and revealed
by the subsequent denudation of the newer overlying strata ;
but if this last agency of denudation had not been called into
play, and the Secondary rocks had not been removed, it would
have been impossible, or at all events unsafe, to have pronounced
a decided opinion as to the positive existence of coal over the
minor areas to the east and south (Map VI.).

The presence of coal beneath London and in the south-
eastern counties of England, assuming the ridge of Palæozoic
rocks to be continuous, and that the Coal Measures were
ever actually deposited along those areas, depends then upon
the amount of disturbance which they have undergone, and the
denudation to which they have been subjected prior to the
deposition of the newer strata.

The safest conclusion that can be arrived at, is that although
coal may possibly exist to some extent over the areas in
question, the probability is that such is not the case to any
great extent. Should this be the case, there is no doubt that
when the time should arrive when coal-mining at such greatly

increased depths offered sufficient inducements to make it a profitable commercial undertaking, modern skill and science would overcome the difficulties which might be opposed to its successful working by the depth at which the coal might occur —a limit fixed by Mr. Hull, in his "Coal-fields of Great Britain," at 4000 feet. This can scarcely be said to be the case now; but with the present enormous and continually-increasing consumption of coal, and the enhanced cost of production consequent on the extension of deeper workings, the time will arrive when the question will demand a positive solution, one way or the other.

As the problem can only be solved by actual experiments, of too costly a nature to be made by private individuals, it is to be hoped that a few borings may be made in the most favourable places at the public expense; for the question is a national one, and where the interests of the country at large are concerned it is but fair that each member of the community should bear his share of the cost.

CHAPTER VI.

SHAFTS AND LEVELS.

Manner of sinking of Shafts and Pits.—Cutting through soft and watery measures.
—Systems of sinking of Kind, Chaudron, and Guibal.—Triger's system of working
in compressed air.—Names of Shafts.—Mining-pumps.—Driving of Levels.—The
animals, plants, and water-courses of the Underground World.—The cost of a
Coal-mine.—The Stakes and the Gain.

THE coal having been found and the extension or prolongation of the Coal Measures ascertained, the preliminary workings must now be commenced by means of shafts and levels.

Before proceeding further it is necessary to explain the
manner in which shafts or pits are sunk. When the ground
is compact and hard, like most limestones and sandstones,
the progress is slow and wearisome, and sometimes the
rock is so hard as to blunt all the tools. In that case the
walls sustain themselves, require neither walling nor timbering, and are allowed to stand open; but when the ground
is soft and fragile, like some sands and most shales, then
an artificial support is required to resist the thrust of the
measures, and the shaft is walled either with brick or stone,
or is secured by timber. This lining consists of a framework
formed of strong segments of wood hooped with rings
called curbs or cribs; or of a continuous wall of masonry
composed of dressed stone, or bricks, resting on a firm foundation. Most of these works, executed according to the
strictest rules of art, and not less difficult than many
connected with railways, deserve to be seen by the full light
of day; but it is the peculiarity of all mining works, from
their very nature, to be comparatively unknown, and very

frequently to have no other witnesses to appreciate their value than the miners themselves.

When water-bearing strata (*nappes*) are traversed, and springs and feeders of water occur, the difficulties are rendered more than usually great. In Belgium, as has been stated, and in many collieries in England, the shaft is fitted with a regular wooden lining or tubbing, resembling a cask placed against the walls, the joints of which are so well fitted as to keep out all the water, and to withstand great external pressure or thrust from the outside.

It frequently happens that the ground is loose—sometimes consisting of a mere stratum of sand saturated with water—assuming the condition of a quicksand. The works are then of a still more delicate nature; but the miner is never disheartened, notwithstanding the desperate struggle he has to maintain against opposing difficulties—it may almost be said against the apparently insurmountable obstacles which the rocks oppose to his labours. Tubbing of cast-iron shields—or cylinders of masonry—are forced down the shaft when it is sunk in comparatively solid ground, but in soft ground they sink by their own weight. As they descend, a fresh capping constructed at the surface is added, and thus the work proceeds until it is completed. It was in this way that the elder Brunel constructed his shafts of $52\frac{1}{2}$ feet (16 mètres) diameter, which form the descent into the Thames Tunnel. In ordinary cases of construction towers rise from their foundations; these, on the contrary, sink bodily into the ground; an inversion of the skill that was displayed by the masons of Babel. Messrs. Kind, Guibal, and Chaudron on the Continent, with numerous others in this country, have distinguished themselves in the useful art of mining-engineering, and have associated their names with undertakings which, from their very peculiarities, and the difficulties under which they have to be carried out, may be at least briefly described here with advantage.

It has been stated that Herr Kind sinks the shafts mechanically, just as he drives bore-holes. Springs of water inundate the pits in the Department of the Moselle, and Herr Kind has carried on his work in the midst of them. The watery ground having been passed through, a circular bed is excavated by means of a special tool at the bottom of the shaft, which is not less than four yards in diameter. This work being completed M. Chaudron presented himself— another engineer as patient and ingenious as the first—and he devised a sliding-piece at the bottom of the cast-iron tubbing (the whole being gradually lowered as the excavation proceeds), the lower flange of which, turned outward, forms a case (*boîte à mousse*) to hold a quantity of moss ; this becomes closely packed and is pressed upon the seat cut for it, by the boring-tool, and upon which it rests, by the weight of the cylinder of tubbing, 98 tons 4 cwts. (100,000 kilogrammes), so as hermetically to shut out all access of water. Concrete or hydraulic cement (*béton*) is placed in the open space, or between the solid ground and the outside of the cylinder, for its entire height, upon the completion of which the water is pumped out by machinery and the work remains water-tight. Then only are the workmen allowed to go down the shaft. The sinking of these shafts, and furnishing them with all the necessary tackle, sometimes costs as much as £80,000. Sums of £40,000, £60,000, and even of £100,000 are stated by Mr. Warington Smyth to have been expended in sinking a single shaft in Durham, in consequence of the difficulty of piercing through the strata overlying the coal of that county (Report on Class 47 in the Paris Exposition of 1867).

The systems just described and that of M. Guibal, which is similar, are of recent invention. They have been especially applied in the Pas-de-Calais and the Moselle, and were preceded by that of M. Triger, conceived in 1841, but which unfortunately is not applicable beyond a certain depth. In the coal-field of the Basse-Loire a curious case presents itself, for submerged sands have to be sunk through, and the shaft established in the very bed of the river. All the ordinary known methods of sinking and draining are insufficient in this instance. It would be easier to empty the cask of the Danaïdes than these shafts, into which the water returns as fast as it is pumped out. It was under these circumstances that M. Triger, a Frenchman, working mines in these localities, conceived the idea of sinking into the ground sheet-iron cylinders of from 5 to 6 feet (1½ to 2 mètres) in diameter, to excavate from them the sand and stones, to divide the apparatus into three airtight compartments, to force compressed

Fig. 31.—Section of Triger's apparatus for sinking shafts under water; after Burat.

H

air into the lower one, and to inclose the workman in that
chamber, the joints of which are made carefully air-tight (fig.
31).   The miner is thus placed in a sort of diving-bell.   The
compressed air, being forced against the bottom of the shaft,
prevents the great mass of the water from filtering through
the sands, and drives back the small quantity which finds its
way in, through a sand-pipe communicating with the surface
from the bottom of the working.   Instead of attempting to
remove the water by pumping, it is kept back and held in
check.   "Imagine an army of mice," the inventor graphically
observed to M. Simonin, "and a cat suddenly to make her
appearance, you would have the picture of the water reaching
the bottom of our shafts by a thousand holes in the ground if
the pressure of the air is lowered, and returning suddenly to
the sands as soon as the air recovers its tension."

The rubbish and running sands are removed in buckets
by hand, or by means of a rope passing over a pulley.   Trap-
doors communicate from one stage to the other, by means of
which an approach to an equilibrium is maintained between
the middle chamber and the one below, while the buckets
pass through them and are emptied outside without any
serious loss of the compressed air.   By this means the shaft
is sunk until solid ground is reached, or the foundations are
laid by means of a lining of wood or masonry, to which it
is necessary to sink from 21·88 to 32·84 yards (20 to 30
mètres)—that being the thickness of the alluvial formations in
the bed of the Loire.

The workmen who carry on their operations in the compressed air, work in it with as much ease as in the open
air; some, however, especially those who have the drum of
the ear very delicate, or who are in the habit of drinking
to excess, can never remain long in this artificial atmosphere; while with others, a slight singing in the ears, a
certain quickening of the pulse, and a nasal tone of voice,

are the only physiological signs which remain. Some, on the other hand, derive a sort of comfort from the air thus made rich in oxygen. The ability to whistle is lost in it, but the deaf recover their hearing for the time, and lamps burn in it with a more brilliant light than usual.

To guard against all the accidents likely to result from an increased atmospheric pressure, the workmen must be made to enter and leave very slowly and carefully ; the air must not be forced in too strongly at the outset ; and in passing from the compressed air to the outer atmosphere, the effects of too sudden a change must be carefully avoided. With this object the middle compartment was contrived, which the inventor (borrowing the term from Hydraulics) terms the sluice or air-lock (*sas à air*). Without the observance of all these precautions, the most serious disorders may be induced : neuralgia, deafness, paralysis, rheumatism, and even rupture of the lungs.

All chances of explosion are avoided by employing three gauges, one fixed near the air-pump, the second at the mouth of the shaft, and the third in the chamber in which the miners work. These indicate exactly the pressure, and whatever derangements may happen, it may be fairly supposed that there will always be one gauge in action, at the very least. M. Triger regulates, moreover, the diameter and the relative speed of the cylinder of the steam-engine, and of the air-pump which it puts in motion, in such a way that the pressure never exceeds a given amount—three or four atmospheres. * Lastly, he also employs safety-valves, and for some short time past he curves the top and bottom of the lock in the form of a hemispherical skull-cap, like that of steam-boilers. When all these precau-

* This limit is that at which the method ceases to be applicable without danger to the workman, and corresponds to a depth of 38·29 to 43¾ yards (35 to 40 mètres).

It is said that the elastic pressure of a gas is 2, 3, 4, . . . atmospheres, when it is twice, thrice, four times . . . greater than the ordinary pressure of the air.

The pressure of one atmosphere sustains a column of water 32 feet (10 mètres) high; in the Triger apparatus, as the column of water is mixed with air, it is calculated that one atmospheric pressure is equivalent to that of a column of water 49·2 feet (15 mètres) high.

tions are observed, no danger is to be apprehended from the adoption of this method, the introduction of which received the encouragement of Arago, and gained for the inventor the greatest honour.

M. Triger's apparatus has been applied in working mines and in great public works. This application of compressed air was used by Brunel, with great success, in constructing the foundation of one of the piers of the Albert Bridge which spans the River Tamar; and it has also been adopted for making the foundations for certain celebrated bridges, especially at Rochester, and more recently at Kehl. Without this ingenious method the construction of a great bridge over the Rhine would have been an impossibility, with its rapid current and shifting bottom; neither could the railway have been carried across the river, uniting France and Germany by one of those bonds of union which do more towards promoting friendly intercourse between nations than all the treaties of peace or of commerce. M. Triger has taken out no patent for his invention; on the other hand, the engineers who have adopted his process for building the bridge at Kehl have notably improved upon it.

The colliery shaft having been sunk and supported, that is to say, timbered or walled, is then often divided into compartments. Sometimes these have a large area given to them, as much as 16·4 feet (5 mètres) in diameter when they are circular. With so large a section the division into compartments is easy, and they then resemble as many distinct or separate shafts. One of the compartments will serve for the passage of the tubs* or cages in which the coal is raised, another for fixing the pumps required to draw off the water; and sometimes, where ladders are used for the miners to go up and down by, a compartment is specially set apart for fixing these. One division

---

* These are strong tubs made of thick wooden staves bound together with stout iron-hoops. Open at the top, they are attached to the drawing-rope by means of chains (fig. 32).

Fig. 32.—Pump-men of Creuzot, in leather dresses, descending the Shaft to examine the Pump.
After a photograph by M. Larcher.

of the shaft serves in all cases as an air-way ; whether the draught be free or whether it be forced, it constitutes the natural chimney through which all the ventilation of the mine is effected. The system of sinking only one shaft and thus dividing it, is happily dying out in this country, and most large collieries have now two or more shafts sunk to the workings.

The most approved arrangements of shafts for a large colliery yielding explosive gas, and where water has to be pumped, is to sink a shaft for pumping, another for raising coals, and a third for ventilation or upcast ; at the bottom of which a large furnace is kept burning. It is in consequence of all the various purposes which single shafts have to serve at the same time, that their dimensions have become so much enlarged. Nevertheless, it is preferable sometimes to adopt smaller sizes and to couple the shafts, as is done at the coal-mines of Blanzy, in the Department of the Saône-et-Loire ; while others prefer them more numerous and scattered, rather than to have them connected with each other.

Little by little, in the continental collieries, the number of shafts increases with the requirements of the works, and then they are named according to the special purposes to which they are applied. These are *winding-pits*, up which the coal is drawn ; *pumping-pits*, in which are fitted the pumps or other hydraulic machines (fig. 32) ; *ladder-shafts*, for the passage of the men up and down ; *air-shafts*, through which the air enters or leaves the mine ; and lastly, *trial-shafts*, for the discovery of the coal. In England they are called *shafts* or *pits ;* in France *puits ;* in Belgium and the north of France a coal-pit is known by the generic name *fosse* or *bure.*

There is also another class of pits, to which it is merely necessary to allude ; these are the pits that have been abandoned—old servants which have had their day, but are no longer wanted. They still exist with dilapidated sides, with buildings about them reduced to ruins, and with machinery dismantled and rusted. Sometimes the air from the workings still circulates to the bottom of the pits; and heated by the breath of the men, the combustion of the lamps and candles, or charged with steam and mephitic gases, it leaves the mine by these openings. Brambles and thorns grow round its mouth and in the pit itself, which thus becomes a dangerous precipice partially concealed by vegetation; except where the shaft has fallen in, and its ruin is the result of neglect. Coal-mining always spreads a certain amount of sterility over the country around; and the desolation of the land, increased by the ruins near the abandoned shafts, ever presents a melancholy picture.

The general appellations of winding-pits, pumping-pits, &c., only denote the special purposes to which the shafts are applied; they would not be sufficiently distinctive in the management of a colliery, and the pits receive, in addition, particular names. The name, in France and Belgium, of the saint inscribed in the calendar on the day when the sinking was begun, when the pick began for the first time to dig in the bowels of the earth, is that generally preferred, and it has the further advantage of fixing a date. The work is, also, willingly placed under the protection of Saint Barbe, the patroness of miners. Sometimes the name of a distinguished member of the management of the colliery is chosen, or else that of a lady more or less interested in the undertaking: a little gallantry does no harm. On the other hand, the dry and severe order of numbers is not despised, and these pits Nos. 1, 2, 3, ... call to mind the mode in which the Americans distinguish their streets. Finally, there are appellations borrowed from the actual names of localities, and others which consecrate the oft-realized illusions of the miners, such as Hope Shaft, Fortune Shaft, Good Success Pit, &c.

In many instances, in Catholic countries, the shaft receives an actual baptism, and the services of the church blend happily with the grand labours of industry. The shaft having been sunk a few yards in depth, is crowned with garlands, and the priest invokes all the blessings of Heaven on this mine, which will be the means of giving work to so many hands.

A steam-engine is almost always erected at the surface, to wind the coal from the pits; and for the last fifty years all possible improvements have been introduced to facilitate the process. A special steam-engine usually pumps out the water, unless the tubs in which the coal is raised by day are employed for drawing the water at night. These pumps are of gigantic dimensions, and persons who have never seen them can scarcely form a conception of their size. They originated in England, where they are known as Cornish or Newcastle engines, after the name of the part of the country where they were first used: in Cornwall, at the copper and tin mines; at Newcastle, at the collieries. Watt applied his ingenuity to devise a steam-engine to supersede the immature fire-engine bequeathed to him by his predecessor Savery, and the atmospheric machine of Newcomen.

The pumping-engine has remained in its principal features as it was contrived by Watt, and it may be said to have been created by that great engineer by a single effort. Imagine an immense cylinder from 80 to 100 inches in diameter, and many yards high, in which the piston is moved by the steam (employed expansively) with a calculated speed, and makes, at most, six or eight strokes a minute. The piston-rod is attached to one of the ends of the enormous main-beam, while at the other end it is connected with the pump-rod, formed of heavy beams bolted together (fig. 32), descending by its own weight, and raised by the engine. A long line of pipes is fixed to the wall of the shaft; these vary much in diameter, and sometimes reach a depth of from $437\frac{1}{2}$ to 547 yards (400 to 500 mètres), equal to six to eight times the height of the

Monument, or ten to twelve times the height of the Column
Vendôme in Paris.

When the water is not discharged at an intermediate level,
the pump raises its contents, by a series of lifts, to the mouth
of the shaft. "Where are you, Academicians of Florence, and
you, Galileo, who reply to the fountain-makers of the Medici,
that water cannot rise more than 32 feet, because nature only
abhors a vacuum up to that height?"

The only defect of Watt's engine is that of occupying so
much room. The Belgian engineers, with a view to economy
of space, have connected the main-rod of the pumps directly
with the piston-rod, working through the bottom of the steam
cylinder, and have placed the cylinder itself immediately over
the centre of the shaft. (This was one of the earliest forms of
steam-pump employed in the mines of the United Kingdom.
Many of those direct-acting engines may be found in ruins in
this country, few, if any, being now in work.) In both cases
the engines have sometimes to raise from the mines actual
rivers of water. The only limit to their work is their capabilities. Some beds of coal extend under the bed of the sea, as
at Whitehaven; and infiltrations of the ocean waters are then
added to the drainage from the land. Some engines are of as
much as eight hundred horse-power. As for the consumption
of coal, it is more than moderate, often scarcely amounting to
three lbs. per hour per horse-power; and those engines are the
best which consume the least fuel. This would naturally cause
astonishment were it not known that these pumps are not only
used in the collieries, but in the metalliferous districts. In
Cornwall, where coal is very dear, the shareholders conceived
the happy idea of publishing monthly reports of the duty or
work done by their engines, the unit of which is the number
of lbs. of water raised one foot high by a bushel of coal. These
reports give the name of the engineer, and the total consumption of coal. It may easily be imagined what emulation such

a system of publicity produces among mining engineers when thoroughly carried out; but at present not one-fourth of the engines at work in Cornwall and Devonshire are to be found in the *Engine Reporter*.

The shafts form part of those excavations which are generally characterized by English and American miners as narrow or *dead* work, because they are not remunerative except indirectly. Amongst the unremunerative works must be included the galleries, drifts, levels, or adits, which are worked into the hill side—as the miners phrase it, "from day." These are usually worked at the same time as the shafts, or rather in place and instead of the latter, for proving the coal, ventilating the mine, the passage of the men, the free drainage of the water, &c. Sometimes the drifts or headings are driven in the seam of coal, and have the same inclination with it, the same thill or floor, and the same roof, in which case they are called *inclines*. When the slope exceeds 45°, or more than half a right angle, the level is called an *up-hill drift:* with such an inclination the works can only be carried on by means of engines or by ladders. When carried horizontally, or with a moderate amount of slope, the drift is called a *level;* but if carried at right angles, or transversely to the seam, it is called a *cross-heading, rise-drift, jinny-road,* or *half-rise road.*

The same custom is followed in naming the levels, as the shafts. The saints of the calendar, numbers, the names of men and places, are taken into contribution. The list, it is clear, is inexhaustible, and there is plenty of choice.

The same difficulties met with in sinking the shafts are encountered in driving the levels, and are even, sometimes, more threatening; for advantage cannot usually be taken, in these latter works, of circular or elliptical forms to resist the pressure of the ground equally and on all sides. Besides, the rocks forming the roof in this case bear with all their weight, and when the shales traversed swell and give way, the floor and roof of the level may have a tendency to unite by one of those movements technically called *crushes* or *creeps* in this country, and in Scotland *sits*. The creep is a protrusion or swelling up of the floor of the mine where the floor is soft and the coal tender, and is produced by the pressure of the roof upon the pillars of coal which are left to support it, and which causes them to sink into the floor. Such an occurrence takes place in many collieries, and there are instances of levels in which a man could stand upright, not being passable eight days afterwards, except on the hands and knees.

The linings adopted in the shafts, such as brick or stone

Figs. 33, 34.—Views of walled levels.

walls, timbering, &c., are likewise employed in the levels; and the means of strengthening them in order to traverse the disturbed, soft, moving, or watery measures are in both cases nearly the same.

When a gate-road or horse-level is intended to last a long time, and the measures it is driven in are insecure, it is arched with masonry, instead of being supported by timber, and the same arrangements are made as in the shafts. Calculations have been made showing that in all cases where a work ought to last longer than from eight to ten years, it should be walled, and not timbered.

The form usually adopted for walling levels is that of a semicircular arch with upright walls, "pack-walls" (fig. 33); or elliptical, when the upper arch sustains the pressure of the measures, and the lower or inverted arch carries off the water (fig. 34). Over the channel a planking is placed for the passage of men and trams or waggons, fitted with the requirements of a railway.

The walling of levels is effected with centerings as in the case of bridges and arches, and offers nothing specially worthy of notice. It is necessary to use good materials, such as dressed stone or brick, and to employ hydraulic mortar or Roman cement.

Figs. 35, 36.—Sectional views of timbered levels.

For timbering the levels, sets of three in a trapezohedral shape are placed at intervals, and between them planks or round pieces (figs. 35, 36, and 37). A complete set consists of four pieces: the cap or *head-piece*, the two *uprights*, *legs*, or *stanchions*, and the *sleeper* or *sill* (*semelle*). Generally there are but three pieces, the latter only being used in cases where the floor is very soft, or when it consists of shales, which are apt to swell up (fig. 36).

The weight from above and the lateral pressure of the measures do not fail to destroy some of the timbers, which first become bent and curved, and then break readily enough.

It is necessary to replace them without loss of time, in order to avoid the danger of a falling in.

The timbers are seldom used squared, but are preferred in the state in which they leave the forest, all that is done being to remove the bark and to cut them into the necessary lengths. The joints are simply notches cut at the ends through half the thickness of the timber. The axe should be the tool used by the mine-carpenter (fig. 38), for wood cut with the saw rots too rapidly, the latter tool cutting through the fibres, which are merely separated or divided from each other laterally, with a cleaner cut, by means of the axe.

Fig. 37.—Longitudinal view of a timbered level.

Larch and oak are commonly used for timbering mines in England; in France fir and chestnut are found to be the most convenient wood for the purpose; in the Alps the beech is employed, and the oak or pine in Italy. It is customary in some places to char the wood to prevent rot and fermentation.

In a hot and damp atmosphere, like that of most coal-mines, the timbering of the levels soon becomes covered with fungi and a vegetation consisting sometimes of light and cottony filaments, mostly of a beautiful snow-white; and occasionally of substances like tawed leather, and of a yellowish colour. The rotting of the timber, also, gives out a particular odour like that of creosote, which is well known to persons conversant with mines, and is by no means a disagreeable smell. Particular insects—moths, flies, and gnats—collect near the parasitic vegetation, some of which may probably be new species, not hitherto noticed by naturalists, and to which, therefore, it is fit that their attention should be directed. As much may be said to botanists with regard to the vegetable productions, because the appearance of species previously unknown having been shown to exist in grottoes and caverns, there is much greater reason to suppose that the same may prove to be the case with reference to mines, since in the latter, as in a world apart, exceptional conditions prevail. It is said that the various forms of life have a development dependent upon the medium or conditions in which they are placed; and what can be stranger than those of mines, where all the normal conditions of temperature, moisture, pressure, and even composition of the air, become so thoroughly modified?

Fig. 38.—Timberer's axes. Scale $\frac{1}{10}$.

The rats, which steal in everywhere, find their way also into these underground places. They frequent the spots where the miners take their meals; and when the latter eat their food above ground, these gnawing animals, for lack of crusts of bread, nibble the candle-ends or the remains of wicks, and even wood. They are met with everywhere, scampering along the levels, or timidly hiding beneath the stays. A colliery explosion occurred in South Wales, at the Mountain Pit, near Aberdare, when one man lost his life and four others were badly injured; the cause of the explosion being a very singular one. In the pit is an old working well known to be full of gas, and the men were warned not to go beyond the danger signals. The colliery is infested with rats, and in the pursuit of one a workman incautiously ran beyond the prescribed limits with a naked light in his hand, and the result was an explosion. Another sort of animal, the bat, finds a congenial home in abandoned workings, which are warm, dark, and quiet; and the *Vespertilio* of the coal-mines, in the full enjoyment of liberty, is the forerunner of the formation of a particular mineral which has been named *guanite*.

The dimensions adopted for the main levels of mines are less than those of shafts, and the largest scarcely exceed, except in special cases, six feet in height and ten feet in width (commonly from five to ten feet), the latter width being only adopted for main roads or rolley-ways, where there is much traffic. There are also levels, such as those for drainage, for which very large dimensions are sometimes necessary. Some of these enormous *drains*, or water-gates, are of considerable length, from three miles to four or even five, like the longest railway tunnels, and discharge "to day" a regular river. In certain English mines these streams of subterranean water have been utilized, and actually converted into canals for the transport of the coal itself.

At the mines of Rocher-Bleu, in the Bouches-du-Rhône, where workings for coal have been carried on for more than a century, there is a water-way above $1\frac{3}{4}$ miles (3 kilomètres) in length, and wide enough to carry a boat, which is fed by cross-levels, like a river by its tributaries. The peculiar nature of the soil, which is fissured and porous, drains away all the water from the surface into its empty cavities; and the streams of the country are exhausted. To compensate for this the water-way redelivers the volume of a river. After the violent storms of which this part of Provence is sometimes the scene, the section of the level is not sufficiently large to carry off all the water which is poured out of its mouth up to the very top of the arch, and is discharged into a large stream, of which it constitutes nearly the entire source. In 1853, when making some delicate experiments at these mines, M. Simonin gauged the volume of water passing through the water-gate after heavy rains several times, and found it to amount to two hundred gallons in each second.

When the shafts and levels have been made and properly secured, and the machinery erected at the surface, the coal is open, and ready for working away. We can now calculate the cost of a colliery. The total sum sunk must be estimated by scores of thousands of pounds. In the first place there are the borings, always a very expensive item, costing from £6 to £8 a yard, which we may suppose to be completed. But there are also shafts of large diameters, the sinking of which costs from £40 to £80, and even more, per yard, and which may have a depth of from 547 to 656 yards * (500 to 600 mètres), that is to say, from four to five times the height of St. Paul's Cathedral, or from eight to ten times the height of the towers of Notre Dame, in Paris. There are levels the driving of which exceeds £20 per yard, and whose length extends to from three to more than four miles. The thousands of pounds spent in these works are completely and for ever sunk. A sinking fund only, that is to say, a sum annually set aside out of profits, will allow of their recovery.

The machinery erected for the different services will come to £40 and more per horse-power; the power of some of the winding and drawing engines is as much as from six hundred to eight hundred horse-power. In Moselle and the Pas-de-Calais, shafts fitted with their engines and all the necessary machinery have cost as much as £80,000. And the profits they are supposed to make will be denied at certain collieries.

---

* The depth of the Monkwearmouth Coal-mine, in North Durham, is 1590 feet, and the temperature ranges from 78° to 80°, and in some parts of the mine even to 89°. The deepest coal-mine in Great Britain, that of Dukinfield, near Manchester, is 2151 feet in depth, and the temperature at the bottom is constantly 75° Fahr.

But with railways no attempt is made to conceal their earnings, however much they may be subsidized by the government; and when certain privileged mines are instanced, no
account is taken of those which have failed, especially with
us, and where millions have been utterly and entirely lost.

A larger calculation is necessary with respect to mineral
industry, that field ever open to the activity of man—to
interest, if it be preferred—and the love of gain, which is often
the main-spring of our actions: the working of mines has
absorbed in our days as much capital as Law's scheme for
colonizing the Mississippi during the period of the Regency.
The entire sum sunk in the coal-mines of France now in work
exceeds 300,000,000 of francs, or £12,000,000 sterling.

On the supposition that every ton of coal got yields a profit
of $2\frac{1}{2}$ francs, or 2s. 1d. (which is nearly the average profit
made in the French collieries), and assuming the entire production to be at the present moment 12,000,000 tons, we
thus arrive at a total profit, for all the collieries of France,
of 30,000,000 (£1,200,000), or ten per cent. on the capital
sunk.   In England and Belgium the calculation furnishes
similar results, but it must be remembered that above
100,000,000 tons of coal are raised in the British Isles.   This
revenue is small, in comparison with all the risks incurred,
and the high rate of interest generally paid for money at the
present day.   Is it not fair that some of the capital employed
in the most precarious of industrial pursuits should yield large
returns, when so much has been sacrificed and irretrievably
lost on worthless or unremunerative undertakings?   When a
coal-mine yields good dividends, it ought to prove a source of
congratulation to think that the favourable results have been
most frequently deserved, and purchased at the cost of prolonged and wearisome efforts, of many years of patience and
perseverance, and by the expenditure of large sums of money.

In certain cases, when success has been attained, it is complete—exceeding even the most sanguine hopes. The mines of the Department of the Nord furnish many instances of this ; as where a certain share which when issued was worth 25,000 francs (£1000), is now worth 700,000 francs (£28,000). There is an instance of a mine, the value of a share in which has risen from 16,000 francs to 70,000 francs (£640 to £2800) ; in another case the value has increased ten-fold, rising from 1000 francs (£40), to 10,000 francs (£400). If such facts did not exist, it would be a cause for regret. It is by the attraction of such large gains that speculators are tempted to join in such undertakings, to risk all their capital, and arrive at the discovery of the coal. Without such allurements few persons would have the courage to explore the ground for coal beyond the limits where it is already known to be ; and all the great works which in France have led to the discovery of the coal-basins of Valenciennes, the Moselle, and the Pas-de-Calais, would never have been undertaken. These remarks are equally applicable, or rather more so, to coal-mining in this country, where large fortunes have been expended—as at Monkwearmouth, and Shire Oaks colliery—before the adventurer was rewarded by reaching the coal.

CHAPTER VII.

HOW THE COAL IS WORKED.

Coal-mines of Turkey and China.—Mode of working thin seams of Coal in England and France ; Holing, or *Travail à col tordu.*—Mode of conveying the Coal to the surface formerly practised in France by Mendits, and by Putters and Bearers in England and Scotland.—Employment of women and children in Mines.—Trappers. —Old system of working Coal in French mines by Falls.—Improved systems of working by Remblais, or Long-work.—Miner's tools.—Transport of Coal under-ground.—Horses in Coal-mines.—The Waggons used underground in Belgium, Somerset, and South Staffordshire.—Modes of descending and ascending the Shafts.—Day and night crews.—Names and duties of men employed at Coal-mines, underground and at the surface.—Novel sensations experienced on first going down a Coal-mine.—Underground workings described.—Exaggerations of Authors. —Salt-mines of Wielliczka and Bochnia.—Mode of making Underground Survey of a Coal-mine.—Magnetic compass and chain.—Anecdote of an old Coal-viewer.— The great importance of correct plans of workings.—Animated aspect of under-ground operations ; in the drawing-shafts, and around the pit-mouth.—Sorting and washing Coal at the surface.

THE methods in use for getting the coal are not less ingenious nor less perfect than those which have been already described for the execution of the preliminary works, involving much ingenuity. The advance in improvement is certain when the necessity for economizing fuel has been recognized, when its value has been learned, and its universal application for all industrial purposes admitted. Formerly the modes of working were rude, as are those still in operation in Turkey or China.

It is stated that at the Turkish mines of Heraclea (in Asia Minor) the works present a confused labyrinth of galleries crossing each other in all directions, and driven in such a way that not merely is drainage rendered impossible, but nearly all the surface-water is allowed to accumulate in them.

The Sultan, in the exercise of his absolute power, granted the privilege of working these mines. Then, by an ingenious combination, and carrying to an extreme the resemblance to the lion in the fable, he divided the collieries into three shares, two of which he reserved for himself, selling the third for a good round sum to pachas with many tails.   These last have caused the works to be carried on, paying the men very irregularly, who in their turn have repaid the compliment by working irregularly also.

Turkey likewise possesses coal-fields in the Libanus, which are also worked in the same unsystematic manner.   It is said that a Caimacan, or Turkish prefect, who was charged with the superintendence of the mine, when some pillars had been left in the levels to support the roof, ordered the men to take away "those great blocks of coal," which he supposed to have been forgotten.   The result was that the ground above fell in, and crushed the Caimacan himself.*

The coal-fields of China are no better worked than those of Turkey, and the son of the Celestial Empire is in that respect a worthy rival of the descendant of Mahomet.   The miners confine themselves to surface-workings, which are abandoned as soon as the water reaches them; otherwise the men are compelled to retreat before the falling in of the ground and explosions of fire-damp.   As there is not a school of mines at Pekin, and as the Chinese have not followed (at least in practical geology) the methods of those whom they style *outer barbarians*, matters could hardly be expected to go better.

It is scarcely necessary to go back a century to find in European coal-mines the same disorder that now prevails in those of China and Turkey.   The nature of certain seams obliges the miner, even of the present day, to adopt what may now seem to be rude and primitive expedients.   In working thin seams of coal in England and France, the miner is obliged to assume a very constrained position, especially if it be an

---

* This is no fiction.   See the remarkable article on Coal (*Houille*) by M. Lamé-Fleury (" Dictionnaire du Commerce "), and the work of Mgr. Mislin, " Les Saints Lieux."

object not to remove the barren measures; and the mode of working recalls to mind that of the ancients as regards metalliferous deposits, when slaves and prisoners were condemned to the mines. The collier, lying on his side, with neck bent, makes a horizontal cut with his pick at the bottom of the seam (fig. 39); then, crouched on his knees, he makes two vertical cuts at the sides, which is called *shearing;* and lastly, the coal thus partially freed is broken down by means of a wedge or a lever. If the coal is very hard it is blasted with gunpowder. It is in consequence of the peculiar attitude which the miner is obliged to assume in *holing* the coal, as this

Fig. 39.—Holing Coal.

mode of working is called, that it has received in France the well-chosen name of *travail à col tordu.* The following was formerly the usual mode of bringing the coal to surface practised in that country :—

When the coal was cut away, the *putter* (*porteur* or *traineur*) or *bearer* came and conveyed it along the face of the workings to the better roads, either in bags, which he carried on his back, or on small sledges (*sleds*) or trams, the rope of which passed between his legs and was fastened to his waist. Half naked, doubled up, and leaning on a staff, the putter went panting along, doing this work of a slave, and often climbed to the surface, with the sack on his back, up an inclined road.

Then he went down again to repeat the same task afresh, like a new Sisyphus, and so on from morning to night.

This system was in force in the coal-mines of the Bouches-du-Rhône, near Aix, ten years ago, having been first adopted towards the middle of the eighteenth century, when the mines in question were originally worked for the use of the soap-works at Marseilles. The fathers taught their children, for youth and vigour were required for such labours as these. The tram, which was very low and of a triangular shape, ran on three little wheels, one being placed at each angle. On the frame, consisting of some rough planks, rush baskets (*couffins**) full of coal were placed. The carriage itself, called the *courriau*,† was drawn by a boy to the main level, where the baskets were emptied into waggons (fig. 40). When these mines were visited by M. Simonin, the bearers sometimes carried their load along inclined galleries to the surface; but afterwards various improvements were introduced into these collieries, and perhaps the *courriau* itself may now have disappeared altogether.

These putters and bearers are called *mendits* in Provence—the mendit, properly speaking, being in the language of the troubadours the servant of the shepherd—and the expression, in its more extended sense, means the assistant or labourer of the miner. These children perform their task cheerfully and quickly; so true it is that use reconciles man to the severest toils.

The bearers at Saint-Etienne formerly performed as hard a task. With bare feet, and supported by a staff, they were obliged, in a day's work, to carry on their backs a certain number of loads (*faix*) up the inclined gallery communicating with the surface, supporting their burthen on the staff while stopping to rest themselves. The roads were slippery and very rough, and the employment was extremely dangerous.

---

* This word, borrowed from the provincial dialect, shows a Greek origin, κοφινος, 'basket.'
† A corruption of the word *chariot.*

In some English and Scotch mines, in which the seams of coal are as thin as those of the Provence collieries, boys, called *putters*, perform a task in some respects similar to that of the *mendits*, drawing four-wheeled trams along the low and narrow

Fig. 40.—The *Mendits*, or Putters of coal in Provence.

levels, which are sometimes not more than a yard high. With belts round their waists they couple themselves by a chain to the tram, and draw it along, occasionally on their hands and feet, over the uneven and often muddy roads (fig. 41). In the higher places they change their position, and push the tram from behind with the help of their head and arms. Like the mendits, they acquire great skill in this work, and draw with ease an average weight of three hundred lbs. of coal at each journey; but it is not a pleasant sight to witness.* Formerly, in some of our collieries, women performed this laborious task, while their companions on the surface turned the windlass by which the coal was raised.

In some few of the Scotch collieries, indeed, as late as 1843, girls were employed in the same duties as the mendits of Provence; but many years have passed away since in any of the mines or collieries of the United Kingdom women—young or old—have been allowed to work underground. Referring to

Fig. 41.—Putters or Trolley-boys in England formerly.

the Parliamentary Report of Mr. Tremenheere on the employment of women and children, a full description will be found in it of the practice thus referred to by M. Simonin as having once prevailed in this country. Happily this degrading labour is now a thing of the past. Girls used to carry on their backs

* In one of these Scotch mines a laughable circumstance happened to Bruce, the celebrated traveller, to whom the mine belonged. On his return from his African expedition he thought he had reason to complain of his manager, and wished to settle the difference on the spot. The descent into the mine was by ladders, at the foot of which low and tortuous galleries extended, which Bruce, then old and stout, could scarcely traverse on all-fours. Every moment he struck his head without being able to help it. At length they came to such a narrow passage that Bruce stuck midway, and could not get out again. What was to be done? Miners were sent for to dig away the ground; but time was required to effect this, and he was kept in the mine for several hours before he could be released The dinner-time came, and at last Bruce was extricated from his unfortunate situation; when he declared that in none of his travels in Abyssinia, or to the sources of the Nile, had he ever met with such an adventure.

a basket, fastened to a leather strap which passed round their forehead. To this strap their lamp was also attached. Equipped in this manner they bore the coal. The load in the basket was often balanced by large lumps which hung around their necks. These girls advanced in parties with their burthens, and climbing up ladders the whole length of the shafts, which sometimes exceed a hundred yards in depth (fig. 42). If a strap broke, or a block of coal fell, the bearers who followed were sometimes grievously hurt, and they have been, though rarely, even killed on the spot. This primitive mode of raising was very barbarous, and was abolished at the earnest recommendation of the commission for inquiring into the employment of women and children in mines. Colliery owners had allowed matters to become so seriously bad that public opinion was roused, and the government, which, especially in England, is averse to any interference

Fig. 42.—Bearers of coal in Scotland.

with private concerns, considered it absolutely necessary then to interfere, and by legislative enactment (5 & 6 Vict.

c. 99) the disgrace was swept away for ever in the year 1843.

In working thick seams of coal in France a mode not less barbarous than those already described has long prevailed; but in this case it is in the working or extraction of the coal, and not in its conveyance underground. In the mines of Commentry, Creuzot, Epinac, Blanzy, &c., it has been especially practised, even within the last few years. It is called working by "falls" (*par éboulement*), a term which sufficiently indicates its nature. The miners, armed with long-handled picks, caused the fall of large masses of coal overhead, at the risk of being crushed. As it was necessary to support the immense empty spaces which were thus produced, whole masses were abandoned in the mine to serve as props or pillars, and two-thirds of the coal remained entirely wasted. As may be conceived, the whole resulted in complete confusion. It was necessary, finally, to escape from the falls, which frequently took place on a large scale, and a considerable quantity of the fallen coal was also abandoned in the mine. Another great disadvantage of this mode of work was that ignition was frequently produced in consequence of the decomposition of the iron pyrites amongst the wastes of small damp coal, by a well-known chemical change developing a large amount of heat. Lastly, immense cracks and enormous subsidences extended upwards to the surface, so that the onward progress of the excavations below could be followed by the movement of the ground above. Buildings cracked and fell in; water found its way into the interior of the mine, and all in consequence of the empty spaces which no precautions had been taken to fill up. Such conditions as those can be seen, in this country, only amongst the ill-worked collieries of the thick seam of South Staffordshire. The more recent workings in this ten-yard coal are systematically conducted; but over a large portion of the South Staffordshire district immense quantities of coal are irreclaimably lost, through the system of wanton carelessness which formerly prevailed. This mode of working bore the expressive name, in most French coal-mines, of *foudroyage*, which explains its effects very well.

Nowadays the necessity of producing coal at the lowest possible price, in order to counteract opposition; the necessity of leaving nothing in the bords or stalls, on account of an exhaustion that has already begun to make itself felt; and lastly, the advantages which have everywhere made themselves felt of well-planned and well-managed works, without which there is neither safety nor profit—all these reasons have led coal-proprietors in France to adopt the working by *remblais*. This mode, in almost universal use, consists in taking out all the coal and filling up ("packing" or "stowing") the spaces with rubbish (*gobs*), out of which the useful mineral has been removed. One of the greatest cares of miners is, in future, not to leave in the mine a single atom of coal.

The improved working of coal-mines presents a certain sort of resemblance to the cultivation of fields or forests. Besides, are not mining operations, as it were, a kind of cultivation of the soil? Thus, the Italians do not say, like the English, the *working*, or like the French, the *exploitation*, but the *cultivation* of mines. The first thing to be done for a field or a wood, from which it is sought to derive the greatest possible amount of profit, is the management of the crops and the felling of the timber, to maintain the roads, and to organize transport, either by means of horses or engines. So it is with a coal-mine. No more (*foudroyage*) square-work or (*éboulement*) falls; no more narrow, winding, and badly-kept roads; no more conveyance on the backs of men: but a methodical setting off of stalls, or a regular system of working away the mineral by long-work, and a rapid conveyance along good roads on underground railways, either by horses or by engines. With all these improvements the cost of winning and raising the coal has been diminished by one-half, and the lives of the men have been less exposed to risks.

Let us turn our attention to this management of collieries, which is well worth the trouble bestowed upon it. It is far more interesting than that of forests and coppices, for the coal when removed does not grow again. In the first place, let us examine the various cases which may occur with regard to the modes of winning or cutting the coal, according as the seam is more or less inclined, and has a greater or a lesser thickness.

If the seam be slightly inclined, levels are driven in it, cutting each other at right angles, leaving pillars between them, by which the mine is divided like a chess-board. These pillars are worked at one of the angles, the roof being artificially supported by rows of props. If the seam be very thick the coal is removed in successive slices, the men standing on the coal already cut, as is practised in getting the Dudley thick, or ten-yard, coal in South Staffordshire.

Sometimes, instead of setting off or dividing the coal regularly, it is considered best (particularly when the seams are thin and slightly inclined) to open a great length of face, the waste of the workings being thrown behind in the *goave*, *gob*, or *waste*. A longitudinal road is sometimes carried through the goave or gob for carrying away the coal, and galleries, or cross-headings, are driven for the circulation of air, the passage of the men, removal of the coal, &c. In many instances, also, the whole of the waste or space behind the men is not filled up, and the pressure of the overlying rock is supported by means of timber props, placed at regular distances apart.

When the seam has a high inclination and is thin, other methods are adopted. Standing on scaffolding or the refuse or coal already cut, the miner gets the mineral by cutting it in steps, which, being always free on two faces, greatly facilitate the work. The blocks intended to be worked in this manner are divided from each other by pairs of horizontal galleries or levels (*étages* or *niveaux*), and by up-hill drifts or *air-heads* (called *cheminées* or *remontées*), carried up the rise of the seam. From these cross-headings the coal usually descends into the main levels, from which it is carried to the drawing shafts.

Lastly, when the seam is highly inclined and of considerable thickness, a longitudinal gallery or drift-way is cut in one of its walls—at the roof for instance—and the coal is worked from the roof back to the wall, by great cuts which are afterwards gobbed. When a face is worked away the men stand on the coal already cut. A drift-way is opened to send down the coal into the main level, and the work advances by a succession of slices.

Fig. 43.—Pick:—St. Chamond.

It is not intended to dwell at any greater length in this place on the various systems of working, it being sufficient for our purpose that the general principle should be well understood. Besides, each mine may be said to present a particular case; and it is especially in the working stall, or *chantier d'abutage*, and by the mode of occurrence of the seam, that a colliery engineer calculates the plan that should be adopted and the modifications that should be made in the theoretical mode of procedure.

The tool specially used by the collier is the pick (*pike, slitter,* or *mandril*), which is of various forms, according as it is intended for working in hard rock, or for cutting coal; it may also have one or two points (figs. 43 to 48).* In the mines of the north of France and in Belgium picks of a peculiar shape

* All the mining tools represented in figs. 43–55 are drawn to a scale of $\frac{1}{10}$ of their actual dimensions.

are used, amongst others a kind of double-pointed pick, or
rivelaine (fig. 49). This tool is employed for making a horizontal under-cut, and the seam, being thus partially freed by
holing, the coal is broken down by means of wedges or bars
driven into the mass (figs. 51 to 54). When the coal is very
hard blasting with gunpowder is used, an operation which will
be described further on.

The removal of the coal is effected underground through
the drift-ways, which are sometimes merely passages (*couloirs*)

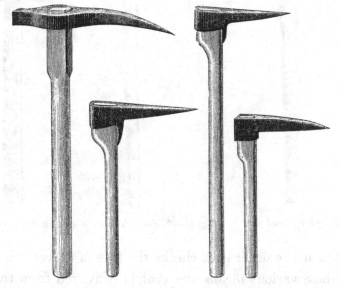

Figs. 44, 45, 46, 47.—Various forms of Picks.

furnished at the bottom with a trap-door, before which the
trams or waggons are brought to be loaded by hand or with a
shovel (fig. 55).

When the size of the drift-way is tolerably large, and the
slope does not exceed 45°, the drift-ways are turned into
inclined planes called *jinny-roads* or *jig-brows*, where the loaded
trams run down by their own weight on the rails which are
laid upon the floor of the incline. The movement is brought
about by the force of gravity; and the descending trains pull
up the empty trams by means of a rope passing over a large

drum or sheave (the old system of *montagnes russes*, or Russian mountains), and regulated by a brake, which, pressing upon

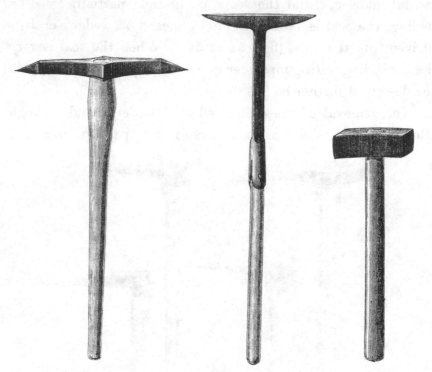

Fig. 48.—Double-pointed pick.   Fig. 49.—Belgian pick or *rivelaine*.   Fig. 50.—Hammer.

the drum at the other end, checks the rate of descent.

By these various means the coal is conveyed from the face of the workings along the main road-ways. These last, driven in the direction of the seam, are furnished with railways. Horses or men (trolley-boys) convey away the trams to the winding shaft, where the coal is capsized into the

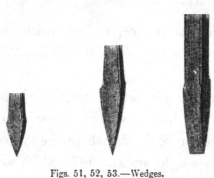

Figs. 51, 52, 53.—Wedges.

buckets which hang from the end of the rope, and it is drawn up the shaft. A more economical mode, and one attended with less injury to the mineral, consists in attaching directly to the rope, or in running into a cage which it carries, the trams which come from the face of the works loaded, while the empty ones detached from the cage or rope go back in their turn to be filled with coal.

The horses which draw the waggons on the underground railways are sometimes sent down into the mine fastened to a rope, but generally in the English collieries on a properly constructed platform and cage,

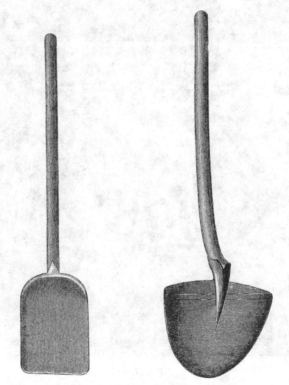

Fig. 54.—Bar.   Fig. 55.—Shovels.

either in nets or baskets (fig. 56). When the former mode is adopted, the horses do not make the slightest movement, being paralyzed with fear, and to all appearance dead, but when they reach the bottom of the pit they gradually recover their senses. These intelligent animals get used to their new mode of life in a very little while, and soon become

acquainted with all the passages, curves, and dangerous
places. It is strange to see how carefully they avoid
running against the trains at the sidings and stopping-
places, and how they know the right distance from the air-
ways at which they ought to stop, so as to allow the driver, or
the boy in charge of the *stoppings*, proper room for opening and
shutting the doors. The horses are cared for like useful ser-

Fig. 57.—Stable in a mine.

vants: the stable is large, well ventilated (fig. 57), and the
litter often renewed. Hay and oats of excellent quality are
supplied at feeding time; the horses become fat and plump,
their coat grows long and glossy, and they seem to prefer
living in the warm and equable atmosphere of the mine to the
great roads or fields in sunshine, wind, rain, or frost. When
they have once entered the mine, in general they never leave

Fig. 56.—Descent of a Horse down a Mine-shaft at Creuzot.

After a photograph by M. Larcher.

Fig. 58.—Working Coal in the thick coal of South Staffordshire.

it, but work there for years, and often end their life in that useful servitude. In some collieries the horses are regularly brought to the surface, and they may be said to be a part of the plant or stock of the colliery, and are entered in a special list with names that might make race-horses jealous.

In some collieries the trains of trams are not only conveyed by horses, but by locomotive engines, or they are drawn by stationary engines—especially when working inclined planes— though the internal traffic of a mine is generally carried on by means of men or animals.

The waggons, carts, or trams used for carrying the coal underground, are usually made of wood or sheet-iron, and run on four wheels. They resemble in shape, but not in size (being on a much smaller scale), the contractor's waggons which are employed in railway constructions. Sometimes, instead of small carriages or trams, tubs or buckets borne on a wheeled framework are made use of. In some levels with a slippery floor, the coal is conveyed by dragging in sleds or sledges. In fine, the carriages differ very little, the form being that of a small waggon, or of a round or oval cask, open at the top, by whatever names they may be called : waggons, trams, berlines, tubs, skips, &c. The capacity varies from 8 to 14 bushels (3 to 5 hectolitres), that is to say, from 5 to 8 cwts. of coal (240 to 400 kilogrammes). In Somersetshire, and in Belgium, great iron buckets called *hudges* are used (*cuffats*), nine feet high, and holding about $1\frac{1}{4}$ to $1\frac{1}{2}$ tons, or 40 to 55 bushels (15 to 20 hectolitres). In the central districts of England, especially South Staffordshire, the coal is raised by stacking it on a skip, or tray, in large blocks which are kept together by broad iron rings ; the rings diminishing in diameter from below upwards (fig. 58). The skip is fastened to the winding rope. This system is both simple and original, but it is only applicable to exceptional qualities of coal, which are only got in large lumps without yielding " middlings."

K

A kind of discipline is adopted in the underground working of coal. The miners are nearly everywhere sent up and down the shafts in cages travelling in guides. They take up their post in the morning, and leave it at night, to make way for others who compose the night shift or set of men. The works are only stopped on Sundays. On the following week the night shift relieves the day crew of the week before; unless each crew has a special duty, one of winning the coal, the other of maintaining the works. In that case there are miners whose duties, like those of the *remblayeurs*, are always the same : to work by night and sleep by day. Paraphrasing the song of the Belgian miner, they can say—

"Ma lampe est mon soleil, tous mes jours sont des nuits."
My lamp is my sun, and all my days are nights.

The men are organized in companies, and most frequently work by contract, that is to say, they are paid so much for every ton of coal got. There are pickmen or hewers (*piqueurs*), who work out, or win the coal ; rolley-men or putters (*rouleurs* or *traîneurs*), who fill it into tubs, corves, or trams, and convey it to the shaft or to the nearest horse-road ; drivers (*conducteurs* or *toucheurs*), who manage the horses ; stablemen ; and hookers-on, or *accrocheurs*, who hook the tubs or waggons to the rope, or run them into the cages. There are also *remblayeurs*, who pack or stow the gob or waste ; *cantonniers*, who keep the roads in order ; timberers (*boiseurs*), who prop up the galleries ; *mineurs au rocher*, who drive the levels or who carry on the labours of discovery in the barren rock ; and trappers, who open and close the doors: altogether it is a little army.

On the surface are the landers (*receveurs*), who unhook the waggons ; weighing men (*basculeurs*) ; banksmen, sorters (*trieurs*) ; washers (*laveurs*) ; engine-men (*machinistes*); stokers (*chauffeurs*) : then pump-men (*pompiers*); smiths (*forgerons*) ; carpenters (*charpentiers*) ; lamp-men (*lampistes*) ; &c.—whose titles explain the nature of their duties plainly enough. All

Fig. 59.—Miners going down a Shaft.   After an engraving by Bonhomme.

these men are superintended by head-miners (*maîtres mineurs*), called *overmen* in England, *gouverneurs* in the mines of Saint-Etienne, *maîtres-porions* in the Belgian mines, and *caporaux*, or corporals, in the Italian and German mines. The engineer who commands and directs all these people, in English coal-mines is called the *viewer*.

The visit to a coal-mine is always extremely interesting, and even exciting, to a novice. The underground workings are reached by the shaft. Mounted on a bucket suspended from a rope in some collieries (fig. 59), but on a well-arranged platform with a cage and shield overhead, the whole travelling in guides, in others, an unpleasant feeling is experienced at starting, in the sensation of vacancy which the going down a shaft always produces. The bucket rubs against the walls; the space is narrow, and appears still more so than it really is, on account of the darkness. It is but dimly lighted by the lamps. Water filters through the rock drop by drop, in a fine rain, and now and then the thought occurs that a stone might fall from the wall and smash your head; that the rope, stretched by the weight it supports, and whose oscillations are perceptible, may also break, or the bottom of the bucket come out. In the middle of the shaft the thought occurs of a collision, or of a possible entanglement; but when the obstacle is escaped you breathe more freely, and soon reach your journey's end, happy to have escaped with so little trouble. Visitors sometimes decline to go down the mine in this way, while others cower down at the bottom of the bucket, where they remain motionless through fear, and on reaching the bottom it is actually necessary to turn the bucket over to get them out, and they only recover their senses with difficulty. The miners, on the contrary, make this journey twice every day without a thought of danger; and they laugh and talk in going down, just as an old soldier goes under fire without shrinking, and gaily faces the shower of grape-shot.

Two or three times every four-and-twenty hours, but usually twice, morning and evening, the fresh turn of hands enters the mine. The sight is a curious one; the men press forward in a body, then, at the sound of the bell, they disappear in crowded groups in the tubs and cages, or down the ladders. They are heard talking on first leaving; but the sound of their voices is soon lost in the shafts, until it becomes merely a hoarse murmur, and only the pale glimmer of their lights is distinguishable.

Prayers are offered up in some continental mines by the miners before going down; in most mines, however, this is neglected, but more than one miner crosses himself on leaving, and whispers an invocation to the Virgin or Saint Barbe, the great patron of miners. When they arrive at the bottom of the shaft they separate, and every one goes to his place of work.

Let us visit these different quarters of the mine, and enter the subterranean labyrinth. In the stalls and working-places where the noise is heard, and where the smell of gunpowder is perceptible, the miners are getting the coal; in the levels the rolley-boys and horses are crowded together, and trains go and come; at the bottom of the shaft it is the noise made in hooking on or unhooking tubs which is heard, and the shouts of the hookers-on to the landers at the pit-mouth. The lamps only shine at certain points, lighting up the faces of the men, the shape of the waggons, and the coal which glistens here and there; the rest is cast in shadow, and yet the whole effect is animated and startling.

The galleries cross each other in all directions, like the streets of a town with many turnings. There are cross-roads and squares; each road has its name and destination, but as there are no sign-posts, a stranger loses his way at first, soon finding it, however, by practice. Some of the galleries, which are long, wide, and well ventilated, form the principal thorough-fares and great streets, constituting the fine quarter of the mine. The others, which are sometimes low, narrow, tortuous, ill supplied with air, kept in bad repair, and liable besides to be only in temporary use, are like the old quarters, which will soon have to disappear. This underground town is inhabited night and day; it is lighted, but with lamps. It has railways, traversed by horses and locomotives. It has streams, canals, and fountains—strong springs of water which, in truth, could be very well dispensed with. There are even certain plants and living creatures which are peculiar to it, and life, as has been said, seems to assume special forms in it. It is the black and deep city, the city of coal, and the lively centre of labour. The inhabitants only live in it part of the day or night to do their work; and the crews or shifts relieve each other two or three times in the course of the four-and-twenty hours. There are not, as might be supposed, either promenades, shops, or houses, and still less resident miners who never see daylight again when they have once entered the works. The horses only, in some districts, as we already know, never leave the mine.

We must here put an end to the dreams of people of the world and artists, and expose once for all the fables, the tissue of errors propagated by excited tourists or by writers ignorant of facts with reference to the work and life of miners. The romantic and dramatic colouring will be lost, but if we know our vocation, truth is still more dear to us, as it was to the disciple of Plato.

Some authors, bearing in mind the slaves condemned to the mines in ancient times, and those still banished to Siberia by the despotic will of the Czar, have spoken of men who spend all their lives underground, who are born and die there, painfully subjected to the labours of the Troglodytes. There are two mines in particular on which the imagination delights to brood—those of Wielliczka and Bochnia, in Austrian Gallicia, where they do not work coal, but a rich mass of rock-salt. At the intersection of the galleries the miners have carved out of the solid rock obelisks, columns, statues, and even a chapel. There was no need to pretend that there were in these salt-mines houses several storeys high, bazaars, theatres, coffee-houses, hotels, springs and streams of fresh water, and even a windmill! It has been stated that the miners never left these dismal abodes, that they were born and died there. All that is pure fiction, invented by the lively imagination of writers who were more poets than miners. It is not the less true that a large mine in active work resembles in some respects in appearance, and by the animation which prevails in the working places and the levels, an actual town, and we have ourselves employed the comparison.

It must now be explained how in this subterranean town and obscure labyrinth people find their way about, how the position of the various stalls and the daily progress is recognized. The sun is absent; and as the map only can show the true state of the localities, the viewer of the mine makes a detailed plan of the works. But if topographical surveys are often difficult to make on the surface, it may be imagined what they must be underground, where the lights shine dimly, where distant points cannot be seen, and where the intersection of the gal-leries forms a sort of apparently inextricable labyrinth. These obstacles have been overcome; and the compass which enables the mariner to find his way across the open sea, also enables the miner to find where he is underground. The collier makes his survey with the compass. The mysterious property possessed by the magnetic needle of always pointing to the pole, an angular variation excepted, which is known for every locality, furnishes a base-line with which all the observed directions are connected. The compass is supported on a tripod stand, and is free to move in gymbals, similar to those used in ships' compasses, which allow it to balance in a horizontal position. The needle oscillates as if it were alive; gradually the movements become more and more feeble, and at last the needle remains almost motionless. The angle of direction is then read off; the angles of inclination, which will afterwards give the differences of level, are taken by means of a graduated semicircle, furnished with a plummet suspended by a string. Lastly, the distances are measured with a chain, furnished with copper or brass links.

Care must be taken to remove all iron objects when working with the compass, and even the rails of the galleries are taken up. If this precaution be neglected, whatever others may be observed, such as that of holding the compass very high or equally distant between the rails, the reading of the number of degrees indicated by the needle will be attended with error, and the plan rendered in consequence faulty; which may lead, in the prosecution of the work, to very serious mistakes.

M. Simonin knew an old viewer who on such an occasion had the rails covered with straw-matting, or with a layer of small coal (slack), in the belief that by those means the compass would be protected against the effects of the iron; in utter forgetfulness of that branch of physics—which he had learnt at school—called magnetism. And yet he never failed to begin by asking his assistants whether they had provided themselves with copper lamps, or had a knife or a key about them.

It is for the purpose of avoiding all the inconveniences resulting from the presence of iron in the use of the compass that many coal-viewers, in making underground plans, adopt nowadays the same instruments which they use for ordinary surveys—the graphometer, the theodolite, the repeating circle —which are furnished with glasses for reading off, and are supported on very low tripods. Lamps, kept as nearly as possible at the same height as the limb or divided circle, serve as points of sight to take observations on. The instrument gives both the angles of direction and inclination, while the distance between two stations is measured with a Gunter's chain or a tape divided into yards (or mètres). The whole process of making the plans underground, as well as at the surface, is conducted on the same principle—to take angles and distances for constructing triangles, the unknown sides and angles of which will give the angles and distances required to be known; whence the name of triangulation, sometimes given to the survey of maps.

In working from place to place, as has been explained, whether with the compass or with more perfect instruments, the complete sketch of one gallery is obtained first, then that of another, and so on. In this way a plan of the mine is constructed, or the vertical projection of all the workings, and the section or horizontal projection, together with the difference of level between all the points.

Making an underground survey is always an interesting operation. When the compass is used the work is usually done by night, because it is necessary to take up the rails in the levels through which the trains pass by day. The viewer, with his assistants, stand round the instrument. He enters in a special note-book the observed angles of direction and inclination, as well as the distances, and the various other observations he may consider it necessary to make with regard to particular points. He alone is all attention. The assistants, who carry and stretch the chain, and the boy who carries the instruments, all do their duty heedlessly and without thought. If there be a mistake in the figures, it is not they to whom it matters, nor is it they who will be blamed. The overmen, only, give their chief constant support, and sometimes hold counsel with him (fig. 60). This survey of the map revives, in fact, the discussion of all the problems connected with the management of the works; the geological accidents which may have affected the measures are interpreted; and the direction in

Fig. 60.—A Consultation in the Mine.

which such and such a level should be driven is determined with precision.

The graphometer or similar instruments ought always to be employed, in preference to the compass, in mapping certain very delicate works: amongst which may be especially mentioned the driving of a level from several points at the same time, like railway tunnels; the driving of a level which ought to meet or intersect another level; and lastly, the deepening of a shaft the use of which cannot be stopped. The operation is carried on, in this case, by an inner shaft sunk below the old one, so that the two works accurately join together afterwards.

There are few mining operations demanding more unremitting attention than those just alluded to. A trifling deviation in the direction or slope may ruin the work; the levels and the shafts may not meet, or may join badly; and an imperfect junction completed afterwards remains as a mute witness but everlasting reproach, showing want of care and vigilance, and often, it may be more truly said, a deficiency of skill. The miners consider themselves as much interested as the viewer that no mishap should occur. They not only adhere faithfully to the orders that are given them, but, before meeting, they knock with the pick and hammer, and judge by the way in which the sound is transmitted through the solid ground between them, whether the two working places are in the right direction for meeting each other.

The importance of underground plans is now thoroughly understood. Without them there can be no regular or systematic workings, nor can any efficient management be possible. In the case of a lawsuit between two adjoining mines, it is often the plan which settles the matter, and that irrevocably and with the infallibility of geometry. In some collieries, also, a special viewer is charged with these duties. He is the geometer, the surveyor of the mine. He has his assistants, his instrument-bearer, his chain-men, a whole bevy of helpers.

At the map-office he puts on paper the observations he made underground.

The surveyor draws on the plan the progress of all the work done in the mine. The various levels or stages, the stalls, the waste, are all shown by different conventional colours; particular signs indicate the direction and dip of the seam, and all accidental occurrences, such as undulations, contortions, heaves, &c. There are general maps and detailed plans; and both are often accompanied by geological sections. The principal points on the surface of the ground are denoted on these plans, so that it can always be ascertained whether buildings are endangered; whether there be any reason to apprehend infiltrations of water-courses; whether it be necessary to begin to reserve the wall of coal which ought to be left intact between the mine and old workings, or between two adjoining setts or properties. On these plans the outcrops of the seams in work and the boundaries of the several formations are also marked; and, finally, the contour lines, by means of which the altitude or height above a given level of any one point on the ground or in the mine, may be instantaneously ascertained.

It is on the plan that the viewer lays down this base of his operations, and that he meditates, like a sailor on his chart, when he has to make an important decision. No work of long continuance is stopped without referring to the plan, for that alone indicates the true state of things; it shows the block of coal already extracted, and that left to be got; it is, in a word, the great revealer of the past and the future. A practised eye discovers the true position of a mine, and it is for that reason that in many collieries the plan is not willingly shown to visitors.

We have already depicted the animated but severe aspect displayed in the workings underground. At the surface, around the drawing shafts, the activity is redoubled, for there is more animation and life there, the labour being done in the open air and by daylight. The waggons filled with coal and piled one above the other, have rapidly ascended, drawn up by the steam-engine. They are emptied by ingenious ways: some have a movable bottom, others are overturned mechanically. Now at the bottom of the shaft the operation is performed in a reverse order, that is to say, the tubs or waggons are loaded and stacked without the loss of a moment. At a given signal the engine-man, who is on the alert, turns on the steam anew, the ropes glide over the pulleys; one going up loaded with full tubs, the other going down with the empty ones. If the shaft is not divided, the engine-man is on the watch to see that no collision takes place midway. Warned by an index which moves before his eyes, along a graduated scale fixed to the wall, he moderates the speed. The critical point passed, the speed is quickened again, and he has scarcely time to think of it before a fresh load is brought to the surface. It is by promptitude and good organization of the service, coupled with rapidity and the power of the engines, that in some cases enormous quantities of coal are drawn from a single pit: from six to eight hundred tons a day in France and Belgium, and one thousand tons in England and the United States.

At the pit-mouth the lander marks the number of tubs brought up, while the banksmen empty them, and, in France and Belgium, the gang of sorters and washers separate all impurities by hand, or on sieves mechanically moved backwards and forwards in water. The unwashed coal is classed by means of rakes, sieves, or riddles into lumps of the same sizes. Lastly come the loaders, who load the mineral into carts and waggons.

The approaches to the pit-mouth, which present so animated a scene during the day, are in the jargon of collieries called banks (*plâtres, carreau, haldes*). Coal is always kept burning at the pit-mouth in a basket of original form, furnished with iron-bars. The miners place their wet clothes on this hearth when they come out of the mine; it is there that the smoker lights his pipe, and that the surface-hands or day-men warm themselves and talk. It is, also, around this fire that the underground-hands assemble before going below.

The fittings which surmount and surround a colliery shaft present a startling picture, before which the wondering visitor stops. An elevated scaffolding of massive and strange forms, and the thousand details of which appear confused, support the heavy cast-iron pulleys over which the ropes pass. Edifices often with architectural pretensions cover the winding and pumping engines. Outside is the group of boilers; the imprisoned fluid boils in the apparatus, while the superfluous steam escapes with a whistling noise through the safety-valves. A lofty chimney, crowned with a thick canopy of smoke, serves for all the furnaces. Then, in continental collieries there are the places for cleaning and picking the coal, where automatic machinery sometimes does all the work. On one side are the vast sheds for storing, measuring, and loading; the factory where (in some of the French and Belgian coal-mines, though to a small extent in this country) the middlings and slack are compressed and formed into brick-shaped blocks; and, finally, the long line of coking-ovens, where the coal is baked—an operation which has for its object the separation of all the volatile matters. The coal, converted into coke, contains a less quantity of sulphur, and is possessed of a higher heating power. It is thus applicable to new purposes, for example, to the smelting of ores and metals, to the heating of locomotives —though coal is often used—&c. For nearly similar reasons wood in forests is often burned into charcoal: coke is only the charcoal of coal.

The forge where the tools are sharpened, the repairing shops, the miners' houses, the offices, the office of administra-

Fig. 61.—Horse Whim formerly used for drawing Coals and Water. After Lançon.

tion itself, surrounded by the inevitable garden, are all established round the principal shafts of the colliery. The railway stretches away with its parallel lines of rails, on which the waggons run, sometimes drawn by the swift locomotive, which conveys to the most distant places the black and useful mineral.

At some mines certain shafts display splendid fittings, of which France, Belgium, and Rhenish Prussia offer examples that may be cited; but the drawing-shafts are not all on such a grand scale. Between the years 1849 and 1852 there were several shafts at Saint-Etienne and Rive-de-Gier working in the old-fashioned way, that is to say, with horse-gins (*baritels*) or whims (*wargues*), kinds of wheel-and-axle machines or wooden drums placed upright, around which was wound a round hempen cable, like those used on board ships.* A quiet horse, with a slow and even pace, and his eyes covered with leather blinkers, turned the whim; and the tubs, as soon as they reached the surface, one by one, were emptied at the pit-mouth (fig. 61). This system of winding dated from the workings of the Middle Ages. Some of the shafts are perhaps still in use in the Saint-Etienne basin; but the greater number of them, become objects of remembrance, only serve now to enliven the stories of the evening, when the old-miners, perched on the axle, relate to their sons stories of past times. In this case, at any rate, the narrators cannot say: *In our time things were better managed than they are nowadays.*

The whim still continues to be the necessary concomitant of mines at their commencement, especially in the sinking of trial-shafts, when it is an object to husband resources, or when the distance from a coal-mine in actual work does not allow of the economical employment of steam-engines or more expensive apparatus.

The English collieries, if they were the first to give up the

* At the present day ropes are made of iron-wire; and instead of being round they are generally flat. Heavy flat ropes made of the fibre of the aloe are much employed in the north of France and in Belgium, where some of the shafts are of greater depth than any in our own country.

use of the whim, are always distinguished by a stamp of the rude simplicity from which they seem finally anxious to emerge. M. Simonin remembers seeing in 1860 near Bilston in South Staffordshire, not far from Birmingham, on a very productive and busy coal-field, where roads, railways, and canals cross each other in all directions, and where factories of all kinds are scattered about, mine-shafts open in the fields without shelter or any buildings around them. The winding-engine, placed between nearly adjacent shafts, sends a rope to each. When all the coal has been worked away within reach of a shaft, another is sunk, and the engine, encamped as it were in an invariable centre of activity, still continues to serve it. In France, where nature has been more niggardly in her distribution of coal, such primitive modes would scarcely be allowed. They are only acceptable in England, where coal is so universally abundant. In some of the collieries of that favoured land the waggons drawn out of the mine may be seen unloading direct into the boats, on a canal, or in a dock, or passing without unloading from the pit or subterranean level on to the railways at the surface. All the advantages which result from this are readily perceived; for coal is a very troublesome article to carry, being heavy, bulky, friable, and liable to many causes of waste.

It may be remarked here that England, in the working of her coal-mines, has not that practical superiority which is generally assigned to her. She has only received a larger share than any other country, by an unexampled accumulation of mineral fuel. A certain mine is mentioned, the daily produce of which exceeds two thousand five hundred tons a day, or seven hundred thousand to eight hundred thousand tons a year, which in France is the produce of whole coal-fields; fifteen English collieries of this importance would take the place of all the French deposits, for they hardly reach twelve million tons a year. The small Newcastle coal-field alone

furnishes double this last amount.   England possesses the
advantage of having coal deposits on the sea-board; and when
they are inland they are in the midst of the most important
industrial centres, which have been created, it is true, by coal,
and where it finds on the spot an almost unlimited consumption.   England exports a part of her coal from the ports which
surround her mines; she dispenses to European countries the
portion which each is in want of, and she supplies the entire
universe.   The coal which she consumes herself, in itself larger
than the consumption of all the rest of the world, is employed
in manufacturing machinery, stuffs, and all those products
of such varied kinds with which she inundates all the foreign
markets.   In that is the secret of her industrial, commercial,
and maritime power; to that she is indebted for part of her
political importance: but to assert that the English work their
mines better than other peoples is a great mistake.   They
may deliver their coal at a much lower rate, it is true, for
reasons which have been already given; but as regards artistic
constructions, modes of working, internal management, and
surface arrangements, everything is with them inferior to what
is seen in many coal-mines in France and Belgium.   When
the treaty of commerce was made in 1860, official or private
inquiries demonstrated in the most convincing manner the
inferiority of the English in this respect; who, themselves well
aware of the fact, are introducing, at this very moment,
numerous improvements into their mines.

The foregoing considerations do not in the slightest degree
diminish the value, the energy, or the aptitude for work of the
Anglo-Saxon race, so highly favoured in all respects; but it is
not the less certain that the chief cause why she excels all
others in the production of coal is that nature has bountifully
endowed her with more extensive coal-fields (so far as the area
of her Coal Measures is concerned), richer and more favourably
situated, than those of any other country.   The seams of coal

are there frequently thicker, are situated at a less depth, are always more compact, and the measures harder, than is the case in most other countries. Lastly, England, surrounded by coasts which are everywhere accessible, is furnished besides with a network of iron roads and navigable ways which are more developed than those of the continent. It must also be said that the absence of all government interference enables her industry to have full play, for no restrictive measures fetter the liberty of labour in England. It is in this last respect that the French should desire to resemble the English; while as regards the able working of coal-mines, the last may in their turn borrow from the French more than one happy inspiration.

Some idea of the importance of the English coal-fields may be formed from the following statement of the production of her collieries and the export of her coal for ten years past :—

	Coal produced. Tons.		Exported. Tons.
1857,	65,274,047	....	6,737,718
1858,	64,887,899	....	6,529,483
1859,	71,859,465	....	7,006,949
1860,	79,923,273	....	7,321,832
1861,	85,512,144	....	7,855,115
1862,	83,510,838	....	8,301,852
1863,	88,165,465	....	8,275,212
1864,	92,662,873	....	8,809,908
1865,	98,150,587	....	9,170,477
1866,	101,630,544	....	9,307,749

CHAPTER VIII.

THE FIELD OF BATTLE.

WE have only till now dealt with the beginning of the struggle. It has been seen how great an expenditure of capital and labour a mine involves; it must now be told what it costs in human life, and with what sacrifice a piece of coal is sometimes purchased. It is not without reason that the art of mining borrows some of its terms from the art of war; that in France a year's work is called a campaign, the different

L

underground working-places posts, a gang of miners a brigade or squadron—in England a crew or shift—while in Cornwall the underground manager is called a captain, and the store-keeper a purser. Is it not said that they attack the coal, and is not the mine itself the collier's field of battle? Is it not there that in his struggles against all dangers he may be said to combat them foot to foot? The four elements of the ancients—earth, air, fire, and water—all conspire against him.

Fire menaces him in blasting, in the firing of the coal, and in explosions of fire-damp; the air, by becoming rarefied, or mixed with mephitic or explosive vapours; the earth, in falls of roofs, &c.; the water, by inundations. The collier opposes to all these (often invisible) enemies, the calm stoicism, the approved courage, and the practical science which tend to make the brave and skilled miner. And the underground soldier is the more meritorious, in that he is encouraged neither by the certainty of advancement, nor by the hope of honourable recompense, in this contest in which he risks his life at every moment. He has only the satisfaction of observing discipline, and of faithfully doing his duty.

Fire is perhaps the most terrible enemy the collier has to encounter, and it is first in the firing of shots in blasting that he ought to guard against its attacks. We know how the work is done with gunpowder.* The miner holds in his left hand the bar, drill, or gad, with a broad and sharp bit (fig. 62) pressed against the rock, and with the other hand a hammer (called in French *massette*, fig. 63). When the work requires

---

* Gunpowder, applied to the throwing of warlike projectiles from the day of its invention, was used many centuries afterwards for the removal of rocks; it being only since the seventeenth century that miners began to make use of it. Its most useful application has been made the last, and that only very lately. This use of powder has given a very great impulse to mining operations in Europe, as well as in America. Powder has, also, been everywhere introduced into the working of quarries. For twenty years, in quarrying marble, in getting limestones for building, clfambers or holes filled with considerable quantities of powder are made, and whole hills are sometimes shattered and thrown down in this way. More recently the employment of an explosive compound, nitro-glycerine, in quarries of slate—especially those of Carnarvonshire—has rendered it possible to detach, at one operation, as much as a hundred cubic yards of rock; but these are exterior operations, and different altogether from purely mining operations.

the hole to be of great size, a single miner is no longer
sufficient, but then a man kneeling takes the bar between his
two hands, while another over him wields the hammer (fig.50).
There are sometimes two strikers, hammering alternately on
the head of the tool, like smiths on an anvil.

The hole or bore is enlarged by degrees. From time to
time it is wetted to prevent the steel from heating, and to
collect the powdered stone. It is cleaned out with the scraper
(*curette*) (fig. 64). More than once the bars have to be changed,

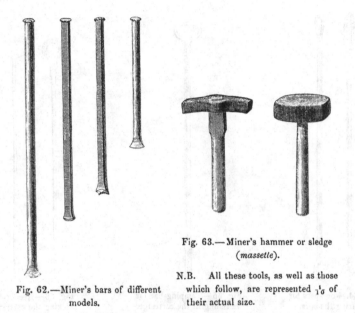

Fig. 63.—Miner's hammer or sledge
(*massetle*).

N.B. All these tools, as well as those
which follow, are represented $\frac{1}{6}$ of
their actual size.

Fig. 62.—Miner's bars of different
models.

for they lose their temper, and soon get blunt, especially when
not made entirely of steel. When the hole has been made to
the required depth, it is dried with old rags placed in the hook
of the scraper, and a cartridge is inserted at the bottom of the
bore. The cartridge is prepared beforehand, and the quantity
of powder employed depends upon the nature of the rock and
the effect that it is desired to produce. Clay, powdered brick,
or earth taken out of the mine, are used for tamping. A round
iron-rod, the tamping-bar (*bourroir*), acts the part of the ram-
rod for guns (fig. 65). With the needle or pricker (*épinglette*),

L 2

fig. 66, a sort of skewer terminating in a point, the cartridge
is at the same time pricked.   The pricker is put down while
the tamping is being done, and on its withdrawal, a channel
is left against the side of the hole, along which are inserted
a series of small tubes called fuses (*cannettes*), made of connected wheaten straws filled with powder, and terminated at
the upper end with a sulphur match, or a small piece of touch-
paper.   The rush or straw is rarely used in England now, the
safety-fuse having entirely superseded it.   This safety-fuse is

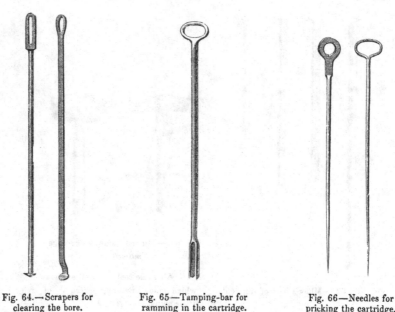

Fig. 64.—Scrapers for   Fig. 65—Tamping-bar for   Fig. 66—Needles for
clearing the bore.   ramming in the cartridge.   pricking the cartridge.

made of hemp twine, twisted into a continuous spiral, forming
a twisted tube, the powder occupying the centre, and being
finally covered and protected by a coating of tar.

A stall, when blasting is being carried on, always offers a
splendid sight.   In order to lose as little time as possible, all
the shots are fired together.   At a signal given by the leading
man, a light is applied to the match, and all retire as fast as
they can.   The shots go off at close intervals; the air is driven
afar; the masses give way, and are shattered; fragments of
rock are blown here and there, and the noise of the explosion

Fig. 67.—Explosion in a Coal-mine.

is re-echoed throughout the mine.   It is a regular cannonade.
A thick white smoke with a nauseous smell, deleterious gases resulting from the combustion of sulphur, charcoal, and saltpetre,
fill the level, and are only slowly dissipated.   Then the men soon
return to their work, and sound the rock with a hammer, in
order to judge of the effects produced by the explosion.

When a shot does not go off, it is only after the utmost
precautions have been taken, and a proper delay, that the
miner ought to approach.   There are instances of shots having
gone off ten minutes after being lighted.   Numerous and
dreadful accidents have arisen in consequence of too great a
haste on the part of the miner in his anxiety to return to his
post, to untamp the hole.   In many districts this is not
allowed; then new holes have to be made.

Accidents from blasting arise, also, from other causes : instead of using cartridges, the powder is merely poured into
the bottom of the hole, under the pretext that it will act more
powerfully.   It is scattered over the sides, and in working the
tamping-bar, or the needle, a spark may be produced by the
friction of the iron against silicious substances, precisely as in
the impact of steel.   The spark is then communicated to the
powder, and the charge explodes in the face of the miner, who
is blinded or disfigured, if not killed on the spot * (fig. 67).

* " In a certain Cornish mine (South Caradon) two miners deep down in the shaft were. engaged
putting in a shot for blasting : they had completed their affair, and were about to give the signal for
being hoisted up; one at a time was all their coadjutor at the top could manage, and the second was
to kindle the match and then mount with all speed.   Now it chanced while they were both still below,
one of them thought the match too long; tried to break it shorter; took a couple of stones, a flat and
a sharp, to cut it shorter; did cut it of the due length, but, horrible-to relate, kindled it at the same
time, and both were still below!   Both shouted vehemently to the coadjutor at the windlass, both
sprang at the basket; the windlass man could not move it with them both.   Here was a moment
for poor miner Verran and miner Roberts!   Instant horrible death hangs over both, when Verran
generously resigns himself : ' Go aloft, Roberts, and sits down ; 'away, in one minute I shall be in
Heaven !'   Roberts bounds aloft, the explosion instantly follows, bruises his face as he looks over; he
is safe aboveground: and poor Verran?   Descending eagerly they find Michel Verran too, as if by.
miracle, buried under rocks which had arched themselves over him, and little injured; he too is
brought up safe, and all ends joyfully, say the newspapers."—Carlyle's "Life of John Sterling," p. 278.

As Verran was anxious after this to work aboveground, and also to gain a little schooling, a few
hearty admirers of his heroic act were glad to subscribe a little sum to enable him to spend some
months at school.   Here he acquired the great arts of reading and writing, then established himself in
a farm and married a schoolmistress, with whom he and his affairs have prospered as they deserved.

To obviate these accidents the notion has been conceived of making the tamping-bars and needle of copper and of bronze; but it has been shown that bronze, when rubbed violently against the hard rock, produces sparks in the same way as iron, and for this reason wooden tamping-bars (*bourroirs*) are used in some collieries. Copper-tipped tamping-bars are now employed with safety in many collieries. The needle, if withdrawn too quickly, may in a like manner fire the powder. As for the needle, the straws, and the sulphur matches, which are also the causes of numerous accidents, they have been almost universally abandoned in favour of the safety-fuse, already described, which is cut in the required lengths and fixed to the cartridge. These fuses burn slowly, and allow the miner time to get away with entire safety. Their defect is a liability to get twisted and even broken during the process of tamping. Lastly, as they form a part of the tamping which is rammed in the hole, it is impossible to withdraw them when once lighted; but this is never the cause of serious accidents. One day at Rive-de-Gier, at the bottom of a shaft which was being sunk, the men had set fire to the charge and gave the signal to be drawn up. On starting, the horse which turned the whim at the surface got entangled in the traces and fell. The miners called out, but received no reply. One of them courageously jumped off the bucket and plucked away the sulphur-match; if this had not been done with that promptitude which belongs to true courage, a frightful accident would have happened.

It is always at the bottom of the shafts that the firing of powder is most attended with danger. The men can only escape to the surface after the charge is lighted, and it is necessary that they should be raised with all speed, at any rate to a certain height. The horses which turn the whim understand this signal very well; and at the first sound of the bell the good animals start off at full speed.

The electric battery has been applied to the firing of charges of powder in mines during the sinking of shafts. M. Simonin has seen it used in some collieries, especially in that of Montsalson, near Saint-Etienne, in 1852, at the Imbert pits, which were then being sunk. In this way the charge can only be fired at the precise moment desired, and from the very mouth of the shaft The electric currents traverse a conductor of thin metal wire, one end of which is fixed in the cartridge; and the electric spark sets fire to the powder. This system does not always succeed in its application to ordinary mining purposes, partly on account of its expense, and perhaps still more from the nicety of the arrangements which it involves.

Gunpowder being absolutely indispensable in all mining operations, attempts have been made to substitute its analogue, gun-cotton, for it; but this new substance, which up to the present time has not succeeded either for sporting guns or for muskets, has not been more successful in blasting rocks: which is the more to be regretted, since its employment in mines might have prevented many accidents.

In 1847 a pupil of M. Pelouze—M. Sobrero—discovered that glycerine, when treated with nitric acid, was converted into a highly explosive substance, which he called nitro-glycerine. It is oily, heavier than water, soluble in alcohol and ether, and acts so powerfully on the nervous system that a single drop placed on the top of the tongue will cause a violent headache, which will last for several hours. M. Noble, a Swedish engineer, has lately succeeded in applying it to a very important branch of his art—namely, that of blasting. From a paper addressed by him to the Academy of Sciences, we learn that the chief advantage which this substance—composed of one part of glycerine and three parts of nitric acid—possesses, is that it requires a much smaller hole or chamber than gunpowder does, the strength of the latter being scarcely one-tenth of the former. Hence the miner's work, which, according to the hardness of the rock, represents from five to twenty times the price of the gunpowder used, is so short that the cost of blasting is often reduced by fifty per cent. The process is very easy; if the chamber of the mine present fissures, it must be lined with clay to make it water-tight; this done the nitro-glycerine is poured in, and water after it, which, being the lighter liquid, remains at the top. A slow match, with a well-charged percussion cap at one end, is then introduced into the nitro-glycerine. The mine may then be sprung by lighting the match, there being no need of tamping. In 1865 M. Noble made a series of experiments with this explosive fluid at Falmouth, and in some of the Cornish mines and granite quarries. Its explosive powers were found to be enormous. Its tendency to change, and spontaneous combustion, from which some most disastrous accidents have arisen, has for the present placed it out of general use.

The firing of powder is not the only source of danger with which miners are menaced by fire. Spontaneous combustion also takes place in coal-mines, produced by the heating of the small coal, from the decomposition of the iron pyrites it contains in contact with moisture. When the small coal of certain seams are left in the mine they speedily undergo this chemical decomposition, especially in a moist and hot atmosphere, which is accompanied by a great development of heat. The coal soon ignites, and the fire, finding in the coal-seam a natural aliment, spreads rapidly through the mine.

In such cases dams (*corrois*) of clay, or walls, are built up to isolate the field of disaster. If the construction be carefully made, the fire, deprived of atmospheric air, gradually goes out. For a certain time a more than Senegalian temperature, 50° to 60° Centigrade (122° to 140° Fahr.), is experienced in these parts of the mine, and the miner can only endure this enormous heat by working in a state of nudity.

Fig. 68.—Mode of using the Fire Extinguisher in a Coal-mine.

The building up of dams by miners to cut off fire is one of their most trying and laborious tasks. Compelled to remain in an impure air of a very high temperature, they may be rendered insensible. By holding under the nostrils and over the mouth a rag soaked in lime-water or ammonia, the effects of mephitic vapours are in some degree neutralized; but this has no effect upon the heat, and therefore the spells of work are of very short duration.

The use of steam and carbonic acid has been suggested for the extinction of fire in coal-mines. This latter gas might be made by combustion of a mass of coke, and projected on to the spot on fire, which being by that means deprived of the element necessary to support combustion, would go out of itself. Steam acts in a similar way, like an inert gas. These means have been used with complete success in English mines. The annihilator (*Extincteur*) might also be adopted, a machine of recent invention, which discharges on to a burning mass water charged with carbonic acid under a very high pressure —a sort of soda-water. A labourer carries the apparatus on his back, and projects the gaseous water by means of a jet like that of a fire-engine (fig. 68).

When the fire gains a height which it sometimes does, all these modes are ineffectual, and it becomes necessary, so to speak, to let the fire take its course. In England and in Scotland it has been found necessary to close certain mines for a long period. Usually it is sufficient to construct around the coal on fire those dams or clay walls (*corrois*) of which mention has been already made. The fire occasionally lasts many years; and there are a few instances of mines where, after an interval of from twenty to thirty years, the colliers have still suffered from the heat prevailing in the vicinity of the *corrois*.

It has sometimes happened that the walls are inefficient, and that the combustion still goes on, in which case extreme measures of safety must be adopted, and the works flooded.

A mine near Charleroy was inundated in this manner in 1851, by turning the waters of the Sambre into it, for a period of three months. In Great Britain, in the same way, conflagrations have been subdued of many years' duration, by which the stalls had been rendered inaccessible. Edmonds' main colliery, near Barnsley, was filled with water for two years. In this country it is always preferred, as has been already stated, to close the burning collieries hermetically, and to wait patiently for the arrival of the time when the works can be reopened with safety.

Fires are not always combated by such radical measures as these. They are frequently allowed to burn on; and there are instances of vast masses of coal which have thus been burning for years. When all communication with the external air is not entirely cut off (and some imperceptible fissures are quite sufficient to prevent this), then the devouring element pursues its course without interruption. It partially burns the coal, which it converts into coke, which has nearly the same chemical composition as the diamond, and a considerable degree of hardness, but without transparency or crystallization. It calcines the sandstones and adjacent schists, changing their colours to a sort of red, and altering their composition. At Brûlé, near Saint-Etienne, there is a coal-mine which has been on fire from time immemorial. The soil at the surface is barren and baked: hot vapours escape from it; sulphur, alum, sal-ammoniac, and various natural products are deposited in it: it might be supposed to be a portion of the accursed cities formerly consumed by the fires of heaven and earth.

Other burning coal-mines are cited in France; for example, those of Decazeville in Aveyron, and of Commentry in the department of Allier. The inhabitants have even, for a long time, kept up these fires for the sake of working the aluminous salts which are given off from the coal, and are deposited on the surface of the soil as whitish efflorescences.

In the carboniferous basins of Saarbrück and Silesia there are likewise coal-mines which have been on fire for a long period. In Belgium, between Namur and Charleroy, at a place called Falizolle, the fire has been alight for many years. The inhabitants, before the concession of this part of the coalfield, were in the habit of working the coal on their own account—an unsystematic labour carried on upon no settled plan. Now it frequently happened that in their venturesome progress two colliers came upon each other suddenly, which caused endless disputes, often degenerating into sanguinary fights. One of the most favourite ways of keeping rivals or competitors at a distance was to throw pieces of old leather on a burning brasier, which gave out an insupportable stench. One day the fire extended also to the coal, since which time it has never ceased burning. The fire, which burns underground, is seen through the fissures at the surface. The sulphur deposits itself round these vents in coatings of a citron-yellow colour, acid gases are evolved, and it is a sort of pseudo-volcano, resembling a miniature Vesuvius.

In England, as at Saint-Etienne, the ignition of the coal, especially in Staffordshire—where, from the peculiar nature of the coal, combustion is not uncommon—has produced surprising effects of alteration in the measures containing the coal. The sandstones have become vitrified, baked, and dilated by the fire, the banks of plastic clay hardened, and changed nearly into porcelain.

In the environs of Dudley there was formerly a coal-mine on fire. The snow melted in the gardens as soon as it touched the ground. They gathered three crops in a year; even tropical plants were cultivated; and, as in the Isle of Calypso, an eternal spring prevailed. It is by somewhat similar means that early fruit and vegetables are grown in the depth of winter in some of the gardens round Paris, where the temperature of the soil and the surrounding air is artificially raised by means of currents of hot water made to circulate in pipes underground.

In another Staffordshire colliery, the firing of which dates many years back, and which is called by the inhabitants *Burning Hill*, it was noticed, as at Dudley, that the snow melted on reaching the ground, and that the grass in the meadows was always green. The people of the country conceived the idea of establishing a school of horticulture on the spot. They imported colonial plants at a heavy expense, and cultivated them in this kind of open-air conservatory. One fine day the fire went out, the soil gradually resumed its usual temperature, the tropical plants died, and the school of horticulture was under the necessity of transferring their gardens elsewhere.

These subterranean ignitions only trouble the miner by the mephitic vapours which they give out, and the high temperature which they cause in the stalls. It is otherwise with fires caused by the explosive gas of coal-mines—that combination of hydrogen and carbon, that brother of illuminating gas— which the miners call fire-damp or stythe when mixed with atmospheric air (*grisou*, otherwise *feu grisou* or *sauvage*). Here it causes a fearful explosion, which sometimes sweeps away hundreds of men at a time.

No meteor, however terrible it may be supposed to be, can be compared to an explosion of fire-damp. Let one of those scourges of heaven be imagined which appear sometimes as if designed for the punishment of human beings—a thunderbolt, a hurricane, a cyclone, or a whirlwind—burning, overthrowing, destroying everything in their course, and the effects produced by them will still be inferior to those caused by an explosion of mine-gas. A discharge of cannon loaded with canister shot, and fired point blank into a crowd; a powder magazine taking fire in the midst of a body of workmen; a gasometer exploding in a factory—can scarcely give

Fig. 69.—Explosion of Fire damp.

an idea of an explosion of fire-damp suddenly overtaking the miner.

The moment the mixed gas comes in contact with the flame of a lamp a tremendous explosion takes place, resulting from the combination of the components of the fire-damp, hydrogen and carbon, with the oxygen of the air. The two former separate to combine with the oxygen, with which they have the greatest affinity. The double phenomenon only takes place at a high temperature; without flame it would not arise. The reaction produces an effect like the most brilliant lightning, and makes itself heard by a clap of thunder. The explosion spreads instantly into all the galleries of the mine; a roaring whirlwind of flaming air destroys everything it encounters, overthrowing trams and bratticing and trap-doors, mounts into the shaft, and lifts from their foundations the staging which covers its mouth, through which it discharges thick clouds of coal, stone, and timber.

The men are blinded, thrown down, scorched, and sometimes burnt to a cinder (fig. 69); often their clothes take fire, and not unfrequently they are buried beneath the ruins of the fallen roofs. When an attempt is made to fly to their assistance, there is not time to rescue them; there are only corpses left which are scarcely recognizable. The calamity spares nobody, even though as many as one or two hundred miners may be at work; death extends over the whole of the mine where the explosive gas was present.

The air-doors are thrown down, the ventilation of the mine is reversed, the underground atmosphere is vitiated by the combustion of the fire-damp, and the stalls are filled with steam and carbonic acid. Sometimes the temperature rises so much that the coal is converted into coke at the sides of the galleries, and the commotion is so great that the dams have to withstand both fire and water, and the wallings, raised for the purpose of resisting the thrust of the measures,

are themselves overthrown.   Then to a scene of already inde-
scribable desolation are added the horrors of inundation, falls
of the ground, and fire, when the explosion has already made
only too many victims!   To add to so many horrors the foul
air, carbonic acid, the after-damp or choke-damp, spreads
through the mine, and suffocation terminates the existence of
those in whom the explosion had left a spark of life.   Do not
suppose this picture to be coloured by fancy.   In 1812 a
terrible explosion took place in a coal-mine near Liége.   Sixty-
eight miners who had escaped the shock of the fire-damp
were suffocated by the gas resulting from its combustion.   At
Jemappes, in 1860, nine colliers, alarmed by the sound of
an explosion which had taken place at some distance from the
stall where they were at work, endeavoured to escape up the
ladder-shaft, where they were overtaken by the choke-damp
developed by the explosion, and were instantly suffocated.

One instance among thousands will be sufficient to give to
those who have never lived near collieries a faithful account
of this most fearful of mining accidents.   To take at random
from all the heart-rending histories which are current in the
French collieries.   In 1835 the mine of Méons, near Saint-
Etienne, was the scene of a fearful explosion.   It was night.
The overman and three other men went down into the mine,
where they found the stableman already looking after the horses,
and the carpenters repairing the woodwork.   Suddenly a frightful report was heard.   The masonry surrounding the mouth of
the shaft, made of large dressed stones, the framework supporting the pulleys, were all blown to a distance even of more than
a hundred yards from the opening.   The very tubs and the
ropes were lifted from the bottom by the devastating hurricane, and hurled into space.   The viewer arrived filled with
alarm; he thought a boiler had burst, but it was an explosion
of fire-damp.   An attempt at rescue was promptly organized,
but the lamps were extinguished in the galleries, the mine

being filled with smoke and bad air. The rescuers fell suffocated, and two of them were dead.

An ambulance was established at the entrance to the mine, and another descent made. The colliers, when a comrade has to be rescued from danger, will sacrifice themselves to the last man. The search was carried on throughout the night, by feeling, for the lamps would not burn. At ten o'clock the next morning none of the victims had been recovered, and a number of asphyxiated searchers filled the ambulance.

An agitated crowd, consisting of the families of the miners, pressed round the pit-mouth. One woman in particular attracted attention, on account of the deep sorrow she displayed; she was young and beautiful; she carried a babe at her bosom, and large tears fell from her eyes. She is the wife of the overman. She requested, as a special favour, to be allowed to enter the mine in order to recover her husband; but no woman being permitted to go into the workings, she anxiously waited outside.

Now they returned into the interior. The wrecks of the timberings, and the coal which had been blown down, formed a desolating sight. From time to time blocks of coal were still heard to fall, as they became detached from the walls. The carpenters who were repairing the timbering had been crushed. All the horses (of which there were six) were found suffocated in the stable, and the groom dead with them, lying in the manger, apparently asleep; the hay had been set on fire and was still burning. At length one of the men who went down with the overman was discovered, and alive! Blown by the explosion to the bottom of a gallery, he had been horribly scorched and nearly blinded! His escape was cut off by falls of coal and rock, and as he had no light, he durst not move. Hearing no noise for fifteen hours, and despairing of ever beholding the light of day again, he awaited death with resignation; consoling himself with the thought (to make use of his own words), "that his wife and children would be supported by the benefit club of the mine."

After this first rescue, another man was soon found, almost buried in the rubbish, but still alive. Like his comrades, he had been driven to a distance by the explosion, and thrown upon the ground. He had seen *a river of fire* pass over him, and to save himself had placed his hands over his eyes. They were frightfully injured, and he himself had become blind!

At seven o'clock in the evening the overman and the third man were at last found, disfigured, charred, and at a great distance from each other. Their hats and lamps had been the sport of the hurricane.

This memorable catastrophe long inspired the miners of the country where it occurred with a sort of superstitious terror. They no longer went to work without safety-lamps, or without first imploring divine protection and commending themselves to Saint Barbe, the miner's patron saint, whose statue was solemnly erected at the entrance to the principal gallery.

The wife of the overman became insane from grief. Her madness was, like herself, of a mild and gentle nature, and she wandered through the villages, asking passers by the way to a distant country where she hoped to find the father of her children. She lived three months in this state, and then she died. Her memory is preserved in the country like that of a legendary personage, and if you go to Saint-Etienne, the old colliers will tell you the story of Marie, the wife of the overman of Méons.*

Need other narrations be added to this? Is it necessary to furnish other examples? They may be found on all sides. Almost every coal-mine has been struck in turn, and the explosion is always accompanied by the same deplorable sacrifice of

---

* Dr. Riembault, in the excellent work on "L'Hygiène des Ouvriers Mineurs," has given in full detail, and after the actual notes of the viewer who then managed the colliery at Méons, an account of the lamentable event which has only been briefly noticed here.

life. In the Department of the Loire, sixteen years ago, M. Simonin witnessed some of these disasters when not a victim escaped, when the explosion was heard even at the surface, and caused alarm there; but still more recent facts can be mentioned.

In 1861, in one of the collieries of Merthyr Tydvil, an explosion caused the death of forty-seven miners. In December, 1865, the gas fired again, and out of forty men at work underground, thirty-two were killed on the spot. Amongst those who were working in the neighbouring stalls, twenty-two were seriously injured by the repercussion of the shock.

Only the two brothers John and Thomas Hall, amongst the eight men who escaped death at the point where the accident occurred, could furnish any particulars. John, the elder brother, was engaged at one of the ends of the stall, when the beginning of the explosion was heard. He rushed towards his brother, who was at work a short distance off; but a second explosion more severe than the first overtook them and threw them down. They gradually recovered their senses, and were able to lift themselves up. The air was heavy and hot; they could only breathe with difficulty. John, remembering that he had his can of tea, bathed his own and his brother's face with it, which gradually revived the two miners. Supporting each other, they tried to reach the entrance to the mine. They crawled on hands and knees, in the midst of darkness, over the bodies of their comrades, some of whom uttered mournful and agonizing cries, while the rest were silent, already cold in death. After encountering a thousand difficulties, these two men regained the light of day.

On descending the mine, and by the light of the lamps, a harrowing sight presented itself: thirty-eight bodies were discovered lying on the ground, six were still breathing, the others insensible, lifeless. In the adjoining galleries twenty-two men were seriously injured. The wounded and dead were

M

carried away, slowly and laboriously, as must needs be the case in a mine. At the mouth of the shaft there was the sad picture which is always observable under such circumstances. Women, old men, and children, bathed in tears, seeking a husband, a son, or a father. What a joy for those who beheld them alive, even though wounded; what grief for those who found only a corpse, when they could recognize it again!

Before the invention of Davy's safety-lamp, the construction and use of which will be explained presently, fire-damp was the great bane of the coal-mines, many of which remained unworked on account of the presence of this invincible enemy. Since naked lights or ordinary lamps could not be used, the idea was conceived towards the end of the last century, by Spedding, of Whitehaven, of lighting the working stalls by means of a steel-wheel, made to revolve rapidly against a sharp piece of flint, by which a stream of sparks was given off. The steel-mill was sometimes hung on to the breast of the miner, and there worked by him.* Sometimes when the fire-damp was very abundant, and was given out at known points, the gas was lighted once for all. A true fountain of fire was thus obtained from these natural gasometers, and the jets of natural gas were called *everlasting lamps.* One of these lamps is mentioned as having been burning for nineteen years in the Newcastle coal-field. The gas, collected in pipes, was carried outside the mine as high as the head-gear, where it was set alight, the ignited stream of gas burning with a flame eight or nine feet in length, with a sound like the roaring of a blast-furnace.

M. Imbert, who has given a vivid description of the salt-wells of Sz'chuen, in China, states that the air which escapes from them is very inflammable, and that if a torch is presented

* "The expense formerly incurred in the article of steel-mills seems almost to exceed belief. Dr. Cranny once stated, that in a single working of a colliery in the neighbourhood of Sunderland, the expense of steel-mills was about £80 every fortnight, so many of them having to be kept at work at a time to give anything like a sufficient light."—"History of Fossil Fuel," p. 241.

to the mouth of the shaft, the gas explodes like gunpowder, and ignites into a great column of fire, from twenty to thirty feet high. This gas is conducted through bamboo tubes to the salt-pans, under which it is burned to effect the evaporation of the brine.

The largest fire-wells are those of Tse'lieoutsing, situated in the mountains, forty leagues from Wutung. In one valley there are four pits which give flame to an amount truly frightful, but no water. These pits, for the most part, have previously afforded salt water; which being drained, pits were sunk to more than 3000 feet in depth, in the hope of meeting with an abundant supply of water. The expected supply of water was not procured; but there suddenly gushed forth an enormous column of air which brought with it large, dark particles, more resembling the vapour of a glowing furnace than smoke. This air escaped with a roaring and frightful rumbling noise, which was heard at a great distance. The mouths of the pits were surmounted by a stone wall six or seven feet high, to prevent any one applying fire to the opening of the shafts either by accident or design. This misfortune happened in August last. As soon as the fire was applied to the surface of the well it caused a frightful explosion, and even something like an earthquake. The flame, which was about two feet high, leaped over the surface of the earth without burning anything. Four men courageously carried an enormous stone to the pit, and placed it over the orifice. It was immediately hurled up into the air; three of the men were scorched, the fourth escaped unhurt: neither water nor dirt would extinguish the fire. At last, after fifteen days of hard work, a quantity of water was brought over the neighbouring mountain, a lake or dam was formed, and the water was suddenly let loose, which extinguished the fire, at an expense of about 30,000 francs (£1200).*

* Imbert, " Annuaire de l'Association pour la Propagation de la Foi," vol. iii. p. 369.

In some collieries it used to be the custom to light the fire-damp every night.   The time is still remembered at Rive-de-Gier, in France, when a man came every evening to set fire to the gas in the mine—to provoke the explosion, in order that the working stalls should be accessible again the next day. Wrapped in a covering of wool or leather, the face protected by a mask, and the head enveloped in a hood like a monk's cowl, he crawled on the ground before firing the explosive mixture, to keep himself as much as possible in the layer of respirable air; for the fire-damp, being lighter than the atmosphere, always ascends to the upper parts of the levels.   In one hand he held a long stick, with a lighted candle fixed at the end of it, and he went alone, lost in this poisoned maze, causing explosions by advancing his lamp, and thus decomposing the noxious gas.   Having fired any mixture of fire-damp, he naturally changed his position and walked upright, since the carbonic acid produced by the explosion rapidly formed the lowest layer of air.   He was called the *penitent*, on account of the resemblance of his dress to that of certain religious orders in the Catholic Church (fig. 70); and this word seemed at the same time to be dictated by a bitter jest, for frequently the penitent, a victim sacrificed beforehand, was blown away by the explosion, and never returned alive.   In other mines this brave collier was called the *cannonier*.   When the fire-damp killed him on the spot, it was said that the cannonier died at his post on the field of honour, and that was all his funeral oration.   The same person in English mines bore the expressive name of *fireman*.

Such were the more or less effectual means employed in coal-mines to counteract the fire-damp, prior to the invention of the safety-lamp.   Sir Humphrey Davy was engaged on a series of researches on flame, and in the course of his investigations he discovered that small metal rings have a very remarkable power in reducing the size and the illuminating

Fig. 70.—The *Penitent*, or Fire-man, igniting the Fire-damp.

power of the flame.   By reducing the size of those rings he found that the passage of the flame was entirely prevented, and that a gauze composed of very fine metal wire, placed around the flame of a lamp, would not allow the flame to pass through. All the heat of the combustion is expended in raising the temperature of the metal, which is a good conductor of heat, and the flame does not retain heat enough to burn on the outside of the gauze.   The explosive gas would pass through the wire gauze freely enough, and be exploded at the flame within it; but the ignited gas could not pass back through the gauze, and hence could not communicate with or explode the gas on the outside (fig. 71).   This observation, which seems insignificant in itself, became a revelation to Davy.   In the year 1815 attention was much directed in England to explosions of fire-damp, and the fatal obstacle which they opposed to the working of coal-mines.   " I surround the candle of the coal-miner with a metal gauze," says Davy ; "the flame will not pass through. Confined in its cage, it will not communicate with the gas, and explosions will not take place.   If they do, they will be merely little partial explosions, in contact with the flame, but they cannot spread."   Experience has confirmed all the predictions of the illustrious chemist, and the human race ought to hail his name as that of one of its greatest benefactors.   It is just to state that George Stephenson contested with Davy the priority of the invention ; having, without any knowledge of the researches of the chemist, devised a lamp in which the glass was protected by wire-gauze, and air was admitted to the flame through small apertures in the body of the lamp. The result, however, of a very careful examination of the question by a meeting of coal-owners, on the 11th October, 1816, was a decision that the merit of discovering a real safety-lamp belonged to Davy.   The two first Davy-lamps used in a colliery are preserved in the Museum of Practical Geology.

It is not merely in coal-mines infected with fire-damp that

the safety-lamp has been found of use. It enables us to approach and enter without fear sewers, the holds of ships, and all the close places where explosive gases are produced or accumulate. By the security and safety which it insures, it deserves its name, and it might also be called, like Aladdin's, the *wonderful lamp.*

Most inventions have at the outset some defect. Thus an objection has been made to the Davy-lamp (fig. 71) on account of its feeble illuminating power, and often if the draft of air is strong in the mine, of its allowing the flame to be driven through the gauze. To obviate this it has been proposed to surround the flame with a glass cylinder, keeping the gauze above the lamp (fig. 75). Other inventors have suggested that the metal cage should be altogether dispensed with, and have merely replaced it by a lamp with a thick glass chimney,

Fig. 71.—The first Davy-lamp.

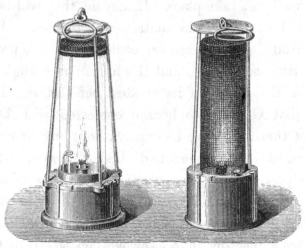

Figs. 72, 73.—English Safety-lamps.

the outer circumference of which is protected by some stout vertical iron wires to prevent the glass from getting broken, and the end of which is closed by means of a wire-covering (figs. 72 and 74). It is not necessary to describe here all these contrivances. In England and France the Davy-lamp is chiefly used. In many of the districts in this country, where the most fiery collieries prevail, the Stephenson, or

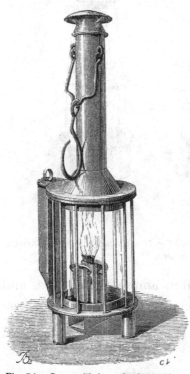

Fig. 74.—Dumesnil's lamp for fixed lights.

N.B. All the figures of lamps are drawn to a scale of $\frac{1}{5}$.

"Georgie," as it is called, has been preferred to the Davy. In Belgium, Mueseler's lamp with a glass cylinder is preferred (fig. 75). Dumesnil's is employed for fixed lights (fig. 74). Mons. Dubrulle, a lamp manufacturer at Lille, has effected some happy modifications of the lamps of Davy and Mueseler (figs. 76, 78, and 79). In other respects, whatever may be the system adopted, the lamps should be made in such a way as to go out if the miner attempts to open them ; or even in such a way that he shall not be able to open them at all, although he may still be at liberty to manage the wick.

In all cases they ought to be carefully examined, one by one, before every descent into the mine.

When accidents happen with safety-lamps, it is almost always owing to the imprudence or recklessness on the part of the miners. In a late catastrophe at Merthyr-Tydvil, where sixty-two men were either killed or injured, only sixty-one lamps were recovered. That of the culpable pitman—doubtless a smoker, who wanted to open his lamp in order to light his pipe—must have been carried away and blown to pieces by the explosion which it had first brought about.

It is well known how habitual contact with danger renders
people indifferent to danger itself.   If the safety-lamps are

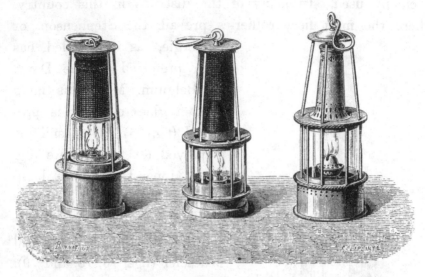

Fig. 75.—Mueseler's lamp with gauze and glass cylinder.   Fig. 76.—Dubrulle's lamp with chimney.
Fig. 77.— Petroleum lamp.

locked with a key, the miners often amuse themselves, and
try their ingenuity, by opening them out of sheer recklessness

Fig. 78, 79. — Dubrulle's lamps with wire-gauze.

in order to get a better light to work by, or perhaps with the
sole object of seeing how the fire-damp burns.   Should the

gas be more abundant than usual, a serious explosion may be, and often is, the result.

In 1852 M. Simonin visited a colliery near Saint-Etienne, over which he was accompanied by the viewer.   He took pleasure in uncovering his lamp from time to time, in order to show the phenomenon of the mixture of fire-damp with the oxygen of the air.   The fire-damp was not very abundant on the day in question.   It burned with a red flame, which was

Fig. 80.—Photo-electric lamp and box.

sometimes of a bluish tinge, and accompanied by a slight detonation.   A mere trifle might increase the quantity of the gas; and both the coolness of the experimenter, and the unaffected way in which he experimented with the fire-damp, were very striking.

In order to avoid the danger of explosions altogether, which lamps with a metal gauze or a glass cylinder do not entirely obviate, electric lamps have been proposed.   One of them, invented by Messrs. Dumas and Benoit, burning in a closed cylinder, in such a way that no fears need be entertained as regards igniting the fire-damp, seems to meet all objections (figs. 80, 81). Its use is also shown by its burning in all

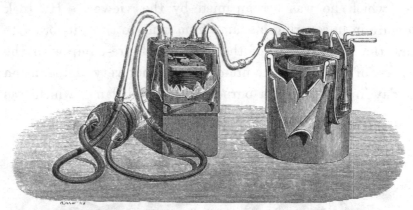

Fig. 81.—Generator of electricity; bobbin and illuminating tube shown out of the box.

atmospheres which are poor in oxygen, especially in cases when other lamps will not act.

In mines free from fire-damp the lamps are of various

Fig. 82.—Lamp used at the mines of Saint-Etienne.

forms. The round iron lamp of classical form is much used at Saint-Etienne (fig. 82); the small lamp made of white-iron (*tin-plate*), and fastened to the hat with a large nail, as at Anzin, and in Belgium (fig. 83); or the tin one in Wales. In some English mines candles are also used: these are held in the hand by a lump of soft clay placed round them, and which serves to attach them to the place whence their light will be best thrown upon the work. In the United States a lamp of tin-plate is used, in the shape of a siphon inkstand, and furnished with a hook at one end for hanging it to the hat.

Lastly, the use of petroleum lamps has been recommended (fig. 77). All these lamps, however, must give place to the safety-lamp, the moment inflammable gas makes its appearance.

Experience has thrown doubts upon the security of the so-called safety-lamps. This has led to actual experiments of a very interesting kind, which have been carried out in the Hetton Colliery, and at Barnsley. The lamps were placed in a rectangular box, and explosive mixtures were forced through at different, measured, rates. When the air was travelling at five miles an hour, the Davy lamp with no shield on the outside exploded in six seconds, and with the shield inside the gauze, in nine seconds; the Belgian lamp exploded in ten seconds, and the small clanny-lamp in seven seconds; the Stephenson in seventy-five seconds. These experiments show that, as we have improved the ventilation and increased the

Fig. 83.—Lamp used at the mines of Anzin, carried in the hat.

current of air, we have at every step impaired the safety of a lamp which under the old conditions of ventilation was absolutely a *safe* lamp. It is, however, satisfactory to know, from the results which have been obtained during the continuation of the experiments alluded to, that modified forms of the original safety-lamps may be used without fear of explosions.

Fire-damp does not exist in all coal-mines (as in those of Somerset and the Forest of Dean, for instance); only certain varieties of coal give it out. These are commonly the fat (*grasses*) or caking varieties, or those which contain much bitumen and volatile matters. The gas is accumulated in the pores and fissures of the coal, often under very great pressure, and bursts out from what are called *blowers* with a slight noise, with a sort of *gurgling* sound, reminding one of that produced by effervescing waters when poured out of the bottles. This noise is a kind of warning call, or appeal, to the vigilance of the miner. It is well known to the colliers, who in England, doubtless from a love of onomatopœia, have given it the name of *puff*.

It has been thought that the density of the air prevents the escape of carburetted hydrogen from the coal, but that is a mistake arising from an imperfect conception of what really takes place. There may be some conditions of the atmosphere more favourable than others to the liberation of carburetted hydrogen, but this is not a case of difference of density. It is well known that a sudden fall of the barometer is frequently followed by explosions in mines. The explanation of this is easy. In the goafs and waste places of the mine the carburetted hydrogen has mixed with air, and formed an explosive compound; by a fall of the barometer—a diminished atmospheric pressure—this expands out into the galleries and working places, and if it meets with a naked light, an explosion ensues.

Another gas, carbonic acid (surfeit or choke-damp of the miners), exudes from the coal by a sort of slow, natural distillation. Carbonic acid is not in itself a poison, it is simply irrespirable; it suffocates by entering the lungs, where it prevents the access of the oxygen of the air. The action or mechanism of breathing is stopped, and with it life ceases.

Carbonic acid does not exist in collieries only; it is one of the most widely spread gases in nature. It is given off in the vats of beer and the casks of wine that are undergoing fermentation, in newly cut hay stacked too hastily and in a damp state; it is the gaseous element of effervescent waters; it is met with in volcanic eruptions and in many grottos; lastly, it is the direct product of the combustion of charcoal and of the respiration of animals. One of the functions of plants is to purify the air by removing this carbonic acid.

Everything, then, concurs to vitiate the air of a coal-mine. Not only does the labyrinth of works sometimes oppose very considerable difficulties to the diffusion of fresh air proceeding from without; but the composition of this air itself soon becomes changed, either, as we have just stated, by mixing with the gases which are spontaneously given off by the coal, or with those resulting from the respiration of men and animals, the combustion of lamps, and the rotting of timber. The firing of gunpowder gives rise to unwholesome emanations —sulphurous acid gas, sulphuretted hydrogen, carbonic oxide, carbonic acid, ammoniacal vapour, &c. Finally, the temperature of the mine is always very high, as much in consequence of the production of these gaseous emanations, as on account of the depth to which the workings have been carried. It is well known that the thermometer rises at the rate of about one degree Fahr. for every 30 mètres, or 60 or 70 feet, of descent beneath the surface.

For all these reasons it may readily be imagined how many difficulties there are in ventilating a coal-mine. The miner has not the less attempted to overcome them, and in most instances with success. He ventilates mines in a way analogous to that adopted in the Houses of Parliament, for the air underground is subject to the same laws as those of buildings; he places doors, which act as air-valves, in the levels, in order to direct the current to take a particular course, to divide it, &c. In general, the mine has at least two shafts, sometimes at different levels; the air enters through one of them and diffuses itself through all the working stalls, and escapes through the other, as it does through the chimneys of ordinary dwellings. The power of the current in the winter will be lower—where natural ventilation is adopted—than what it is in the summer. The strength of the draught depends on the difference between the external and internal temperatures, the latter varying little, while the former follows all the changes of the season. This double movement of the air of mines is what is, as we have said, called natural ventilation.

As the works are developed and deepened, or when the mine only affords one outlet, the spontaneous or natural ventilation is no longer sufficient, and the quantity of air passing through the colliery is increased by means of a furnace constructed at the bottom of the shaft, in which a fire is kept burning day and night. The draught which takes place in this lofty chimney carries away the foul air, while the fresh air which enters from the surface through the downcast shaft circulates through the mine, and is discharged in its turn through the upcast shaft or ventilating furnace.

At other times the gas of the colliery is drawn out by means of an enormous mechanical ventilator, placed at the mouth of the shaft; the fans or vanes of which, put in motion by steam like those of a ship, draw up the air from the mine in their rapid revolution. The velocity is so great that the fans are not individually distinguishable; only a loud roaring is heard, which agitates the air at a distance. It is the noise of a hurricane let loose. The internal air is drawn off by this kind of suction, while the external air naturally passes downwards through another channel to fill up the vacuum, as Aristotle would have expressed it. This fresh air is distributed through the mine, by means of the air-ways, to the most distant workings. Ventilators are also sometimes established in the galleries themselves. These are generally worked by boys, being of much less power than those placed at the pit-mouth. Besides, they have to serve another purpose, to supply air to the galleries where it is most wanted.

Instead of ventilators placed at the surface, air-pumps are sometimes employed, composed of two such machines worked by a steam-engine, and of such a size that those used in experiments in Physics are mere children's toys by comparison.

Not only is pure air supplied to a mine by these powerful means, but all the irrespirable gases are got rid of which accumulate in it, and which often form as many distinct layers in the atmosphere of a gallery as those liquids of different specific gravity which separate from each other in a flask, in the order of their respective densities. The carbonic acid, being the heaviest, occupies the lower part of the gallery. If a dog were carried in it would fall down in a state of suffocation, as in the celebrated Grotto del Cane of Puteoli, near Naples. The lamp of the miner would be extinguished on coming in contact with the carbonic acid. Above this last, forming the middle layer, is the more or less pure air ; and lastly, occupying the higher levels, the light carburetted hydrogen, the inflammable gas, which when mixed with air forms the terrible fire-damp with which we have become acquainted. Sometimes all these gases are intermingled by reason of the operation of the diffusion of gases, forming then one homogeneous whole. It is then that explosions and suffocations are especially to be apprehended. A good ventilation, either by natural or by artificial means, and the use of safety-lamps, are the best means of avoiding accidents from these sources.

On entering a mine where the air is bad, the lights gradually grow pale, and are finally extinguished. A sort of pressure on the temples is felt, a faintness, a sense of extreme lassitude and fatigue in the limbs, the action of the heart fails, and suffocation approaches. Instant retreat then becomes necessary. With what delight a fresher and purer air is breathed : that which was just left being, in the language of the miners, *dead*. The lights either not burning in the bad air, or becoming extinguished on the slightest motion, a better place must be gained without a moment's delay. This accomplished, the lighting of the first match is hailed with joy ; and, less inquisitive than Lot's wife, one walks straight onwards, without casting a look behind.

It is sometimes required to enter working places that are filled with deleterious gases, either for the purpose of carrying aid to suffocated men, or exploring old workings. It has lately been proposed to provide the miners with a respiratory apparatus, carrying a reservoir of pure air. A sort of bag made of leather or metal is carried on the back, and into which air has been driven under a sufficient amount of pressure. An india-rubber tube, proceeding from the bag, spreads out at the other end, and fastens to the mouth against the teeth. It is furnished with two valves, one opening inwards to carry the air to supply the lungs, the other outwards to carry off the air after it has been breathed. The nostrils are closed with a spring. Sometimes the end of the tube opens in such a manner that it can be applied simultaneously to the nose and mouth. For long journeys additional bags filled with air may be taken, which are carried on a wheelbarrow.

This apparatus, devised by Pilastre des Rosiers, whose name has been so sadly associated with aeronautical ascents, has been improved by Humboldt. M. Rouquayrol, engineer to the collieries of Aveyron, invented, some years ago, a more perfect apparatus. The reservoir is made of sheet-iron, capable of resisting pressures of from twenty-five to forty atmospheres. Air is forced into it by means of ingeniously contrived pumps, the pistons of which are fixed, while the cylinders are movable. *Nous avons changé tout cela*, "We have changed all that," the inventor may exclaim with Molière ; for until now it was the pistons which moved, and the body of the pump that remained fixed. The apparatus is fastened on to the back, like a soldier's knapsack (fig. 84). A contrivance consisting of a kind of bellows, placed above the reservoir, allows the air (although very greatly compressed) to enter the lungs at the ordinary pressure only. A small external valve, formed of two leaves

Fig. 84.—Saving Life by means of Rouquayrol's Apparatus.

Fig. 85.—Rescue with Galibert's Apparatus.

of india-rubber, which are kept in contact with each other by the pressure of the atmosphere, opens to allow the escape of the air respired.

In this apparatus, at the same time as it is breathed, the air is sometimes supplied to a lamp especially constructed to burn in an atmosphere in which a common lamp or a Davy could not be used; but in such a case the electric light, burning in a closed tube, would appear to be a still safer and better mode of light (figs. 80 and 81).

Provided with Rouquayrol's apparatus, a man can breathe with ease even under water; in proof of which some conclusive experiments have been made in the beds of rivers and at the bottom of the sea. By means of this invention, submarine operations, pearl-fishing, diving for coral and sponges, repairs to sunken ships, quarrying rocks under water, the raising of ships from the bottom of the sea, all become practicable henceforth on a very large scale and without any danger. The Rouquayrol apparatus might sometimes be found useful in mines for working under water, at the bottom of sumps,* for instance, for repairing broken pumps.

The system invented by M. Galibert for enabling persons to enter an atmosphere charged with mephitic gas must not be passed over in silence, even by the side of the preceding. The simplicity of the mechanism is extreme; it contains no part capable of derangement, and it can be made ready for use in a moment. With a small cylindrical leather bellows, having precisely the shape of a transparent paper-lanthorn, the air is injected into an air-tight goat-skin. The man carries this upon his back, where it is kept in its place by means of straps which go round his shoulders and waist (fig. 85). Two india-rubber tubes are fitted to the goat-skin bag, terminated together by a piece of horn or ivory, hollowed out round its circumference,

---

* The reservoir of water at the bottom of a mine-shaft is generally called a sump (*puisard*, in French). In South Staffordshire the term is applied to a heading.

and held between the teeth ; this is the mouth-piece. The nostrils are compressed by a small nose-pipe made of wood and closing with a spring. Lastly, when necessary, in a very hot atmosphere or in case of fire, the eyes are protected by a pair of spectacles with glasses mounted in leather.

Thus clad, the man is free and unfettered in all his movements, breathing is carried on without difficulty, the air which has been respired is returned into the reservoir, and may be used twice over without inconvenience ;* but at the end of a quarter of an hour it becomes necessary to renew the air-bag. The tube through which the air passes outwards opens at the upper part of the bag, while the tube by which it is returned, on the contrary, goes to the bottom. Besides, the air which has been breathed being the heaviest, naturally occupies the lower part of the bag. At the end of some seconds, a movement of the tongue is instinctively and unconsciously produced, which stops up or opens in the mouth-piece, sometimes one of the tubes, and sometimes the other, according as the lungs inhale or respire the air.

The apparatus of M. Galibert, as it has just been described, is especially used amongst pump-men and well-sinkers, repairers of sewers, &c. It has been employed by the men engaged in purifying hospitals and holds of ships. For mines, the inventor has had reservoirs of tin-plate specially constructed, which are not liable to burst like goat-skin, and which allow the air to be forced in under a greater pressure.

---

* Common air contains 21 per cent. of oxygen, and 79 nitrogen. In the process of respiration three parts of oxygen are used and converted into carbonic acid. The expired air contains, then, 18 per cent. of oxygen. In this state it may be used a second time. With 15 per cent. of oxygen the air is irrespirable.

Fig. 86.—Falling in of a Mine.

CHAPTER IX.

FALLS OF ROOF AND INUNDATIONS.

New enemies of the Collier : Falls of roof.—The well-sinker Giraud.—Cochet the
miner.—Irruptions of water.—Inundations at Heaton Colliery, near Newcastle ;
at Workington Colliery ; at a colliery in Durham.—Inundation of a mine near
Liége ; of a colliery of the Loire.—M. Duhaut and his pupils.—Song of the
students at the Mining School of Saint Etienne.—Terrible inundation of the
mines of Lalle.—Bravery of Auberto, the Piedmontese timberman.—Affecting
details.—Irruption of water in the mines of Beaujonc.—History of Hubert Goffin ;
his heroic conduct ; his death.—Inundation of a mine at Charleroy.—Adventures
of Evrard, the collier.

THE contest is far from being ended. We are about to see
the collier contending against obstacles scarcely less terrible
than those which have been already described ; viz., falls of
roof and irruptions of water. The rock is split, and bulges
out, or else the ground is loose and running. We know how
it is resisted by ingenious supports of timber or masonry.
Nevertheless, these works sometimes give way under the
enormous pressure of the earth, and the result is a fall which
buries the miner in the ruins (fig. 86).

Falls are the kind of danger which always enter into the
imagination when underground works are thought of. Nobody
in France has forgotten the story of Giraud, who was excavating a well near Lyons in 1854. The poor fellow, dashed to
the bottom of the hole by a fall of the ground from above,
which had been perhaps insufficiently propped, beheld a sort
of vault suddenly form above his head, which crushed him
under its weight, and kept him prisoner together with his
fellow workman. The question then was how to save these
poor fellows. It was necessary to dig a new shaft near the
first, and then to connect the two by means of a drift-way,

which should strike it at the point where the accident had taken place.   In spite of all the exertions which were made, a whole month was spent in bringing the operation to a close, for fresh falls occurred in the new workings themselves. Giraud and his comrade heard the noise of the picks, and replied to the workmen, thinking every moment that the hour of deliverance was at hand.   Vain hope! The second man died.   Hunger added its horrors to the sufferings of the survivor, as in the sad story of Ugolino.

Giraud, a person of greater energy than his companion, bore up.   The corpse of his friend, which lay near him, poisoned the little air he had to breathe ; but the desire to live sustained him.   Neither hunger, nor this unpleasant proximity, cast down this man : he wished not to die.   He carried on the contest for an entire month.   Every moment it was expected that he would be reached, when some fresh accident happened, which rendered it necessary to begin the work anew.   Giraud did not succumb ; he replied distinctly to all the questions that were put to him.   France, indeed, all Europe, watched the contest day after day, and a bulletin was published every evening of the day's progress.   On the thirtieth day victory was achieved, and Giraud was saved. Pale, wan, and reduced to a skeleton, his body was a mass of sores.   Gangrene had attacked all his limbs, caused by the corpse, which for three weeks had been rotting by his side.   The unfortunate well-digger was carried to the hospital at Lyons, where, after lingering on for some time, he expired.

Falls of ground may be ranked amongst the greatest perils which the miner has to guard against.   If the shock be direct, the man is crushed on the spot ; or if he escape, it is at the cost of a limb.   Masses of rock from the roof, bell-moulds (*cloches, culs de chaudron*). as lumps of ironstone are called in the figurative language of the colliers, sometimes become suddenly detached without the least warning, from the shales or friable coal forming the roof. These lumps, frequently of great size, falling on the head of the miner, often kill him outright. *

In other cases, the wallings and timberings give way under the enormous pressure of the ground. Woe betide the workmen who do not flee in time : they are crushed in the mass of ruins (fig. 86). There are, however, instances of persons having escaped with their lives, of whom that brave collier Cochet is one. Overtaken by a fall of coal in the mine of Creuzot in 1864, he had sufficient strength to call for help. His comrade, who had left him for a moment, returned and gave the alarm. The most skilfully devised means of rescuing him were immediately commenced. One part of the coal having been removed from around the sufferer, his head and one hand became visible. Cochet was lying under a mass of broken timber, thrown upon the floor of the gallery on his right side, with his legs doubled under him. To move was impossible ; but fortunately his chest was not compressed. Air was supplied by means of a ventilator and a tube. The rails were sawn, as well as the cross-pieces and other timbering in which the miner was entangled. When the gallery was cleared so as to reach him from below, his legs were first set free. As for himself, he did not lose courage ; he retained all his composure, and gave his preservers more than one useful suggestion. Finally, after six hours of horrible suffering, he was literally withdrawn from this tomb, where he was upon the point of being buried alive. All the workmen had vied with each other in ardour and skill to save their comrade from

---

* The globular nodules, oblated at the bottom, and on that account locally called *bell-moulds* in the Bristol coal-field, occur in the sandstone overlying the coal, and sometimes in the heart of the rock, but separated from it by a thick coating of oxide of iron. The nodules of bituminous shale met with in the soft roofs of some Scotch collieries, are called by the miners *creeshy* (greasy) *bleas*, from the sort of unctuous smoothness which causes them to fall out when the coal is worked away from under them. What are sometimes called *cauldron-bottoms*, and render the roof of the coal in some mines highly dangerous, occur in the roof of the Bensham coal-seam in the Wallsend collieries. They are often so large as to require six or eight men to lift them.—" History of Fossil Fuel," p. 248.

death.   In such cases, zeal and energy are never wanting in the miner, and he is never deficient in those sentiments of strict confraternity which ought to subsist between all who are liable to the same accidents.

The danger of underground inundations is as formidable as that of falls of ground.   We have seen the coal-miner oppose wooden or metal dykes to the liquid element which bursts in on all sides, enormous pumps and tunnels which he transforms into canals, thus turning the evil to good account.   The water accumulates in the mine, in a body, in basins, in actual lakes. He keeps it there by dams made of cement or clay—by wooden framework, the different pieces of which are geometrically put together like the stones of a wall or a vault.   We have seen equally cleverly devised masonry built up in the shafts; and yet the pressure of the water is sometimes so great as to overcome all these obstacles.   An old English collier, who believed the earth was alive, compared the veins of water met with in mines to the veins and arteries of the human body. "When the water breaks into our working-stalls, it is the earth which revenges itself upon us for having cut one of his veins." *

Inundations may stop the working of coal, equally with fires.   The mines of Rive-de-Gier were on the point of being irretrievably ruined in 1838.   The works had been badly carried on in all the concessions, for at that time the art of mining was not so advanced as it is now.   Old workings of many centuries standing served as channels for the passage of some of the surface-waters into the interior of the mine; the River Gier itself broke in.   The persons working the mines, nearly all of which were connected together, at least in a geological sense, not being able to arrive at an understanding,

---

* Perhaps it is in consequence of some such belief that the Belgian miners call the water which flows out of the coal "the blood of the vein" (*le sang de la vein*).   But here vein is synonymous with seam of coal.

MAP IX

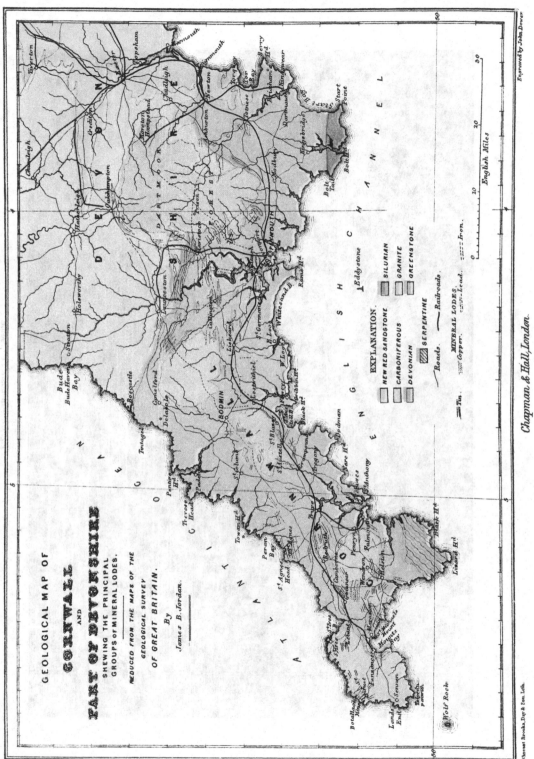

GEOLOGICAL MAP OF

CORNWALL

AND

PART OF DEVONSHIRE

SHEWING THE PRINCIPAL
GROUPS OF MINERAL LODES.

REDUCED FROM THE MAPS OF THE
GEOLOGICAL SURVEY
OF GREAT BRITAIN.

By
James B. Jordan.

EXPLANATION.

NEW RED SANDSTONE
CARBONIFEROUS
DEVONIAN
SILURIAN
GRANITE
GREENSTONE
SERPENTINE

Roads —— Railroads

MINERAL LODES.
Tin. Copper. Lead. Iron.

English Miles

Engraved by John Dower.

Chapman & Hall, London.

Vincent Brooks, Day & Son, Lith.

Fig. 87.—Inundation of a Mine.

it became necessary that a law should be passed uniting them into a sort of syndicate, in order to extricate them from the danger.

In this instance the admission of the water was known, and also the way to prevent it; but there are some kinds of underground inundations the cause of which may be said to be unexpected. The rains of heaven, descending like a deluge on to the surface, and flooding the ravines, sometimes find their way into the mines through the galleries, or the natural fissures in the ground, and lay them waste. A whole river breaks into the works, and carries away and drowns men, horses, &c. (fig. 87).*

The irruption of water from old workings is, perhaps, most to be apprehended. These irregular openings and ancient excavations exist in most mines, and in the absence of proper plans their extent is often unknown. Some are many centuries old, and date from the earliest mining times; forming the worst neighbours a colliery can have, by becoming the receptacle of the rainfalls which accumulate in them, magazines of irrespirable gas, and the scene of subterranean fires. All these reasons prevent them from being visited, and the plan being carried on to make it agree with that of the mine. When the requirements of the workings lead in the direction of these old places, they must only be approached with the utmost precaution, especially when bodies of water are suspected. It is usual, in such cases, to sound the ground carefully in advance, by means of a borer worked horizontally in front of the face of the heading. The tool, which is screwed to a long rod, varies with the nature of the rock which has to be perforated (figs. 88 and 89). If the water is not met with, the whole distance examined is safely traversed, and then a fresh boring is made.

* Wheal Rose Lead Mine, Newlyn, Cornwall, was flooded and lost by a sudden and enormous fall of rain; and in the year 1856 the South Tamar Consols, near Beerferris in Devonshire, was destroyed by inundation, owing to the giving way of the bed of the River Tamar, under which the workings were carried.

If water is met with it is merely a jet like that of a fountain, and is allowed to run off quietly.

In some cases, however, the quantity of water entering through these borings is sufficient to drown the mine, an instance of which occurred in 1825 in a colliery near Liége. The water, driving out the boring-tool, rushed in with impetuosity. The workmen made several attempts to plug

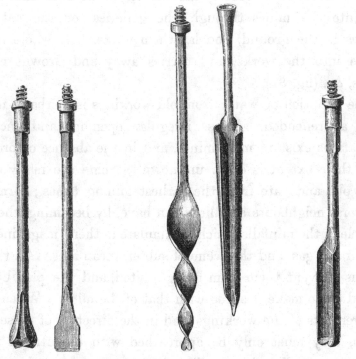

Hand boring-tools for the interior of mines.   Scale ⅛.

Fig. 88.—Chisels for hard rocks.   Fig. 89.—Augers for soft rocks.

the hole, but could not succeed in doing so, and in a short time their only safety was in flight. Some hours afterwards all the working-stalls were flooded, and became altogether inaccessible. The old workings extended under the very bed of the Meuse, and the river may be said to have been turned into the mine.

The proprietors were not discouraged by so great a disaster; but although pumping machinery was put up at four shafts at once, it was not till the end of seven years that the water could be kept under. It was then only that a descent could be made, when every inlet for the water was closed by means of banking and brattices put up with the greatest care, and the works were renewed. This example of energetic perseverance is one of the most remarkable in the history of coal-mining.

One of the most fatal cases of inundation took place at Heaton Colliery, near Newcastle, on the 3rd of May, 1815, owing to a failure in exploring through a fault, where the ordinary thickness of coal between the workings of the colliery and the inundated wastes of the adjoining Old Heaton had been reduced till it gave way under the extreme pressure. A dreadful inundation was the consequence, which immersed ninety human beings in the upper part of the workings, from which there was no possibility of relief. Some miraculous escapes were made by some of the persons who happened to be close to the shaft, but the place where the water burst in being many hundred yards distant, and forming a descending plane of one in ten, the tremendous force of the current bore down every obstacle, and hurled in the most awful confusion men, horses, carriages, and all that stood in its way towards the shaft, where some fortunate beings escaped when the water was nearly breast high. Many months elapsed before the waters were sufficiently drained to render the workings again accessible. On the 19th of February in the following year fifty-four bodies were brought to the surface, and were all recognized with one exception, principally by their clothes. Twenty-nine were buried at Wallsend, and the mournful procession filled the road for a mile in length.

The bodies of these poor people were without smell, but so perfect and clay-like, that upon pressing the flesh, the indentation remained. No opinion can be formed whether they had suffered much bodily pain; but their mental agonies must have

been of the most maddening description, since no earthly hope
could be entertained that it was in the power of their fellow-
men to afford them any succour. *

Workington Colliery furnishes another sad instance of the
destruction of a coal-mine by inundation. This colliery was
worked by means of two pits situated on the sea-shore and
ninety fathoms deep, to the distance of 1500 yards under the
Irish Sea; and the workings were driven considerably to the
rise, so that they had been brought at length within fifteen
fathoms of the bottom of the sea.

Although the pillars of coal which had been left to support
the overlying strata were barely large enough to afford the
necessary support to the roof, the manager, in his anxiety
to produce a large quantity of coal, proceeded in a reckless
manner to reduce them still further in size by working them
partly away. Warning of approaching danger was soon given
by some heavy falls of the roof, accompanied by discharges of
salt water, and several of the men left the colliery through
dread of the consequences.

Matters had gone on in this way for some weeks, till on
the 30th July, 1837, the whole neighbourhood was appalled
by the breaking in of the sea; and so extensive was the
commotion, that many persons, at the distance of hundreds of
yards, observed the swirl of the waters directly over where
the fracture took place. A few of the pitmen escaped by
groping their way to the day-hole in the rise workings; but
thirty-six men and boys, and as many horses, together with
all the stock underground, were irrevocably destroyed, the
waters having filled up the whole of the extensive workings
in this thick seam to the level of the sea in a few short hours.

The case was the more deplorable since the deluge, before
it occurred, was not only the subject of common conversation,
but, also, after repeated warnings had been given.

* See "On the Winning and Working of Collieries," by Matthias Dunn, p. 224.

This colliery is yet, and ever must remain, under water, as also the bodies of the people; from the magnitude of the breach of the strata under the bed of the sea assuredly no engine power can ever be made effective in the draining of those workings. *

Recently, at a colliery in Durham, an inundation which threatened to be fatal to a large body of men has occurred. In this case two Cornish miners are said to have holed through to the old workings.

In these irruptions of subterranean waters, the lives of the men are almost always in jeopardy. Thirty years ago, in a colliery of the Loire, they were driving in the direction of old workings. The miners drove the levels carelessly and without properly securing them, and, in fact, they neglected to take any precautions; when, all at once, the water burst the narrow wall which yet divided it from the men at work, and descended upon them like an avalanche. The terrified men fled up an inclined gallery, in which they took refuge; but this gallery had no outlet, and the water rose to the place where they were. The people above ground were most anxious to save them, but what was to be done? how could they be rescued? where were they? were they not already suffocated or drowned? But there was no hesitation, everything being done which humanity suggests in such a case. The men were supposed to be still alive, so their comrades resolutely set to work to find them. The surveys of the mine were well executed, and furnished a ground plan and vertical section of the workings; the former giving the precise position of the stall in which the miners were working, and consequently of the shelter they might have sought. The section, on the other hand, made known the vertical depth of this point from the surface, and therefore the degree of inclination between a given spot and the place of refuge. A gallery was

* "On the Winning and Working of Collieries," by Matthias Dunn, p. 230.

driven on this slope in the direction of the supposed point. The blows of the picks struck upon the rock as a signal remained unanswered at first; afterwards a sort of faint answer was heard in reply to these repeated knockings. It is a well-known fact that sounds are transmitted through rocks for very considerable distances; and the American Indians avail themselves of it, when, with ear placed on the ground, they listen for the sound of the approaching horseman. "Come, boys, courage, and on with the gallery." In a little while sounds were transmitted more distinctly, and it no longer became necessary to tap with the pick in order to attract attention; even the sound of voices was heard. Quick with the borer! God be praised! the colliers were there, all living. They were communicated with, and asked what they wanted most. "Light before all things," they exclaimed; and many days had elapsed since they had eaten anything. Lights were handed to them, and afterwards broth was poured down a tin pipe passed through the bore-hole. At length the last blow of the pick was struck, the sufferers were out of danger, and the prisoners saw daylight again. What must they not have endured during those long hours of waiting! They had eaten their candles, and devoured even their leather straps; and yet so distressing is prolonged darkness, that their desire was to see before they asked for food.

M. Simonin's excellent master, M. Duhaut, who was tutor at the Mining School of Saint Etienne in 1851, related to his pupils this accident in all its details. He had assisted at the rescue. The professor drew a moral from this story, after the fashion of the worthy Æsop in his fables. He attempted to prove from it the utility of adding a vertical projection to mining plans, connecting each point in the interior by true geometrical co-ordinates with the surface. And the boys, in order that nothing might be wanting to the lessons of the master, composed a ballad which began thus :—

> Mineurs, écoutez l'histoire
> De trois malheureux ouvriers,
> Restés sans manger ni boire
> Pendant six grands jours entiers.
> Au fond d'une galerie
> Serrés comme en un bocal,
> Ils auraient perdu la vie
> Sans la coupe verticale.

This was sung at the beginning and end of lessons, when the master was not present, to all sorts of tunes, with admirable ardour.

One of the most terrible inundations within the memory of colliers, is that which took place in 1862 at the mine of Lalle, near Bességes in the Department of the Gard.

On the 11th October, between three and four o'clock in the afternoon, a violent storm visited the country, and it is even averred by some that a water-spout burst at this place. The waters of the river Cèze, as well as those of a stream and of a ravine which is dry at ordinary times, and both of which are tributaries of the Cèze, rose higher than had ever been seen before, even at the time of the great floods of 1840 and 1855. It was a vast inundation—a regular deluge. All at once the water made a whirl at one point, and then rushed tumultuously into the mine, through a large opening which formed itself over the outcrop of one of the coal-seams.* The plane of the seam itself served as a channel for the water. An alarming rumbling noise reverberated from gallery to gallery; all the hands were at the moment at work in the mine; the catastrophe was grievous! A gang of startled colliers, nevertheless, were able to escape in time by one of the ladders; while some others hastily mounted a shaft closely followed by the water. A noble act of devotion took place. A brave timberman, a Piedmontese, Auberto by name, also escaped up a shaft, but not before giving the alarm to his

---

* A year afterwards, when passing through Lalle, an inhabitant showed M. Simonin the spot where the fissure was formed, and gave him an account of the accident.

comrade, who was working at a lower level. He ran to another opening, fastened the tub to a rope, descended, and called out, the water falling the while in torrents. Five men made their appearance ; four of whom got into the tub, while the fifth, hesitating a moment, was lost. They had scarcely reached the surface before Auberto caused himself to be let down again, and perceiving a young man entangled amongst the timbering of the lower gallery, he drew him towards him, threw him into the tub, and re-ascended, at the very moment when the water took possession of that part of the mine. Auberto had saved six lives, and would have saved more, but no other point was accessible, the whole mine being then under water. One outlet remained open, which had been formed by a foundering of the ground close by the opening through which the waters found their way in. Lights were seen shining there, and ropes were thrown in lashed to trees. Alas ! the violence of the water increased, the ground fell in afresh, this last outlet also became closed in its turn, and all the men were drowned ! Half an hour sufficed to convert this mine into a lake. The air and gas of the interior, violently compressed by the weight of such a vast body of water, found their way out even through fissures in the ground, producing the effect of an explosion of gunpowder, throwing the earth to a distance, and even overturning houses.

At the earliest intimation of the accident, M. Courroux the engineer, and the overman Martin Degasso, are on the spot, where they are soon joined by the engineers of the neighbouring collieries or foundries, all of whom might be named, for they all came, accompanied by their overmen. An engineer was also summoned in haste from Saint-Etienne ; for there were not too many people for conducting all the operations. Lastly, M. Parran, of the Imperial Corps of Mines, is dispatched from Alais to direct the salvage, in which all the authorities equally take a part.

No immediate succour is possible; the colliery is, perchance, only one vast tomb, for out of a hundred and thirty-nine lamps which had been given out in the morning, only twenty-nine were accounted for; or, in other words, there were a hundred and ten miners still in the mine, and amongst them the four butties—the captains with the soldiers. All these persons were scattered, one at one point, another at another, at different depths, as occasion required. How were these poor fellows to be found? was it even certain that there was a single one of them still living?

Whilst a dyke was being made at the surface to keep off the water, and the promptest and surest means of preservation were being studied on the plan, a young rolley-boy, who had previously been employed as a hooker-on in the underground winding, entered into a gallery on Saturday afternoon, the 12th October, twenty-four hours after the accident took place. He knocked on the walls, and after listening for some time, thought he could distinguish sounds answering to his own. Having called his comrades, he repeated the experiment, which was followed by the same result. The engineers were informed; everybody hastened to the spot. M. Parran sent some persons to insure the utmost silence, and made a signal by knocking with a pick at equal intervals of time. He has left an exciting account of these operations. "With ears resting on the coal," he says, "and holding our breath, we soon heard, with profound emotion, extremely faint but distinct and timed blows—in a word, the miners' signal—which could not be the repetition of our own, because we had only knocked at equal intervals."

A solid wall more than twenty yards thick intervened between the prisoners and their rescuers, which had first to be cut through; but the greater part of the miners were shut up in the mine. Who would remove the rock? The neighbouring companies generously lent their hands, and the first blows of the pick, which were soon heard, bore hope to the hearts of the prisoners. From six o'clock in the evening the work was carried on. Operations were commenced at five different points, by means of inclined drift-ways driven in the direction of the places where the victims were supposed to have taken refuge; the starting point of these drift-ways being in the very gallery where the signals were heard. One pickman only, relieved when his strength was exhausted, worked at each heading with all the energy he was capable of. The coal was carried away in baskets as fast as it was hewn down, and passed from hand to hand, along a chain of men who were ranged one above another in the drift-ways. The labour proved more difficult in proportion to the depth, in consequence of want of air, and it became necessary to put up ventilators, and sometimes the lamps would only burn in front of the air-pipe. On Monday, the 14th October, at 2 o'clock in the morning, the captives were communicated with. They said: "We are three"—"There are three of us," and gave their names. The efforts were redoubled, but as though by a sort of fatality, the coal increased in hardness. On Tuesday the work was carried on to the utmost; the air failed, and the heat became unbearable. The best pickmen were to the front, hewing the coal with all their strength, the prisoners the while making themselves continually heard. Finally, the same day, at midnight, one of the drift-ways reached the hiding place of the prisoners, two of whom were still alive—the youngest sobbing, the other in a high state of fever; the third, an old man, was unable to survive the trying ordeal, and lay dead not far from his companions. His body was found at the bottom of the gallery, in the hole where the three miners had taken shelter.

Covered with warm wrappers, refreshed with a cordial, and lying on beds, the two surviving men were carried gently to the mine-infirmary, where they received medical attention; and the next day they were already in a satisfactory state.

The work of preservation had continued without intermission from Saturday evening at six o'clock to midnight of the 15th October; that is to say, for seventy hours. On calculating the quantity of solid rock removed in driving one of the drift-ways, which had not been interrupted for a single instant, it was found that a full *month* would have been required, under ordinary circumstances, to do the work which had been accomplished in *three days*. This simple calculation proves the activity and the energy which had been exerted, as well as the discipline which had been unceasingly maintained, throughout this difficult operation.

The most precise details of the circumstances which marked their confinement were taken from the actual mouths of the two rescued colliers. They were at work in a heading when the water was heard coming in upon them, upon which they ran to the upper end of the gallery where they were discovered —a narrow place, with a considerable slope, and very slippery. They dug a little place in the shale to sit down in, with their hands and the hook of their lamps. The water reached to their feet, and they were in a sort of bell, in which the air was highly compressed. * They felt a singing noise in their ears, and they lost their voices. Their lamps went out for want of oil. They tapped with the heels of their shoes on the walls of the gallery to summon assistance, which was the sound that was heard, but only after they had been knocking for twenty-four hours! Convinced that help would arrive, the oldest of the three, he who would never behold the light of day again, shed tears of joy. Another, mad with thirst, descended into the level, with the water up to his arm-pits, in a vain search for a way through the rubbish; but he afterwards regained

---

* It has been calculated that this pressure must have risen to four atmospheres; that is to say, that it was four times as much as that of the external air.

O

his place, guided by the voices of his companions.   The youngest, seventeen years of age, frequently fell asleep, and would have fallen into the water but for the aid of his neighbour, who held him in his arms like a child, and thus saved him from death.   At one instant, the noise of the ventilator connected with the operations of their preservers reached their ears more distinctly, when they imagined that a new influx of water was about to take place, and became discouraged.   The old man especially was in constant activity.   Overcome by his efforts, he lost his resting place, slid over the inclined plane, and fell into the water and was drowned, without moving a single limb, or uttering a single cry.   Frozen with horror, held motionless in their places, the two others dared not go down to his assistance, even to raise his head.   Later, they even refrained from announcing this sad accident to those above: "There are three of us," they cried.   He who was devoured with thirst nevertheless determined to move, but touching a dead body while drinking, he soon climbed back again.   Fatigue, bad air, and this fearful vicinity to a corpse rendered him delirious, and he said to his comrade: "Come; let us leave this."   The other was frightened, and in order to divert his attention suggested that he should go again to drink; but he did not drown himself; his friend will have been saved twice.   He strikes against the dead body again in passing; desolating spectacle which no light comes to illuminate!

In the meanwhile the water got lower in the level, but it was cold there, and the two captives remained in their places, where the air was dry and warm.   At last they are recovered, and are carried into the light by the arms of their comrades. By a strange phenomenon, these two men had lost all notion of time, and thought they had not been in the mine longer than twenty-four hours—they had not even felt any hunger. *

* Other instances of similar facts are mentioned.   Some miners of Hainault, in the seventeenth century, who lived twenty-five days shut up in a rise, after an irruption of water, thought they had only been there eight or nine hours.

While the operations for saving the lives of these two victims, who were so miraculously preserved from death, were being carried on in this part of the colliery, other works were undertaken with the view of penetrating the interior at other points. Pits were dug where the miners were suspended from ropes for fear of explosions, while other workings were repaired which had been injured by the flood, and by means of these they proceeded to get rid of the water. One of the old shafts was undergoing repair at the time of the accident. In ordinary times fifteen days at least were required to refit the engine, put up the ropes, &c.; now everything was done in four days: the pumping began on the 15th October, and was not again interrupted.

Nevertheless, during the night between the 21st and 22nd, a fissure which communicated with the underground workings was pierced by a bore-hole; but the rock being fissured, the air could not accumulate at this point, and the colliers, if they took refuge there, must have been drowned. The gas which escaped through the bore-hole had a bad odour; it was the vitiated air of the mines. The workmen shouted; they rang a bell; but there was no reply.

They continued to bore and to dig shafts. On the 24th October, thirteen days after the occurrence of the accident, a depth of eleven yards had been reached, and the bore-hole was carried down twenty-five yards. All at once the men who were working at the bottom of the shaft heard shouts. Three men were still alive, only separated by rubbish and a vacant space of ground from the point where the workings were carried on. "We have been there a very long while," they cried. Disputes arose as to who should save them, but one of the overmen present was allowed the honourable favour of going down to them first. He met two men, who clung to him and entreated him to extricate them. He encouraged them, and covered them with his clothes. In the meanwhile the timbermen made

o 2

the ground secure, and in a short space of time they hastened to set the captives at liberty.

The third prisoner, a child, was still left. His comrades described the place where they had buried him in the coal in order to keep him warmer. The engineer, M. Courroux, flew to the spot, without even taking a lamp, seized on the child, who embraced him weeping. The three new victims who had been rescued alive from the colliery of Lalle soon found themselves in company with the two others in the hall of the hospital attached to the mine : an affecting neighbourhood in suffering and recovery.

Like their comrades, whose dramatic story has already been related, the three last colliers had fled before the water from the first moment of its breaking in, and finding a rubbish-passage stopped up, they despairingly made an opening into it. They afterwards clambered to the heading of a gallery as a last refuge. Their lamps were out, but they heard the water rise, and retreated before it. The noise occasioned by falls, and the breaking of timber, as well as of explosions caused by compressed air, all reached their ears distinctly, like a frightful tumult which seemed to announce to them the last hours they had to live. One of them had a repeating watch which he made to strike several times ; but it stopped on Saturday morning the 12th October, at a quarter to three o'clock. They remained closely huddled together for the sake of keeping warm. The noise of the tubs was heard plunging into the water in two adjacent shafts. They conceived the idea of reckoning the progress of time by means of the short intervals of rest caused by changing the setts, and thus formed a very near guess as to the duration of their captivity, which they calculated at fifteen days instead of thirteen. To assuage their hunger they eat the rotten wood of the struts, which they crumbled in the water, having previously devoured their leather belts ; but they could quench their thirst at will, and that sustained them.* Afterwards the water rose to where they were, wetting their feet. Subsequently the level of the fluid fell, and then they thought of fastening one of their boots to a string, and of drinking out of this improvised vessel.

Seeing the water gradually retiring, the child resolved to go in search of an outlet. Swimming, or holding on by the walls, he groped his way along the level, but soon fell into a hole and laid hold of a rail. Exhausted and chilled with cold he returned to his comrades, who lay close to him to warm him, and then covered him with small coal, in which position he was found.

These men were liberated after undergoing a confinement of thirteen days; the temperature, the pressure, and the composition of the air in which they were found shut up were favourable to life, and moreover they had the means of quenching their thirst. Under such conditions it might be possible to live for a month. What will not our poor human nature endure when it is compelled, and when the energy of life exists! The pitman Giraud himself, whose harrowing story has been related, did he not live all that time without water, almost without air; or what is still worse than that, in a poisoned air, with a dead body lying close beside him?

Only five men were saved in the catastrophe which happened at the mines of Lalle. All the others, to the number of one hundred and five, perished. Draining off the water by the engines was followed up with ardour, in the midst of accidents without number, the breaking of ropes, toothed-wheels, &c. The colliery was not entirely drained before the 4th January,

---

* M. Chalmeton, manager of the mine of Bességes, adjoining that of Lalle, has followed all the events of this memorable drama. He recently wrote to M. Simonin: " . . . . I have preserved the leather belt of one of the men who was saved, still bearing the mark of teeth. The poor devil had eaten five or six centimètres of it.

" It would be well, if you speak of this catastrophe, to say a few words about Dumas' electric lamp which only came into use a year after the accident, and has been of great service to us. We have been able to work more actively in the galleries, where the oil-lamps burned with difficulty."

Dumas' lamp is represented at pp. 169, 170, figs. 80 and 81.

1863 ; it had held in its interior 200,000,000 quarts of water (50,000,000 gallons). During the interval the bodies were slowly discovered, and heart-rending was the spectacle which the mouth of the shaft presented when the victims were drawn up; relatives and friends pressing forward and endeavouring to recognise or guess at some well-known face. We will not dwell upon those details ; they are too heart-rending. All France read them day by day, in the newspapers of the time, and still preserves the remembrance of them. Let us rather bestow a glance on the gallant men who did their duty, from the managers down to the humblest workman. Every man vied with his neighbour in doing what was needed, however difficult it might be. All the directors of mineral works in the Department of the Gard assembled, or dispatched their overmen, their surveyors, and workmen, who to a man gave proofs of a courage and abnegation of self which never failed for a single moment. The Government has bestowed crosses and medals, which is well ; but a modest column should also be raised on the very site of the accident, to perpetuate the memory of the unfortunate victims, and their brave preservers.

While alluding to the subject of subterranean inundations, that which occurred in one of the Liége collieries in 1812 must on no account be omitted. Its history is as dramatic as that which has just been related, and it remains engraven in the memory of all the old miners of the district in which it happened.

On the 28th of February a sudden irruption of the water, which had been dammed back in the old workings, surprised the colliers in the mine of Beaujonc, some of whom had just time enough to make their escape by means of the shaft, while others in their hasty flight were drowned ; the rest remained close prisoners. The overman, Hubert Goffin, could have gone up in the tub, but would not do so, and he even kept his son, a boy of twelve years of age, near him. Like the captain who ought not to abandon his ship in the moment of danger, he meant to remain in the mine, displaying the most heroic devotion, and the noblest resignation. "I will save all my men," he said, "or I will perish with them." Firm at his post, he encouraged and sustained everybody, striving to revive the courage of those who were on the point of yielding.

Scenes took place such as the pen cannot describe. Two men were engaged in a quarrel, and, while Goffin tried to separate them, some one exclaimed, "Let them fight, we will eat the one who is beaten." At another time all these men were seized with despair. The work that Goffin had caused them to begin, with the object of finding, if possible, a way out, having produced disengagements of fire-damp, they cried to their chief, "Do not close the communication; let us take the lights there and blow ourselves up." Some exhausted miners seemed to be nearly dying; their comrades, as they afterwards acknowledged, watched for the instant in order to devour their bodies.

All the lamps were extinguished for want of air; the weakest and most timid became delirious, complaining that somebody wanted to kill them, by leaving them without food or light. They imperiously demanded something to eat, and inveighed against Goffin.

They contended for the candles, which they devoured. Some went creeping along to quench their thirst. "It seemed," said they, "as though we were drinking the blood of our drowned comrades."

However, help from without came to the colliers. The engineer of the mines, a Frenchman, M. Migneron by name, who died Inspector-General some years ago, and the Prefect of Liége, Baron Menhoud (Belgium belonged to France at that time), superintended, with ardour, the work of preservation.

At the end of five days they were able to rejoin the captives. All were miraculously preserved, twenty-four in number, amongst whom were fifteen boys. Nineteen miners were drowned in the first moments of confusion which succeeded the irruption of the water.

Goffin, preserving his inflexibility to the end, went out the last. "If I had abandoned my men, I should never have dared to see daylight," he replied to those who enquired why he had not saved himself in the first instance in order to rejoin his wife and six children. As a reward for his admirable behaviour, he was nominated a member of the Legion of Honour, and was allowed a pension. Everybody vied in celebrating the courage of the brave collier, and Millevoye, concurrently with the prize offered on this occasion by the French Academy, wrote a piece of poetry which may be seen in the collection of his works, entitled *Goffin ou le héros Liégeois*, which was crowned. It is composed in the mannered and stately style of the period, and is no longer readable; but the intention was laudable and Millevoye must be forgiven.

The proverb says "The pitcher goes to the well till it gets broken." The Liégeois hero, faithful to his original calling, continued to work as a collier. Valiant miner! he fell on the field of battle, and was killed in 1821 by an explosion. The enemy, over whom he had triumphed so often, beat him at last! *

In the mines of Charleroy, where inundations are not less frequent than at Liege, a collier, Jean-Baptiste Evrard, passed in the last century for a legendary type, as Goffin does in the present. His adventure was slightly different; but it created such a noise that the Academy of Sciences

---

* The descendants of Goffin are still in existence. One Sunday in the month of March, 1866, M. Simonin arranged a conference of workmen at Montmartre, in the name of the Polytechnic Association. He related the history of coal, and described the struggles of the collier. Incidentally, a few words were said about Goffin. The meeting being over, a good-looking man stepped up to M. Simonin and begged to congratulate the professor. "You are one of us," said he, "allow a Belgian overman to press your hand." M. Simonin shook a hand hardened by underground work. "Everybody down there remembers Goffin," the miner went on to say, "his daughter is in Paris here; ah! if she had but heard you speak of her brave father!"

M. Simonin made an appointment with this colleague to take him to see Goffin's daughter; but he never saw him again.

of Paris was informed of it, and have mentioned it in their memoirs. *

On the 17th December, 1760, the water which had accumulated in the middle of some old workings suddenly broke into a bord where nine miners were at work. Two had time to escape through the shaft; seven others, amongst whom was Evrard, were carried away by the flood. In the midst of falling ground, and separated from his comrades, Evrard gained a sloping road, and thence a gallery communicating with the shaft; but these last were fallen in. The poor collier, with clothes dripping wet, his body covered with wounds, and suffering from the effects of the bad air, called and shouted for a long time, striking the rock with a pick which he had found on the way. No answer was returned to his signal. He then regained the ascent where he had taken shelter in the first instance, and, overcome with fatigue, slept soundly. When he awoke his clothes were dry.

Dying of hunger, he tried to eat the candles which he had with him, but could not overcome the repugnance which this unusual fare occasioned ; he, however, quenched his thirst by drinking the water in the mine. Nevertheless, he only drank three times during the whole of his captivity, and remained nearly always either in a drowsy state or buried in the soundest sleep ; in addition to this, he did not despair of his ultimate safety.

Nine days after the accident, on the 26th December, the colliery having become accessible, the workmen from without entered to recover the dead bodies. Evrard heard them consulting about raising one by placing a rope round his neck, or by fastening it to his shoulders. He shouted again, knocking with his pick. The terrified workmen imagined that he was a ghost, or the bad genius of the mine, of whom there is still a

* Morand, the *doctor*, who had known Evrard, also tells the story at length, and in the most curious terms. See " L'Art d'Exploiter les Mines de Charbon de Terre," 1768-74.

lingering belief in many collieries; notwithstanding which, however, they did not run away, but knocked on their side. Evrard
replied. They repeated the signal with the like result. Mus-
tering courage to draw near in a body, and hearing the spirit
pronounce its own name and call them by theirs, they did not
know what to make of it all. At length a body of workmen
came, who, fortunately for the captive, had brought some drink,
and determined to set to work. Scarcely had an opening for
escape been made, when Evrard, impatient to leave the hole
in which he had been confined, threw himself upon the first
miner who presented himself. This man, seized by the head
with the grasp of despair, thought he would die of fright, and
was more firmly convinced than ever that he had to deal with
the genius of the mine. The bad air having extinguished all
the lamps, the rescue was effected in darkness. Evrard,
fastened to a rope, was carried to the bottom of the winding-
shaft, and was sent up the first in the corve, accompanied by
the miner, from whom he would not be parted.

The curé (who had come in case there should be any need
of his ministry), and upwards of a hundred persons assembled
at the shaft-mouth, hailed the miner who had been saved with
acclamation. Daylight, the sudden glare of which might have
blinded him, produced no effect upon him. Without troubling
himself about the welcome given him by the crowd, which
grew louder from moment to moment, he sought to assuage
his hunger, and perceiving, at last, three apples which were
baking in the engine-furnace, he seized hold of them and
devoured them. They made him drink a little white wine,
and then handed him over to the doctors, who brought him
back by degrees to his ordinary regimen. He did not recover
sleep before the seventh day; having, perhaps, slept too much
in the mine.

At the end of three weeks he was entirely cured, and went
back to the colliery, where he was only afterwards employed

in work connected with the surface. His comrades remained as firmly convinced as before. They believed, indeed, more strongly than ever, that a spirit inhabited the mine, and some pretended to have recognized in the ghost one of the miners who had perished.

The example of Evrard furnishes another proof, after so many others which have been already mentioned, that the preservation of workmen in danger should never be despaired of, and that prompt and resolute measures should be taken to insure their deliverance, whatever length of time may have elapsed since the first moment of their confinement underground.

CHAPTER X.

THE PERILOUS PASSAGE.

Accidents which occur in the Shafts. —Tubs protected with a roof.—Mode of preventing the Collision of Tubs.—Safety-cages, or *Parachutes*. Advantages arising from the use of Guide-rods.—Visit of M. Simonin to the Mines of Roche-la-Molière, and the Bouches-du-Rhône.—M. Graud, the Manager.—Falls down Pits almost always prove fatal.—Accidents at the Coal-mines of Monte-Massi and Saint-Etienne.—Fatal Accident to a Mining Director in Algeria.—Collision in a Shaft at the Mine of Méons.—Accident to an Engineer of Méons.—Mode of descent into the Salt-mines of Wielliczka ; into some English Mines.—Descent in Baskets or Corves ; Circulation by means of Ladders ; its Inconveniences. Man-engines ; Machine with two Movable Ladders described ; Warocqué's Machine ; Simple Movable Ladders with fixed Landing Stages.—Accidents arising from obstructions of Shafts.—Accident at Hartley Colliery in 1862 ; at Poder-Nuovo ; and at Marles. —Statistics of Mining Accidents.

AFTER all the accidents caused by fire, fire-damp, want of air, falls of ground, and water, those have yet to be spoken of which have the shaft for their scene. These proceed from the breaking of ropes, timber, and pieces of machinery set up over the pit-mouth, from the falling of stones, tools, the fouling of tubs and buckets, &c. In consequence of the variety and persistence of the causes by which they are produced, these kind of accidents make many sufferers.

The shaft, the most dangerous of ways, is, as it were, the miner's tomb, and it is said that the Belgians have named this abyss the grave (*la fosse*) intentionally. In Cornwall the old open workings on the lode were called coffins. In some mines the shaft is the theatre of so many sad events, that the men never enter it without a sort of superstitious terror. I have seen some who never went down it without crossing themselves beforehand, and offering up a brief prayer, although they knew that every precaution had been taken to guard against the fresh dangers of death which are for ever impending over the head of the poor coal-miner.

Thus, the tubs have been covered with a roof or bonnet* (figs. 32 and 59), and the heads of the men are protected against falling materials by a hat made of sheet-iron or of the stoutest leather. A double indicator traversing a graduated plate before the eyes of the engine-man, shows constantly the relative position of the tubs at each point of their course; a bell is rung by the engine at fixed intervals, which gives warning that the tubs are passing mid-way in the shaft, or that they have reached the bottom or arrived at the pit-mouth. By means of particular contrivances signals can also be made from the mine to the surface. The attention of the engine-man is constantly kept on the alert, and the fouling of tubs is thus prevented; their immersion in the sump is checked, unless they go to bring up water, and their sudden fall to the bottom of the shaft retarded; and, lastly, their violent shock against the pulleys overhead obviated. In the better collieries of the North of France, in Belgium, and in England, many of the cages are two-decked, to carry two tubs or waggons on one floor; others, for narrow shafts, having four tiers or decks each holding one tub. Some of the Belgian and English cages are made for six tubs, carried in two tiers of three tubs each, the weight of which, when made of rolled iron, as usual, is 3960 lbs.

By degrees other improvements have been introduced. The shaft has been fitted with guides or conductors; that is to say, with a double vertical wooden-way, along which the cages or chairs are conveyed up and down the shafts, carrying the tubs or waggons, and also the miners. If a stone, a piece of timber, a tool, or a block of coal falls, the bonnet of the cage protects the men and the machine. If a rope breaks, a spring

* Commonly styled in French *parapluie*, or "umbrella;" but which might with more correctness be called *parapierre*.

placed above the cage, and kept compressed by the tension of
the rope, is set free, and acts upon a double clutch made of the
toughest and best-tempered steel. This catch or wedge falls
instantly between the wooden guide and a part of the cage, and
brings the latter immediately to a standstill, even before the
commencement of the descent takes place. The cage remains
suspended with its load, and time is afforded for proceeding to
the rescue (figs. 90 and 91). These safety-cages (called in
French *parachutes*) rarely fail to act or lose their effect. How
many accidents have been prevented by this contrivance, how
many miners have been saved in this way! Thus the passage
of the men through the shaft, which the government had nearly
everywhere forbidden in France, is now authorized at all the
mines which have adopted the use of guide-rods, and provided
the cages with bonnets or covers (*parachutes*)*

Another advantage derived from the use of guide-rods is
to prevent the meeting of the tubs, which is the cause of
numerous accidents. Formerly, where the workmen, in spite
of administrative orders to the contrary, or for want of being
able to do otherwise, traversed the shafts, it was necessary, in
the mid-passage, to be prepared to guard against a collision.
Frequently the space was so confined that the two tubs had
scarcely room to pass each other. As the contempt of danger
was sometimes carried so far as to descend outside on the edge
of the tub (fig. 59), and sometimes it was even considered

* This feeling is not shared by the English coal-miner. The following quotations from the Report
of the Jurors of the International Exhibition of 1862 will show this:—

"The Jury gave careful attention to all the varieties of this apparatus, and were strongly impressed
with the merits of several of them, and the desirableness of enlisting in this cause the interest of
intelligent mechanicians. But they share in the repugnance of colliery viewers to trust to the action
of a spring, on which most of them depend, and which, of whatever substance it is made, is sure by
degrees to lose its elasticity, and is thus liable, unless frequently looked after, to fail at the moment
when required. They are also aware that a great inconvenience, not to say danger, has been introduced
by all those hitherto employed, in consequence of the apparatus being brought into play by a plunge
during the rapid descent of the cage, and that hence several of these inventions, after being fairly tried
for one, two, or three years, have been ultimately removed.

"Nor is it too much to say, although an insufficient argument if taken alone, that the employment of
this apparatus has a tendency to make people careless about the examination and renewal of ropes."

amusement to make it oscillate, it will be understood that a collision, a mere contact, or grazing against the wall of the shaft, might throw a man down.

In 1851 M. Simonin visited, with some comrades, the mine

Fig. 90.—Fontaine's Safety-cage (*parachute*). Scale $\frac{1}{50}$
The cage is being drawn up with the rope; the spring is clutched and holds the levers against the two sides of the guide-rods of the shaft.

of Roche-la-Molière. The system of shafts furnished with guides had scarcely then made its appearance in the basin of Saint-Etienne. The machinery was still of the most primitive kind. Four iron ropes were stretched from the mouth to the pit-bottom; the tubs were furnished at the sides with rings which kept them between the guides, along which they ran. Although this method only guarded against the meeting of

Fig. 91.—Fontaine's Safety-cage (*parachute*).   Scale $\frac{1}{30}$.
The rope is broken; the spring being set free has thrust the levers into the guide-rods.
The apparatus remains suspended in the shaft, awaiting succour.

the tubs, the miners, nevertheless, went to see these fittings for many miles round.

Our shaft was like nothing else. We were in a little tub without a cover, narrow besides, and somewhat dilapidated. "They allowed us to go off by ourselves," says M. Simonin;

" hanging from the rope, we chatted and went our way without even dreaming of taking care of ourselves, when all at once, in the middle of the passage, the tub that was going up ran foul of us. The frail shell in which we were was all but upset, and our first impulse was to hold on tight by the iron chains which supported it. We shouted ; they heard us, let down slowly the tub which had come into collision with ours, and then resumed the first movement. Becoming careful from that time, we avoided the enemy in the passage ; and gained the bottom without meeting any further mishaps.

" This adventure gave me a lesson ; and some years afterwards, visiting at various intervals the collieries of the Bouches-du-Rhône, I was not surprised that the manager, M. Grand, an old mining-wolf (as he was jocosely styled by his colleagues), with as much skill as prudence, always preferred to go into his stalls by the inclined road, or *visette*, as it was called in the language of the troubadours. The shaft was of no great depth, a hundred-and-fifty yards at the most ; the drift-way was not less than a thousand yards, or more than half a mile in length, and what a drift-way it was ! It was a very inclined gallery, narrow and low, while M. Grand, worthy of the name, is more than six feet high. He had furnished it with steps and a ladder, like regular stairs. The steps were notches cut in the rock, worn, reduced.to nothing, smooth as glass, where each person vied with his neighbour in slipping. The ladder was an old wire-rope, worn by the drum, untwisted, and tearing the fingers. This inclined gallery was still used for the descent, but the ascent was effected by means of the upright and confined way, furnished with notches."

A fall down a pit is almost always fatal, but yet there are instances of men having been saved in such a case either by clinging to the timbering, or by having fallen direct into the sump or reservoir of water, and when there swimming, shouting, and waiting until help arrived.

P

In Tuscany two men fell one day to the bottom of the pit, at the coal-mines of Monte-Massi. One was killed on the spot; the second was grievously injured, but had nevertheless the courage and strength to reascend by climbing perpendicularly up by means of the timbering of the pit, which was more than four hundred yards deep.

At Saint-Etienne, in 1863, a young engineer of mines was quitting a colliery through an old pit, in company with the son of the overman, when a block of stone fell out of the wall into the tub, and crushed his companion. He himself reached the surface in a nearly lifeless condition, the block having struck his ribs; and in spite of the hopes of recovery entertained by the doctor he very soon afterwards expired. At the pit-bottom was the overman, who may almost be said to have had a hand in his son's death; since in spite of strict orders to the contrary, and to oblige the engineer, who was fatigued, he had permitted them to go up in the tub; little supposing that they were all to be so terribly punished!

The pit-mouth, which is nearly always open, is a new source of danger. A mining director in Algeria, being engaged in making a plan, had thrown the measuring-tape to the bottom of an open chasm, for the purpose of ascertaining its depth. All at once he disappeared, as if the abyss had swallowed him up, and only his lifeless body was recovered, his skull having been fractured against the wall. Poor fellow! he had grown old in his profession, and had denied his bones to the quicksilver mines and the fevers of the Tuscan Maremma and of Africa. Who would have thought that, after successfully braving so many perils, this sad end was reserved for him?

Accidents in pits were until lately very common. At Saint-Etienne, previously to the adoption of guides, two tubs of coal were sometimes hooked on to the rope, and not always placed one above the other, but frequently side by side. One day, at the mine of Meóns, where the latter system was in use,

Fig. 92.—Collision of Tubs in a Shaft.

as the engineer and overman were going down the shaft together, a violent shock was experienced midway in consequence of a collision. The two men were standing up, holding the lamp in one hand, while the other held on to the chains. The shock unhooked their tub, and they were left holding on merely by the rope. As the colliers, having had no warning, had piled up the ascending tubs with coal, the orders in such a case being to leave them partially empty, large blocks of coal were thrown out by the concussion, and fell down the shaft. By a singular chance neither the engineer nor the overman was injured. Their presence of mind never left them for an instant, and they reached the termination of this rough journey still hanging on by the rope, which they held with a convulsive grasp (fig. 92).

The same engineer of Méons met with another accident which was attended with no more serious results, but which in the case of a man of less courage might have ended in a most lamentable manner. When about to go up through the shaft one day, the engine-man, at starting, raised the rope too suddenly, which had the effect of overturning the tub, and the engineer, hanging on by a foot, and with his head downwards, was hoisted a height of forty yards. The alarm having been raised, they managed to stop the engine and to send help to the person in danger (fig. 93).

This unusual manner of going up recalls to mind that of the aeronauts, who suspend themselves by the feet to the trapeze of their balloon, and perform a thousand dangerous feats in the air; but, at any rate, they are not taken unawares, like the brave engineer of Méons. Formerly there were few colliery managers who had not met with similar experiences; but it is no longer so now that the art of mining has been somewhat improved.

At Liége, as well as at Saint-Etienne and most other collieries, it was formerly the custom to go down to the working places sitting astride or outside the tub. On one occasion an engineer had his clothes caught by a strut which projected out of the side of the pit, and hung suspended from it. The position was critical alike to him as well as to the other

Fig. 93.—Critical situation of an Engineer of Méons (Loire).

persons who had accompanied him and continued to descend; for in case of his falling they would have been crushed by him. Their shouts were not heard, the man whose business it was to watch at the pit-mouth being no longer there, and the alarm could only be given on their reaching the bottom of the

Fig. 94.—Descending the Shaft at Wielliczka.

pit. After twenty mortal minutes the engineer was at length set free. He had vainly felt with his hands for a support—some crevice opposite the stays—by means of which he might sustain himself and ease the strain upon his frail garment, which threatened to give way beneath his weight. He had no sooner got into the tub again than he fell insensible into the arms of his preservers. He had a long illness, after recovering from which he continued his business, but preferred ever after to go to his work down the ladders in preference to the tubs.

There are certain mines in which the descent is made in a very primitive and simple manner, by means of the rope alone, without a tub. Beudant, the mineralogist, relates in his "Travels in Hungary" (*Voyage en Hongrie*), that on paying a visit to the salt-mines of Wielliczka, in 1818, they took him down the pit in a somewhat unexpected manner. The extremity of a rope, brought to the surface, carried round a knot five or six ropes'-ends looped up like a swing, and furnished with a couple of transverse bands, one of which served for a seat, the other for a support to the back. Seated in this aerial chair, the visitors were launched into vacancy, where they formed a sort of living chandelier, the resemblance being rendered in this instance the more close by each one holding a lighted candle in his hand. If there were too many persons, they were divided into two parties, forming two clusters placed one above the other, and they descended in that manner to the pit-bottom, where the miners set them free (fig. 94). If the accounts of travellers are to be relied on, the curious Gallician salt-mines are always entered in this fashion.

A similar mode of descent was at one time practised in some English mines; only the groups consisted of but two men, each of whom passed one leg through an iron chain which was fastened to the rope, and which formed the seat. One man besides could hold on standing, with his foot in a loop. Accidents were of

Fig. 95.—A fall in going down a shaft.

frequent occurrence by this method of descent, and more than one workman, coming in contact with some unexpected obstacle on the way, has been thus precipitated into the abyss before the eyes of his terrified comrades (fig. 95). Excepting in shallow pits this plan is never used now.

When baskets or corves are employed for descending in France, the men get inside, and the boys hold on outside by the suspending chains, often in two ranks one above the other. Sometimes the rope alone is relied upon, the end being doubled back upon itself so as to form a loop, through which a leg is passed, or to carry a transverse bar of wood which serves as a seat. With such primitive methods of circulation, which are gradually being abandoned to make way for the adoption of safety-cages, there is no protection against the fall of stones, timber, &c. (fig. 96); in addition to which a careless engineman may send you into the sump to take a plunge-bath at an unseasonable moment.

The circulation by ladders, like the use of cages with guides, obviates most accidents; but it is attended, in its turn, with certain inconveniences. The men not only lose a great deal of time in going up and down; but this exertion, repeated twice a day, causes fatigue and weakness. It is of little use to provide landing-stages for resting, for the miner does not traverse with impunity hundreds of fathoms of ladders; in time this violent exercise induces diseases of the lungs and heart, asthma, and consumptions, which often terminate fatally, or render him unfit for work after thirty-five years of age. These remarks apply, more especially, to the deep metalliferous

Fig. 96.—A fall in the Shaft

mines. There are no collieries of any extent, in this country, in which ladders are now used for descending into or ascending from the depths. The age of the coal-miner is quite equal to the average life of his brethren; but in the metal mines it is as M. Simonin has stated.

In order to obviate these grievous inconveniences resulting from the use of fixed ladders (sometimes nearly as long as those seen by Jacob in his dream), movable ladders have been fitted in some mines, which are set in motion by the engines. These ladders have been employed for thirty years in the deep metalliferous mines of the Harz in Germany, whence they have passed, in a modified form, into those of Cornwall, in which county six of them are now at work, and subsequently into the French and Belgian collieries. Now they are in common use at many mines on the Continent, though never adopted at British collieries. They are called *Man-engines* in English, *Fahrkunst* (artificial or mechanical ways) in German, and *Warocquères*, after the name of the Belgian engineer, Warocqué, by whom they were improved. The French workmen, who despise German and English names, call them *machines à monter*.

The principle on which these machines are constructed, whether intended for use in an inclined or a vertical shaft, is as follows :—

If the apparatus is double—that is to say, with two rods —imagine two strong parallel rods, furnished at equal distances with steps (fig. 97). By the action of the steam-engine at the pit-mouth, with a crank movement, one of the rods is raised a certain height, while the other is lowered for the same distance; generally from eight to fourteen feet at each stroke. No actual interval of rest occurs; but during the movement of the crank over its turning point the miner passes, from the step on which he is standing, to the opposite step. Another stroke of the engine is made, and the rod moves in a contrary direction, and is followed by a fresh movement of the miner. It is understood by this mode, that whether he goes up or down, the miner rises or descends at each oscillation without fatigue; the only movement which he has to make being a lateral motion from one step to the other at every resting point. He soon gains the bottom or the mouth of the shaft—in twelve minutes, for instance—if the pit be three hundred yards deep, and the engine make fifteen strokes a minute. Of the two double-rod man-engines which have been built in Cornwall, at Tresavean in 1843, and at the United Mines in 1845, only the latter is now at work. At Tresavean the steam-engine made fifteen strokes a minute, which was reduced by spur-gearing to three strokes a minute on the rods. The rate at which the men were lifted was seventy-two feet per minute, twenty-four minutes being requisite for the entire journey of two hundred and ninety fathoms. At the United Mines the rods make three oscillations to eighteen strokes of the steam-engine; and the time taken up in travelling the whole distance of two hundred fathoms is seventeen and a half minutes.* A little less time is

Fig. 97.—The Man-engine, with double movable ladder.

required to perform the journey by means of the rope; but by fixed ladders twice or thrice longer would be consumed, without stopping to rest at any landing, and the men arrive at

* "Descriptive Catalogue of the Mining Models in the Museum of Practical Geology," by Hilary Bauerman.

Fig. 98.—Simple movable ladder, with fixed
landing-stages.

their journey's end greatly fatigued.*

In M. Warocqué's machine, which is the most convenient, and which has passed from Belgium into France, especially at Rive-de-Gier, the steps are replaced by fixed platforms or landing-places with balustrades, capable of receiving two men. The length of stroke is three yards, and the number of strokes from twelve to fifteen per minute. Twenty men can be distributed over the several landing-places without inconvenience, a number which is sometimes accommodated in the two-decker safety-cages furnished with guide-rods; but by the tubs only the half or fourth part of this number, that is, from five to six men, is seldom exceeded, according to the capacity of the tub.

* Engines equally well suited for extraction have been constructed on this principle of the movable ladders, by Méhu, for the mines of Anzin. By means of an automatic movement the tub travels successively from one step to another. It is to such machines that recourse must evidently be had when coal-mines are worked at greater depths than is the case at the present time, when the vertical distance shall be so great that the ropes shall no longer be able to sustain even their own weight.

We have described the mechanism of the double reciprocating man-engine. When the machine is single (fig. 98) as is mostly the case, and especially in inclined shafts, the men pass from the movable ladder on to a step or landing-stage opposite, which is fixed to a vertical wooden rail on the wall of the shaft, where they wait for a new stroke to gain the step of the ladder. There must be no hesitation. If the place should happen to be already occupied on the ladder or on the stage fixed against the shaft, by one miner who is going up as another is going down, he should remain quiet in his place, and wait for a second movement. The slightest embarrassment may cause the most serious accident, and the brutal engine, by its sudden return-motion, may kill the traveller on the spot, or break a limb. It is in this way that the precautions which are adopted to guard against accidents sometimes give rise to fresh dangers.

A single-rod man-engine was built in 1851, at Fowey Consols mine in Cornwall, which extends from the surface to a depth of two hundred and eighty fathoms. There are other single-rod engines in Cornwall: at Levant, two hundred fathoms long, making four strokes per minute; at Dolcoath, two hundred and twenty fathoms long, making three-and-a half strokes per minute; the length of the stroke being, as is the case in all the Cornish engines, two fathoms. There are also other single-rod man-engines at Carn Brea and Huel Reath. *

All the accidents connected with shafts which have been enumerated, are surpassed in intensity by those arising from the closing of the shafts themselves when the mine is only provided with one means of egress. The breaking of struts, and the falls to which it gives rise of some portions of the winding or pumping machinery, are the origin of those new accidents which carry desolation and death into a mine.

* H. Bauerman's Catalogue of Models in the Museum of Practical Geology.

On the 10th January, 1862, the beam of the pumping-engine broke at the Hartley Colliery in the Newcastle coal-field, and in falling through the shaft killed five out of eight men who were being raised in the cage, the other three being miraculously saved.

The mine was in full work, and all the men and boys, 199 in number, were underground. By the shock of the enormous cast-iron beam weighing more than forty tons, striking the walls in its descent, the shaft was damaged in several places; the rubbish and broken timbers were accumulated at the depth of 138 yards from the surface, and an impenetrable vault closed the only mode of egress by means of which the captives could escape. Two hundred and four colliers, inclusive of the five just mentioned, and forty-three horses, met with their deaths by this accident.

As the massive beam, pump-shears, and timberings, had intercepted all communication between the interior of the mine and the outside world, and as the mine was furnished with ventilating furnaces in which a large quantity of fuel was burning, it appears that in the day succeeding the accident the victims died of suffocation. However, they did not experience the horrors of starvation, for a dead pony was found by the side of the miners. Some men, in a moment of extreme despair, tried to force an outlet; timber had been cut and sawed; and the rescuers had heard from outside these desperate attempts, which were, unfortunately, as brief as they were useless. The crowd of relatives and friends who stood around the pit-mouth, finding that the attempts at rescue were not carried on fast enough in their opinion, threatened to undertake the duty themselves, and boldly demanded the bodies which they were waiting for. But all this impatience was calmed, and then the dead, brought up one by one, were solemnly buried. Was ever a longer or more mournful procession beheld in time of war or pestilence?

Amongst the various suggestions made by experienced coal-viewers and by the government commissioner and inspector, in connection with this frightful catastrophe, the opinion that all mines should be furnished with two shafts strongly presented itself. The Hartley shaft was divided by a partition, and sufficed for the wants of the colliery. No doubt, the existence of a second shaft would have afforded the means of escape to the men who remained shut up in the mine. Is it to be understood that coal-owners are to be compelled to sink at least two shafts? As the engineer who conducted the official inquiry remarked, it would be both impossible and unjust, in the majority of cases, to require such an additional work; for, in the first place, the right of the state to interfere in the operations of private industry does not seem to go to that extent; and, secondly, the compulsory outlay upon a second shaft or on an inclined gallery, in certain collieries where the ground is very hard, would not allow of the works being carried on with a profit, and a numerous and important population would thus themselves be deprived of a certain source of work. But the English legislature paid no attention to these observations, for the obligation of providing two different outlets is now imposed by Act of Parliament on all the colliery-owners of Great Britain. It might be said, then, that while the French endeavour to imitate the English in their independent ways, the English have a tendency to borrow from the French, in turn, some of their restrictive measures.

The coal-mine of Poder-Nuovo, near Volterra in Tuscany, became, in 1864, the scene of a sad accident which reminds us of that which took place at Hartley. In the former case, a fall which occurred in the shaft closed every outlet for the miners. The details of this catastrophe were communicated to M. Simonin by his old friend, M. F. Blanchard, Director of Mines in Italy, who obtained the information on the actual spot where the event occurred.

On Holy Thursday, the 24th March, 1864, three men went down the mine.   While they were going down they noticed a certain derangement in the timbering of the shaft, which was sunk through yielding clays ; but they, nevertheless, continued their way.   At the pit-bottom they found a young man already employed in loading stuff into the tubs, and feeling uneasy, inasmuch as he had heard cracking sounds.   Some minutes afterwards stones began to fall, and he was drawn up in haste, having first called to the miners to follow ; but instead of taking his advice, they laughed at him and went on with their work.   On reaching the surface, the lad informed the timberman, the captain (or *caporal*) being unwell.

The timberman descended as far as the dangerous place, about forty yards from the surface,* where, in his endeavour to secure the struts, formed of beams and planks, he caused a first fall.   The timbering, in falling down the shaft, got entangled midway, where it formed a sort of scaffolding.   The timberer caused himself to be drawn up in great haste, calling to the other men below to follow him in the tub, which was going down as he was going up.   He had hardly reached the surface before a second fall took place ; the clays gave way in a mass and completely blocked up the shaft.   In the meanwhile the three miners who remained shut up in the mine had run forwards, shouting lustily for help ; adding that there was an open space above them for a height of thirty yards.

What was to be done in such a case ?   An inclined gallery or drift-way might be driven to the level of the fall to reach the miners ; or the actual scene of disaster having been limited by means of fascines, an attempt might be made to traverse it in order to supply the imprisoned men, in the first instance, with air and food.   But to do this there was need of an engineer, or at least of a cool-headed miner, to direct operations, and to animate the workmen by the influence of his

---

* The whole depth of the shaft was a little more than 100 yards.

example. Instead of setting to work in a methodical manner, everything was begun, nothing was finished, and they lost their heads. The *caporal* himself, sick of a fever and of no use, thought proper to leave; and in order to escape the pursuit of the people of the locality, relatives or friends of the captives, he was not ashamed cowardly to desert his post, and rode away on an ass.

On Monday the 28th, four days after the accident happened, Mr. George Brown, an Englishman, who was engineer of some neighbouring works, from humane motives and actuated by a feeling of confraternity, arrived at the mine. He resolutely undertook the direction of the necessary operations, and perceiving the hesitation of the miners, he descended to the scene of the accident and himself worked alone, day and night, at the clearing out of the shaft. On Thursday the 31st March he effected an opening through the fallen rubbish for the supply of air; he then ordered near him the timberman, the same man who had in the first instance endeavoured to repair the disaster. "We two will go down, underneath the fall," said he, "by means of the rope; on reaching the bottom we will release the victims one by one, living or dead, and will get them out." The timberer had previously called out, and heard one of the men in reply entreat him to be quick, because he was in a dying state.

There was no room for hesitation, and not a moment was to be lost. Mr. Brown substituted for the tub, which was too large, an ordinary bucket in which he placed his foot, ordering the timberer to do the same and follow him. The descent had scarcely begun through the interlaced timbering, than some planks which had been badly secured fell, and struck the Englishman on the chest and face. He wished to go on; but his companion, trembling with fear, uttered cries of terror, and they were both drawn up to day.

Before the brave engineer was able to fix upon some other

man to accompany him, a fresh fall took place, and then began a series of operations which were as long as they were painful. It was only on the twentieth day after the accident that all damages could be repaired, and when Mr. Brown descended to the bottom of the shaft it was only to meet with three bodies, which were already in a state of decomposition. With the help of a lad and a miner, he wrapped the bodies in a cloth and sent them to the surface, where coffins had been got ready in advance, so little hope was there of the success of operations which had been marred by so many accidents.

Each of the three miners was found lying on a plank. They were stripped of their clothing, which they had rolled up under their heads in the form of a pillow, and then fallen into their last sleep. Two were found in the gallery, the third had left his place and was discovered near the shaft; and he it doubtless was who had been heard to give signs of life on the seventh day. The poor fellows must have perished with hunger, for they only took one loaf of bread with them into the mine. Shut up in a confined space, deprived of respirable air, they must have long awaited the moment of preservation with a holy resignation; then aware that the light of day would never shine upon them more, they had themselves arranged their own funeral, and in their lingering agony had in some measure studied to die decently. Their lamps were found, still full of oil, hooked on to a waggon. More than once the miner's lamp, which in some localities reminds one of the sepulchral lamp of the ancients, has in this way done justice to its form.

In the deplorable series of accidents of which the shafts are but too often the scene, the fall in the shaft of Marles in the Pas-de-Calais, of quite recent occurrence, ought not to be omitted before closing so long a list of harrowing incidents.

On the 28th of April, 1866, M. Micha, the engineer of the mine, noticed that the cages could no longer work in their guides. The wood-tubbing with which the shaft was lined showed at the depth of fifty-six yards an appreciable movement of torsion, and a tendency, like the leaning tower of Pisa, to deviate from the perpendicular. The joints opened, and displacements showed themselves at several points.

All the staff, consisting of three hundred workmen, were made to leave the mine; the horses only, to the number of twenty-seven, being left underground. Resolute men went down the shaft with the object of covering the slips, but the result was only to create fresh ones. During a struggle, which lasted two days, the tubbing was heard to crack, the planks to break one by one, the measures to fall, and the water to rush impetuously into the workings, which were of considerable extent, and two hundred and fifty yards in depth.

In the meanwhile a communication had been made to M. Glépin, of Grand-Hornu in Belgium, the consulting engineer of the mine. On his arrival he descended with the overman, or maître-porion, Louis Wyns, to the very midst of the fall; and, believing that he was going to certain death, he embraced his assistants, saying — "I am fifty years of age, and have a wife and children; but I go where duty calls me."

The lamps went out while they were descending, and M. Glépin and the overman were only preceded in the chasm by a lanthorn which they had hung outside to the bottom of the tub in which they descended. By the uncertain and feeble light of this oscillating lamp the enormous opening was perceived that had taken place in the middle of the tubbing, displaced portions of which still continued to fall, and through which the water rushed in torrents. "Let us go up again," exclaimed M. Glépin; "the enemy is master of the situation, and all hopes of saving these workings are lost."

The brave engineer, in relating this fearful incident, added, "I have lived ten years in a quarter of an hour; my hair turned white in this perilous descent, which I shall never

Q

forget as long as I live." At night the falls from the middle of the shaft extended from top to bottom. At the pit-mouth an immense opening had formed—a crater thirty-five yards in diameter, and ten yards deep. Scaffolding, engines, boilers, and buildings, had all by degrees gradually fallen into the opening. At each movement in the ground a fresh engulfment took place. The sky was dark, and covered with clouds. The timbering of the shaft gave out sparks under the enormous friction which was developed by the sudden fracture of the wood. A peacock, shut up in the neighbouring court-yard, uttered lamentable cries at every movement which took place in the ground and at every fresh fall. " No poet could describe, nor painter represent, the desolating spectacles which we witnessed," said M. Glépin in concluding this account, during the recital of which he could scarcely restrain his emotion.

The pit of Marles, which had been in existence for ten years, was one of the most productive in the Pas-de-Calais. The sinking and securing of this important work had occupied the attention of those who were intrusted with the duty of carrying it out day and night, and they had successfully overcome all the difficulties caused by the water. According to the opinion of all practical men, a work of greater difficulty had seldom been undertaken or more ably managed. The capital expended on these works amounted to £60,000, the whole of which was dissipated and lost in two days, notwithstanding the efforts of a whole population of workmen, and the assembled engineers of all the neighbouring mines; constituting one of the greatest calamities, although there was no loss of human life, that coal-mining industry has ever had to deplore. The Company of Marles, however, has not lost courage. With that indomitable perseverance which characterizes most mining associations, they are about to re-open the works by means of another pit, and are taking measures to reclaim the shafts which so unfortunately fell in.

And now that all the kinds of accidents to which the collier is liable have been passed in review, let us resume this long chronicle of martyrs. The statistics of the numbers of killed and injured have been carefully taken in different countries. The study of these documents, which are no less interesting than those drawn up by Dr. Chenu in reference to the French army in the Crimea, shows that in France colliery accidents affect, on an average, two per cent. of the men every year, and that the proportion of deaths to injuries is in the relation of one to five, or four in every thousand. In some years the proportion was still more fearful, one man being killed for every twenty thousand tons of coal extracted. In the year 1864 the total number of lives lost from different causes in the British collieries and ironstone-mines of the Coal Measures appears, from the official reports, to have been 963, showing a decrease compared with the numbers for 1863 and previous years.

In Great Britain, as also in France, the greatest number of accidents are occasioned by falls of the roof and coal, nearly one-half; then a third take place in the shafts, from breakage of engines, ropes or chains, upsetting or fouling of skips, boxes, tubs, &c. The remaining number, or one-sixth of the casualties, occur from blasting, explosions of fire-damp, suffocation—caused by defective ventilation, or for want of air, or by poisonous gases, carbonic acid and after-damp—and, finally, inundations. It is true, then, that a piece of coal often costs more than is supposed, and that the mine is to the collier a real field of battle.

The total quantity of coal raised from 3192 mines by 320,663 miners being 101,630,544 tons, and the proportion of deaths having been about one in every 216 persons employed in 1866, according to the last Annual Report of Her Majesty's Inspectors of Mines, the result is that on an average one accident takes place for every 147,925·5 tons of coal raised, and

Q 2

one life is lost for every 68,484 tons. Even in the Derby, Nottingham, Leicester, and Warwickshire coal-fields, the most favoured districts in England, one man is killed for every 131,034 tons of coal raised, while in Yorkshire the proportion is no less than one in 22,235 tons. In West Lancashire and North Wales it is one in 55,666, and in South Wales one in 78,137; but the credit of minimum destructiveness belongs to Scotland, in the western district of which 131,880 tons are raised for every life lost, while in the eastern part of that country the proportion is only one to 190,625 tons.

In the year 1866, 651 lives were lost from explosions of fire-damp, as against 168 in 1865, giving the enormous increase of 483 out of the gross increase for the year from all causes, which is exactly 500. The number of lives lost in the in-spected ironstone-mines of Great Britain in 1866 amounted to 81. Taking the several groups of Inspectors' districts into which the coal-fields of the country are divided, the returns show the following results for each of the years 1865 and 1866 respectively :—To one death, the number of miners em-ployed was 636 and 83 in Yorkshire; 403 and 112 in North Stafford, Chester, and Salop; 238 and 200 in West Lancaster and North Wales; 182 and 243 in South Wales; 296 and 248 in South Stafford and Worcester; 258 and 259 in North-umberland, Cumberland, and North Durham; 414 and 310 in South Durham; 325 and 321 in Monmouth, Gloucester, Somerset, and Devon; 356 and 368 in North and East Lan-caster; 330 and 467 in Derby, Nottingham, Leicester, and Warwick; 340 and 445 in the western districts of Scotland; and 450 and 662 in the eastern districts of Scotland. Out of the 1484 deaths in 1866, 651 occurred from explosions of fire-damp. The deaths from this cause alone in Great Britain in the ten years 1856 to 1865 were 2019. The total number of deaths from all violent causes in the ten years was 9916, about twenty per cent. of which was caused by fire-damp

explosions.   The number of deaths from falls in mines in 1866
was 361 ; from accidents in shafts, 162 ; from accidents underground, 203 ; and from accidents on the surface, 107.   Of the
deaths from fire-damp, 361 occurred in the Oaks Colliery, 91
at Talk-o'-th'-Hill Colliery, and 38 in the Victoria Colliery in
Dukinfield.

The following list will convey an idea of the terrific nature
of the explosions of fire-damp in the English mines :—

Year.		Men Perished.
1812 at Felen's Colliery,	. . . .	92
1835 at Wallsend Colliery,	. . . .	102
1844 at Haswell Colliery,	. . . .	95
1856 at Cymmer, Rhonddu Colliery,	. . . . . .	114
1857 at Lundhill Colliery,	. . . .	189
1860 at Risca Colliery,	. . . .	142
1866 at Oaks Colliery,	. . . .	361
1866 at Talk-o'-th'-Hill Colliery,	. . . .	91

From the Oaks Colliery they are only now (October, 1867)
beginning the terrible task of recovering the bodies of the
dead.   Of the accidents in which less than ninety were killed,
no notice has been taken in this list.

THE PROPORTION OF ACCIDENTS AND LIVES LOST TO THE NUMBER OF PERSONS
EMPLOYED, AND THE TONS OF COAL RAISED, IN 1866 :—

The number of accidents involving loss of life in the collieries of the United Kingdom are,	837
The number of lives lost by those accidents are, . . .	1484
The number of collieries in the United Kingdom are, . .	3192
The present quantity of coal raised, ...	101,630,544 tons.
A life is lost for of coal raised, ...	68,484 tons.
Tons of coal raised per separate fatal accident, . .	117,537
Persons employed per separate fatal accident, . .	374
Persons employed per life lost, ...	216
The number of coal-miners in Great Britain, . . .	320,663

The accidents, when analyzed, are annually as follows :—

By explosion of fire-damp,	. . . . 169·6
By falls of roof,	. . . . 446·6
By falls, &c., in shafts,	. . . . 197·0
By sundry causes,	. . . . 196·1
	1009·3

CHAPTER XI.

THE SOLDIER OF THE ABYSS.

Miner and Pioneer.—Qualities of the Coal-miner; his intelligence.—Sobriety of French Colliers.—St. Barbe.—Diseases of Coal miners.—Sick-funds and Benefit Clubs ; Churches and Schools.—Humanity of French Coal-proprietors.—Cheap Dwelling-houses.—The Cottage and its Garden.—Predilection of Colliers for detached house.—Failure in France of the system of lodging-houses —Cost of Colliers' houses in France.—English Colliers' cottages.—The unmarried Miner ; furnished rooms and canteens in England, France, Belgium, and the United States ; cost of board and lodging.—The Collier's daily life.—Sunday amusements.—Wages. —Types of French, Belgian, English, American, German, Spanish, and Italian Coalminers.—Coal-mining in the Tuscan Maremma —Daily life of the Miners ; Old Niccolini ; Agostino the Corsican ; the Pistojan mining-captains.—Pierre Lhôte and Father Garnier.—The legion of Coal-miners.

THE daily struggle against what we have called, with the great poet Hugo, αναγκη (*ananke*), the fatality of the elements, has converted the coal-miner into a kind of disciplined and energetic workman-soldier. In this army of labour the old instruct the young, and these last soon acquire, in the assiduous pursuit of their duties in the working-places, patience, reflection, coolness, and a host of other solid qualities, without which he cannot be a good miner. The body must also be inured to the severest fatigue, and brave continual dangers, to become accustomed to life underground. See those men who emerge from the shaft in the dusk of the evening, walking with heavy tread, with blackened faces, with clothes and hats wet and covered with mud. Where are they going? They rejoin their families in calmness and silence. Recognize in them the obscure and manly combatants of the abyss, the pioneers of the subterranean world.

The coal-miner is by nature full of courage and devotion. Ever ready to sacrifice his own life to save that of his comrades, we have seen with what stoical resignation he supports the most painful trials. Fond of his mine, in the neighbourhood of which he has been, in many cases, born, he used seldom to migrate formerly, but now under the excitement of the oratory of paid delegates, strikes and subsequent migrations are unfortunately too frequent. The French coal-miner entertains no ill-feeling towards his employers, for he feels that he has his share in their success, and works with pleasure. On one occasion only, in 1848, he failed in his duty in some of the French collieries, but for this he speedily made amends. This is not, however, strictly correct as it regards the relations of master and man in the English collieries.

Contrary to what is generally supposed, and with some exceptions which have been pointed out, the work of the coal-miner in no respect reminds us of the labour of the slave ; he much rather exercises the moral and physical qualities of the mechanic. The miner accustoms himself to exactitude and obedience ; he executes the orders of his chiefs with docility ; while the daily labour developes all his corporeal faculties, and endues them with a vigorous stamp.

The intelligence of the miner is unceasingly called into play in the execution of the work in which he takes part. The applications of geometry and underground geology become familiar to him, and he eagerly follows the vein or seam, as he calls the stratum of coal, and studies its various phases with all the ardour of a gambler. The disappearance or loss of the seam distresses him ; he tries to overcome the accident caused by this ill-omened fault, which has suddenly come in, to cut off and interrupt the continuity of the coal, to derange the regular mode of occurrence, and to carry it goodness knows where. He experiences no joy until he has recovered the seam ; and if it will but yield a combustible of superior quality, he feels nearly as much pleasure as though the mine were his own. Incessantly on the watch, he studies all the peculiarities of the measures, and follows the track of the deposit precisely as the trapper of the prairies or the untrodden forests follows the trail of the bison or the beaver.

It is a lamentable fact that, in this country, our coal-miners are far from being, as a rule, the most temperate of men, and that too much of their hard-earned wages are spent in drink; but in France, and on the Continent generally, the French collier is described by M. Simonin to be, as a rule, habitually sober, and is seldom addicted to intemperance; he comes out of the mine tired, and seeks rest in sleep. If he frequents the public-house, the café, or the wine-shop, it is only on pay-days, that is, every fortnight, or once a month, and sometimes on Sundays. It must, however, be said that, in certain countries, all the miners have not such exemplary habits, and that the temperance societies would have more than one convert to make amongst them. In Belgium the public-houses, as in many places nearer home, are a perfect curse to the working collier, and keep him in poverty notwithstanding his good wages.

The frequent use of gunpowder has made of the miner a brother, in some sort, of the artilleryman and the marine; and it is for this reason that he celebrates the festival of Saint Barbe. He fêtes her by suspending her image enshrined and illuminated from one of the stagings of the shaft, or even at one of the intersections of the galleries; he drinks to her, glass in hand, on the 4th of December, the anniversary of his great patroness. On that occasion drinking is excusable, and licence is the order of the day at all the mines. The directors give a grand dinner to all the officers, at which the curé, the doctor, the overmen, and other head men are present; while the rest receive double pay. Guns and petards are fired off. The next day everything is in order again, and the mine resumes its ordinary aspect, its usual regularity.

In the confined and sunless atmosphere in which the coal-miner lives half his time, he contracts few special maladies;

nevertheless, in the course of time the bad air impoverishes his blood and causes anæmia, while the dust arising from the coal produces dangerous affections of the chest and lungs. On the other hand, the miner is sheltered from the inclemency of the weather, from cold, wind, and rain, and is more favoured in that respect than the out-door labourer. He has, nevertheless, to be careful not to take cold on leaving the mine, and to observe certain precautions when he has to work in water. Now that the long ladders have been nearly all done away with in coal-mines—though not so in metallic mines—and with them the serious lung-diseases which they induced, it may be said that, accidents excepted, the most formidable enemy of the miner is rheumatism. It is accidents only which expose the miner to a thousand perils, and which cause death, as it were, to march incessantly by his side.

The managers of collieries generally watch over the safety of their men with a paternal solicitude. Sick-funds are everywhere established, and supported by subscriptions, and by the produce of fines, which are imposed at every mine whenever any of the standing rules are broken.

The sick-clubs, which are now generally being introduced into the collieries of this country, are under the management of the miners themselves. Medical attendance and medicine are afforded gratuitously to every sick man, in addition to which he receives a daily allowance, which on the Continent is usually a franc, and which varies in England and Wales according to the state of the fund. When the injury is of so serious a nature as to render the amputation of a limb necessary, or to incapacitate the sufferer from further work, he is in France pensioned for life. In case of death arising from accident, the French company takes charge of the children, and also pensions the widow; and we have many examples of this also in England. Moreover, the old and infirm workmen are not forgotten, and the sick-fund, in this way, is often converted into a provision for retirement from active duty. From this it may be perceived that the generality of French coal-miners have exercised a considerable degree of foresight in providing for themselves and in insuring their lives.

The anxiety of the French mining-interests has not been confined to these humane measures. In most cases they have made equal efforts to provide for the wants of the mind and of the soul, by improving the moral and material condition of their workpeople. At their own expense they have built churches and founded schools, in which gratuitous instruction is given to children and adults, and to which even books are added. In this way the coal-proprietors have worthily recognized their duties. Labour is the lot of man here below; light and help descend from on high. Assistance so worthily rendered carries with it nothing degrading to those by whom it is received; and it is thus, only, that in this age benevolence and patronage should be rendered. Neither charity nor alms should be guaranteed to the working classes, but the most extended and liberal protection, and above all education.

The companies have gone yet further, and in their anxiety for the welfare of their miners have erected cheap lodging-houses at their own expense. These buildings consist of a house (detached when possible) with yard and garden. In many parts of France, among others in the Department of the Saône-et-Loire, various patterns of these houses may be seen, substantially built of stone or brick, and well laid out. The doorway opens into a spacious apartment, provided with a fire-place, and which serves as kitchen, dwelling-room, and bed-room. By its side is a smaller chamber to hold the linen and clothes-presses, and for the children to sleep in. The windows are adapted for the admittance of plenty of air and light; the floors are paved with tiles, and the walls white-washed. The principal chamber is at least 15 or 16 feet in length and breadth, and 9 or 10 feet high—exceeding in this respect the cellars and basement-storeys of many a Parisian house.

Over the ground-floor are the attics; but some houses which have only room below have a first-floor, to which access is gained by a staircase.

Under the house, or at the side, is a penthouse, intended for a cellar for keeping wine and provisions; behind is the garden, where the miner grows vegetables for the use of the family, and flowers to adorn his dwelling. Besides this, he may appropriate a portion for a pig-stye, or a poultry-house, or for keeping rabbits, and try to solve the famous problem of deriving an income of three thousand francs a year by breeding rabbits.

Each of these houses costs nearly £80, inclusive of the value of the land. They are let to the workman at a rent equivalent to from two and a half to five per cent. on the outlay. There are few owners of town-houses who would be content with such a small return for their investments.

It has been stated that it was not only necessary to surround the houses with gardens, but to isolate them as well. The miner, by some of his peculiarities, from the way in which he is brought up and lives, as well as from the nature of his work, allies himself to the agricultural population, and loves to cultivate the ground after he has finished hewing coal. In most cases, in his native village he has worked in the fields in his youth, and experiences in the cultivation of his garden-produce some of his earlier habits. At harvest-time it is very difficult to restrain him, and he gladly escapes from the mine to go harvesting, or to take a part in the vintage; but that is the only way in which he plays truant.

The decided predilection on the part of the miner for an independent house is less easily explained. It may be said that, after having worked all day with his comrades, he feels the want of being alone with his family in the evening; and the excessive fatigue he experiences on leaving the mine may also induce him to desire quiet and repose.

At any rate, this isolation must be provided for the collier, and is one of the indispensable conditions for insuring the success of dwellings for the working miner, with whom towns and large barracks have never become popular. The miner feels a sort of repugnance to that kind of living in common in which morality, tranquillity, and health are sometimes sacrificed. The systems of Fourier, Considérant, and Cabet—the *phalanstères*—and living in common, have, in this instance, signally and completely failed. Philosophers do not sufficiently reckon with poor human nature, and they only see in the featherless biped (according to the definition of Plato) a sociable and not too perfect animal. Imagine the case of a workman in a town where a hundred or even two hundred families live together; the women apt to quarrel in the absence of their husbands; the men, on returning from work, bringing into the house their share of trouble and noise; the children, above all, adding their perpetual cries and noisy games (other disagreeables may be passed over)—soon render the house or barracks uninhabitable, and turn it into a pandemonium. On the other hand, in detached houses everybody is at liberty to do as he likes; nobody troubles his neighbour, and the old French saying, *Charbonnier est maître chez lui*, or "the coal-man is master in his own house," is the order of the day.

The preference for isolation has been so greatly felt in some collieries that the attempt to introduce the system of lodging-houses, in use amongst the working people of Mulhouse, in which four families can be accommodated under one roof, but in separate apartments, has been tried without success at the collieries in the Saône-et-Loire; and at Blanzy, Creuzot, Epinac, and elsewhere, they have reverted to the use of detached houses. The groups of houses which have been built in this style, with roofs covered with red tiles or slates, with white-washed walls, and provided with yards and gardens, compose highly picturesque villages. The streets are wide, regular, open, and often planted with trees.

The houses are sold at cost price to workmen who may wish to buy them. Every kind of facility is afforded for payment, which may be extended over a certain number of years without interest, or by a series of deductions from their wages. As there are some who would rather build a house according to their own fancy, and would prefer not to pay the Company rent of any sort, the latter then sell plots of ground and building materials—lime, stone, bricks, &c.—at cost price, and even go so far as to make advances in money. In this way the dream of the operative house-owner, so highly lauded by some economists, is realized without pressure and without any sacrifice of independence.

The French colliers, from the moment they thus become owners of houses on the very spot where they work, become still more attached to their mine, and feel increased interest in its prosperity. Opposed to strikes, avoiding the public-house and wine-shop, they contract a habit of saving, and become a class which is especially appreciated by proprietors, for the calling is laborious and difficult, requiring long practice and a sort of apprenticeship. Miners are not made to order. They must be inured at an early age to hard work, and to the almost military discipline which prevails in underground operations. The best way to accomplish this result is to establish the honest family of the workman on the mine itself, in the manner which has just been explained.

The measures and philanthropic institutions which we have described are now in use at most French collieries. By means of the establishment of an annual prize in some localities for the best-kept, the tidiest, and most cleanly house, a taste for comfort has been developed amongst French operatives which they do not usually possess. By means of a praiseworthy emulation a kind of studied refinement of living, and even an elegance, has been brought into existence amongst them, involuntarily reminding one of England.

It is in this country, above all others, that the dwellings of colliers should be visited. The cottages in the best districts are ornamental, neat, in many instances detached, and even wide apart. The wife is at home, attentive and thrifty, making the husband's tea or the traditional pudding, and ready to do all that is required in the household. The furniture shines with a bright polish; order prevails everywhere; articles of luxury are seen, books, and a newspaper, which is so often found to be wanting elsewhere. The English miners have even journals of their own, which the French have never yet possessed. The children, of whom there are frequently a troop, are peaceably disposed, and are carefully dressed. It is not the house of a workman, it is rather that of a citizen; it is the cherished home of the Englishman, the sacred and inviolable domestic hearth, such as has given rise to the proverbial saying—*Every Englishman's home is his castle.*

The working-men's houses, in all countries where they have been built, only accommodate the families of miners. The unmarried generally lives with the married miner, who finds him board and lodging for a certain fixed sum. Like the soldier on the march who is quartered on the shopkeeper, he has a place by the fireside, and on Sundays he accompanies his hosts in their walks. In a word, he shares all the pleasures of the family without having any of its cares.

In order to obviate the incessant complaints to which the lodging system gives rise, furnished rooms for unmarried men have been established at several mines in France, and also canteens, where for a moderate sum, less than a franc per day, the workman may obtain two excellent meals, consisting of bread, wine, broth, beef or mutton, vegetables, and cheese. A good appetite, such as the work in the mines provokes, distin-

guishes the guests.   In these establishments everything is
conducted with order and decency, under the superintendence
of a person employed on the mine; and there is nothing there
to call to mind the canteens of migratory workmen employed
on railways in course of construction.

In England and the United States these establishments are
commonly known at the mines as *boarding-houses.* The meals
are more substantial and better supplied with meat in those
countries than in France.   Potatoes and roast-beef, fruit-pies
and puddings, are the standing dishes.   Beer, tea, coffee, milk,
and often pure water, take the place of wine.   The usual cost
of board in the United States is a dollar per day.   Throughout
Great Britain beef is not less indispensable than tea, and this
regimen is easily accounted for in a northerly climate subject
to damp and fogs.   In Belgium the miners have almost
exclusively adopted the use of coffee, and have observed that
this beverage combines hygienic qualities with very nutritious
properties; an observation which has been verified by science.

The collier's daily life has a sort of calmness and regularity
about it, which is attributable to the discipline which regulates
his duties.   Here are no noisy meetings, no shouts or songs,
no nights given up to gambling or drinking.   The work
proceeds uninterruptedly, day and night, by means of constant
relays every eight or twelve hours.   They only cease from
labour on Sundays, on which day the miner makes himself
smart.   After the stains which his skin has acquired, from the
black coal-dust and his working dress, have been got rid of by
repeated ablutions, he puts on his best clothes and goes for a
walk.   The country is close at hand, and he goes there with
his wife and children and friends, talking as they stroll along
the roads, or by the brook-side, through the fields, and under
the tall trees.   Not but that the wine-shop receives also
sundry visitors, who, in the evening, may in more than one
instance succumb to their too liberal potations; but the topers

may be excused on the plea that "once in a way is not a habit" —*Une fois n'est pas coutume.*

"The wages of the English coal-miner are, as a rule, very good, nay in some cases exceedingly high, if men only choose to work, and have acquired the degree of skill which we find, even in coal-cutting, will greatly distinguish certain hewers above others. The colliers of the north can commonly make their 6s. per day, and have often houses free of rent, and coals and schooling at a nominal charge; while the Welsh colliers, in a good stall of the rich Aberdare coal, will get their 8s. and even 10s. a day.

"At the Navigation colliery, Mountain Ash, the highest earnings made in March, 1866, was no less than 12s. 8d. per day for twenty-three days."[*]

From Newcastle down to South Wales the pay for good hewers varies from 5s. to 12s. per day.

The average wages of colliers in South Wales are about from 22s. to 25s. per week, and the same or 22s. per week in the Forest of Dean. In the Forest of Dean and throughout the whole of South Wales the colliers are paid by the ton, the price for which varies in the former coal-field in all the pits, but is uniform in South Wales in nearly every seam, according to the name of the vein.

In Lancashire, where they generally work by the score of pit-carts, and pay for their own hauling or waggons, the average weekly earnings vary from 24s. to 28s.[†]

In South Staffordshire the wages of a hewer are 4s. 10½d. per day, with a daily allowance of two quarts of beer, and a ton of coals per month.

In the Somersetshire coal-field the average wages of a pit-man are about 15s. a week, with an allowance of six cwts. of

---

[*] "Treatise on Coal and Coal Mining," by Warington W. Smyth, M.A., F.R.S., p. 236.

[†] From information furnished by Mr. Simeon Holmes, manager of the Blakeney and Forest of Dean Coal Co. (Limited).

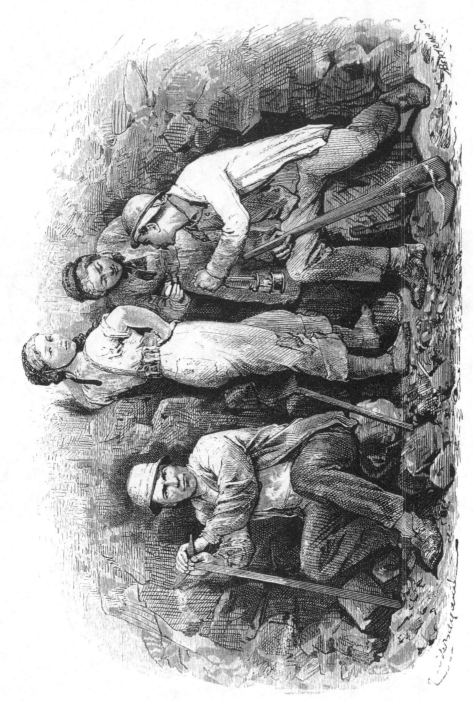

Fig. 99.—Pitmen and their Wives, in their working dresses, at the Collieries of Charleroy, in Belgium.
After a Photograph.

coals a week. The wages of boys vary from 10s. to 12s. a week, according to age and capabilities.

In France and Belgium the wages for the best hewers is on an average from four to five francs (3s. 4d. to 4s. 2d.) a day, and may exceed that amount when the work is done by contract. The other men earn about three francs, and the women and boys from one to two francs (10d. to 1s. 8d.). In the United States the amount, as in England, is considerably greater, frequently half as much again. In France women no longer work in the mines; in England they work only on the surface, and in Belgium they are still employed (figs. 99 and 100); but such cases become every day more and more rare. Boys are nowhere allowed to be employed underground or at the surface under a certain age, and only on certain conditions. The employment of women at the surface consists in sorting, sifting, and washing the coal, a task that is not very laborious, and one for which they possess great aptitude. Underground in Belgium they load the coal in the corves, or draw them to the bottom of the shaft. The duties of the boys are nearly the same as those of the women.

In all coal-producing countries the miner's life is sensibly the same, with the exception of certain habits which are peculiar to each country. The French miner has already been described. In Belgium, where the occupation may be said to have originated in the province of Liége, the miner is distinguished by a still more marked type. Belgium is a true coal country, and many of the terms in use amongst the French coal-miners are borrowed from her. From the time of the Middle Ages, the occupation of miner at Liége carried honour with it, and in some degree ennobled the person who exercised it. The guild of colliers possessed their charter, and peculiar privileges. The arms were *two picks or on a field azure;* and the azure or blue was doubtless the sign of honesty and force, while the *or,* or gold, was emblematic of the wealth derived from coal.

R

The prominent part which the British coal-miner occupies in the industrial prosperity of the United Kingdom must be

Fig. 100.—Woman and young miner of Pontypool.

fully recognized, besides which the English miner applies to his underground labour the vigorous qualities which characterize his race. He is industrious, punctual, reserved, and cheerfully dedicates himself for ever to the same task without evincing any desire either to change his abode or his calling. He does more work than the French miner, because he is better fed. Obedient, zealous, impassible, executing the orders of his superiors without dispute, he contributes in a great degree to the success of the operations by the exercise of all these valuable qualities.

The United States miner displays a resemblance to the English miner in more ways than one. On both shores of the Atlantic the people and the language are the same, only the political constitution is different; and the greater freedom and the more complete equality which prevail in the States of America react on the habits of the citizen and the working-man. The terms employer and workman are almost unknown; the miner labours at the mine as though he could fill any other office; he is paid liberally a specified sum. He is your equal, remember; for his part, he will not forget the respect which is due to each; he begins by respecting himself, and on leaving the mine he dons his black clothes and hat. The levelling of class-distinctions is complete. We are in the land where Lincoln the rail-splitter, and Johnson the tailor, have attained the office of President in succession. While at work every one is attentive and serious; there is no singing, there are no disputes. There is no smoking, but every one is content to chew his quid of tobacco in silence. Away from his work-place, however, the conduct (at least at the liquor-shop or public house) is not always so free from reproach.

There is one trait which seems almost peculiar to the American workman; the fondness he seems to have for his tools, which he always endeavours to keep in good order. In the country in question, where the mechanical arts are so highly cultivated, each tool assumes the most elegant form, as well as that best suited for the special purpose for which it is intended.

R 2

All that has been said is applicable to the free workman in the United States. In the coal-producing states of the centre and south, in Virginia, for example, negroes have been long employed in mining, whether they were owned by the colliery proprietors or were hired for the special purpose. The negro has proved himself to be there, as he is everywhere else, a great infant, with little capacity for long-continued efforts or systematic labour; a good workman, but a bad miner. Let us wish him more success now that he is free, and content ourselves with having mentioned him amongst the great family of the soldiers of industry.

The German coal-miner, whether belonging to Prussia Proper, Saxony, or Austria, preserves the distinctive traits of the Germanic race. Less impulsive and more phlegmatic even than the Anglo-Saxon miner, he has faithfully carried into coal-mining the habits of discipline of the metallic mines, which in Germany date from many centuries, and have been everywhere preserved with a religous and jealous care. The earliest coal-miners of Central Europe came from these mines, where they are regimented in the strict and figurative sense of the term, and form a distinct class of the population. They constitute something more than a mere guild or corporation, and are a sort of caste with peculiar costume, dress, and manners, as well as free-masonry, traditions, and even superstitions. Certain particular terms are handed down from one generation to another, and when on the road or in the country you are saluted with *glück auf,** you may be sure that you are addressed by miners. In the metallic mines we shall find the German with still more strongly marked traits, justifying in all respects the Teutonic saying, " Proud as a miner."

At the western extremity of Europe, the Spanish coal-miner of the Asturias and Old Castile, as also in Andalusia, is distinguished by special characteristics. Sober, living on next to

---

* Literally, *happy out*; meaning happy going out—good journey to the mine and back again.

nothing, and perpetually smoking cigarettes, he is at the same time an energetic, courageous, and intrepid worker, but of a restless and changeable temperament. He is cold, taciturn, or talks little, and on his return home notably clad in his brown cloak, with his sombrero drawn over his face, he looks much more like a country hidalgo than a coal-miner. The Asturian grafts on to the qualities of the Castilian others of a still austerer kind; and he is, also, a better workman and more inured to fatigue. His habits seem to borrow their tone from the ruggedness of his native mountains. In the south of Spain, the less energetic Andalusian has preserved somewhat of the costume and nonchalance of the Moor, who formerly possessed the country; but still he is a capital miner.

In southern Europe there is another type full of character, which must be mentioned; the Italian coal-miner, especially the migratory miner of the Tuscan Maremma, a district once flourishing and populous, and one of the granaries of ancient Rome, but now, at certain seasons of the year, a prey to fever and miasma.

From July till October it would decimate the miners, and the colliery is, therefore, closed during those four months, the works being carried on without interruption during the remaining interval.

The migratory labourers employed in the mountains of Pistoja, on the healthy flanks of the Apennines, leave their country by hundreds after autumn, when the weather begins to get cold and snow may soon be expected. Old Niccolini, the mine-agent, goes to hire them beforehand, and to pay the earnest-money. The exodus begins soon after the middle of October; some going on foot, while others ride, seeking places in the *calessini* or stage-coaches. They come alone, leaving wives and children at home, for too many victims must not be exposed to the *malaria*, or deadly air of the Maremma.

The opening operations, which are soon begun, are impeded

at the outset by the last rains of autumn. The dilapidations which have happened during the stoppage of the works are made good: timbering has broken, the goaf has given way; water has made numerous devastations at the surface, as well as underground; but all is promptly set on foot, and the works are re-opened and carried on with vigour. The coal raised is sent to Leghorn for the use of the local factories and the steam-boats. It has also been taken to Rome, and the Vatican is lighted with gas made from this coal.

The two hundred miners employed on the mine live in lodging-houses or barracks, sleeping on camp-beds, covered with a woollen blanket and with merely a sack of straw for a mattress. They live as they like; but the fare is little varied. Niccolini keeps a shop, where he dispenses to the miners dried cod, salt pork, rice, coffee, wine, rum, and especially maize-flour, with which a thick soup (*polenta*) is made, the favourite dish of the Italian workman from the northern to the southern extremity of the peninsula. The polenta affords little nourishment, and as the work is very hard, twelve hours a day on the average, the continuous labour exhausts the miner and induces fever. But this is not all; for in addition to excessive bodily fatigue is added the daily use of salt provisions and an abode in a hot and humid valley; and scurvy attacks those whom the fever has spared.

A story is told that an Englishman inquiring of a peasant in the Pontine marshes, how he could live in such an unhealthy country, received for answer from the *contadino*, " We do not live, we die." The miners may say the same, and yet every year those who have been spared by the fever and the scurvy return, for bread must be earned, and no work is done at home, where, in the winter, the Apennines are covered with snow.

On Sundays the works are stopped, and the miners amuse themselves all day in the chase, or rather in playing, singing, and with *morra*. They go to the wine-shop to drink the national *poncino*, a mixture like the English punch, made of hot water, sugar, lemon, and rum, which is considered to be a good specific against the fever. These potations are continued, they become excited, and sing, alternately improvising verses like those of Tasso. The poets challenge each other at the outset, like the shepherds in the Bucolics and Idylls. Nothing has changed in this respect for the last two thousand years, and Virgil and Theocritus, if they could revisit Italy, would recognize their pupils. They ever invoke Apollo, and will go and gather at the capitol the laurel crown wherewith to reward the victor. Everybody makes verses in the beautiful Tuscan dialect, which is spoken here with an elegance and correctness which might make the Academicians of La Crusca jealous. The poncino has this effect on Agostino, a Corsican bandit who has taken refuge in this mine, and who improvises too at leisure moments. His language is not of the purest; he hesitates and cannot think of a rhyme, which gives an easy victory to his rival.

To make up for his deficiencies as a poet, Agostino does his duty as a watchman at the mine with a savage fidelity, and, armed to the teeth, daily traverses on foot the twenty-five kilomètres of railway which connect the colliery with the sea. He also goes to fetch the money for the men's pay every fortnight. The diligence from Leghorn tosses to him, as it passes him on the road, the bags full of crown-pieces. It is night, and Agostino has made known on the previous evening, all over the neighbourhood, the object of his journey, just to see whether anybody would venture to attack him. It must not be forgotten that the Maremma is the classical land of bandits (*birbanti*). These gentry frequently make their appearance at the mine, but in a friendly way, in consequence of the numbers of workmen there, who, in case of need, would give chase to the brigands.

Agostino fled from Corsica after a *vendetta*. His countenance bears the traces of past contests, his nose having been split in two by the blow of a stiletto, and his face seared by an enormous scar. Having been found guilty of contumacy, he came to seek service under the Grand Duke, who, honouring his unfortunate courage, allowed the Corsican to take the oaths, and appointed him watchman on the railway of the Monte-Bamboli colliery. Proud of being able to deck himself out with knife, pistol, and gun, this time without opposition, the Corsican performs his service with strictness, and none of the ignorant or savage peasantry whose lands adjoin the line of railway would dare to pull up a rail or place any obstacle on the road. It is said that even the buffaloes, who roam at liberty in the marshes, hesitate to break through the fences when Agostino is going his rounds.

By the side of this savage Corsican the mining captains or caporals present a singular contrast. They are all Pistojans, named Luigi, Sandro, Geremia, and Beppo. Luigi and Sandro, two brothers, who from mere labouring men have become captains, have charge of one of the districts of the mine, one by day, the other by night. Of lofty stature, dignified in their deportment, wearing large beards, calm, reserved, and ever foremost in duty, they remind one of those crack artillerymen or soldiers of engineers who have grown old in the discipline of camps.

Geremia pleases in other ways. Lively, a good talker, alert, and courageous, he is daunted by no danger; wherever there is a difficulty to overcome he is the foremost, and he sets an example everywhere. In case of need, he knows how to face death. His brother, who, like him, was ever ready to lead the van, was killed in the mine by an explosion of gas. As to Beppo, he is apparently the opposite to Geremia, being cold, silent, and reserved; but he has a more observant mind, and has much greater intelligence than any of the others. He endeavours to penetrate the secrets of subterranean geology, tries to find out why the beds affect such and such a position, questions the engineer, and inquires whether some working book has not been written for the use of mining captains.

Pierre Lhôte, the bold overman of Epinac (fig. 101), must not

Fig. 101.—Pierre Lhôte and Father Garnier, overmen of Epinac (Saône-et-Loire), after a photograph. Pierre Lhôte, who is standing up, wears the cloak (*limousine*), the wooden shoes (*sabots*), the leather hat, and the skull-cap of the coal-miners of Central France.

be altogether forgotten. He was at the siege of Constantine, and helped to take the city under Lamoricière. Pierre Lhôte brings into his service all the steadiness of the soldier, and he doubtless became a collier that he might still burn powder.

If the engineer goes to visit a working-place, he says to his men, "Come, my lads, fall in ; here is Monsieur the engineer going by." He is very near putting them in line at *present arms*—the pick raised in his right hand, the left resting on the handle of the shovel. One day the miners at the bottom of the workings threatened to strike, a dispute having arisen about the price to be paid for the work. "Wait for me here, my lads," said Pierre Lhôte, "while I go and speak to· the engineer." On his return, he says, "Listen, my lads, to Monsieur the engineer's answer :—'Pierre Lhôte, go and tell your men that the day shift begins at six in the morning, with an hour's rest at noon for dinner, and ends at four o'clock in the afternoon.' There, my lads, that's what Monsieur the engineer said to me. He is our chief; *I* obey *him, you* obey *me;* come, my lads, to work !" There was no more talk after that about striking.

Side by side with Pierre Lhôte, the underground foreman, walks Father Garnier, the day foreman (fig. 101). He superintends the drawing, washing, and coking of the coal, and also its transport ; in fact, all the details of the upper works fall under his charge. He has an eye to everything, and is ready to satisfy everybody: "Yes, Monsieur Director ! Yes, Monsieur Engineer !"

Such is the colliery population, deserving among all others. Each one modestly executes his task, it may be said his useful duty, the last labourer equally with the most skilful miner, the overlooker as the engineer and the manager of the mine. Soldiers, caporals, captains, all march side by side in the path of duty. What energy, courage, and devotion are displayed by this new legion of labourers, which has been formed amongst the French half a century ago ! It is a world in itself, which has been little studied till now. The public has passed by the collier with too much indifference ; the philosopher, the savant, the artist, the romance writer, have not sufficiently examined

him.   The soldier of the lower regions deserves more than a
mere passing notice, a momentary curiosity, when a hurried
visit is paid to a mine, or a shocking accident happens to
alarm a whole district, and throw hundreds of families into
mourning.   The patient and enduring labours of the coal-
miner are deserving of serious examination on the part of all.
If the multitude could but see them, they would be full of
curiosity and eagerness; but the ground covers the greater
number of these bold and grand colliery undertakings, and the
world at large knows little or nothing about them.   Subterranean architecture and engineering have no other beholders
than those who practise it, and its creations can scarcely
enjoy any great or extended celebrity.

CHAPTER XII.

TO-DAY AND TO-MORROW.

Quantity of Coal annually exported from Great Britain.—British Coaling-stations abroad.—Number of persons directly employed in raising Coal in Great Britain, France, Belgium, Prussia, and other countries.—Quantity of Coal raised in 1866, by Great Britain, Prussia, France, Belgium, Austria, Saxony, United States and other Coal-producing countries.—Average price of Coal in England, France, and the money value of all the Coal annually raised.—Great importance of Coal at the present day.—Production of Coal in United Kingdom, Belgium, Prussia, and France, per head, in 1866.—Consumption of Coal in those countries, per head, in 1866.—Superficial areas of the chief productive Coal-fields of the Globe.—Production of Coal in Europe per square mile of Coal-field in 1866.—Proportion between the areas of the principal Coal-producing countries and the areas of their respective Coal-fields.—Proportion of Coal-fields in each Coal-producing country in Europe. —Increasing production of Coal.—Increase in quantity of Coal raised in the chief Coal-producing countries from 1831 to 1865.—Sir W. Armstrong on the duration of the Northern Coal-fields.—Sir Roderick Murchison on the Coal-bearing areas of the United Kingdom.—Mr. Gladstone.—Appointment of a Royal Commission.— Probable duration of European Coal-fields.—Mr. E. Hull's estimate of quantity of Coal in Great Britain, and its duration.—Mr. W. S. Jevons on the Coal Question; his conclusions.—Mr. W. W. Smyth's estimate of the duration of the British Coal-fields.—Duration of Coal fields in North America and other countries.—Difficulty of finding a substitute for the Steam-engine and Coal.—The combustible of the future.—Bottling sunbeams.

WHEN the collier has wrought the black subterranean domain, and has drawn the combustible from the bowels of the earth, when he has brought it to day, cleansed it, and placed it on the routes for transport, the useful mineral is borne in a thousand different places to diffuse light, heat, and force in all directions. It is a production which is indispensable at the present day to the existence of all civilized nations, and everybody foresees the troubles which the world would experience if the supply of coal were suddenly to fail. There would be no more light in the towns, no more fire in the factories or in most houses, and all the railways would be stopped. Textile and other manufactories, nearly all the workshops, machinery of almost every kind, and a great number of ships, would be laid up in enforced idleness for want of their essential aliment. Material life, as well as much of intellectual life, would be extinguished, in the same way as the life of the body becomes extinct for want of nourishment.

The most polished nations cannot for the future dispense with coal, and the degree of a country's civilization may almost be estimated by the quantity of this combustible which it consumes.

All countries burn coal when they can get it, but very few comparatively are producers of it, and rare are those which nature has endowed with abundance of the black diamond.* Only those will be mentioned the production of which is of importance. In the first place, then, Great Britain holds pre-eminence on the list of coal-producers, leaving all other countries behind; for it raises more than all the others put together, having raised more than 101,630,000 tons in 1866.

First after Great Britain is Prussia, the Zollverein States (which include Westphalia and Bavaria, Map X.) furnishing upwards of 17,000,000 tons; then follows France (Maps II. and III.), producing 11,300,000 French tons of 2200 lbs. avoirdupois; and Belgium (Map II.), which yields about 12,000,000 English tons. The coal production of Austria (including Bohemia, Map X.) is a little above 4,000,000 tons, of which more than 1,500,000 is brown coal; and Saxony gives about 2,000,000 tons of coal, and 500,000 tons of lignite, respectively.

All the other countries of Europe—Germany (with the exception of the above-mentioned states), Hesse-Cassel (Map X.), Hanover, Spain, Italy, Hungary, &c.—scarcely reach a total of 4,000,000; and all the remaining coal-producing countries in the world together as India, China, Japan, Australia (Map XIV.), &c.—very likely do not raise 3,000,000 tons per annum.

* At the Great Exhibition of 1851 the Koh-i-noor created a great sensation, and some English coal-owners, playing upon the name, which signifies "the mountain of light," and making allusion to the application of coal, placed this inscription on a block of the latter—"This is the real Koh-i-noor!"

America, with her immense coal-fields, is destined to become eventually the great coal-producer of the world, but at present the total produce of the United States is under 15,000,000 tons. The quantity of coal raised and sold in the province of Nova Scotia during the year ended 30th September, 1866, was 601,302·2 tons, and during the year ended 30th September, 1867, 482,078 tons; showing a deficiency of 119,224·2 tons during the latter period. The other coal-fields of the New World are insignificant.

By adding up all these totals we arrive at a grand total of 169,000,000 tons, of which Great Britain furnishes much the largest half, while the shares of North America and Prussia amount to a tenth, respectively; that of France and Belgium to a fourteenth each; and, lastly, that of all the other coal-producing countries together to a twelfth.

This statement of the coal-production of 1866 is recapitulated in the following table :—

COAL PRODUCTION OF THE GLOBE IN 1866.

Name of Country.		Tons.
United Kingdom (England, Ireland, and Scotland),.		101,630,544
Other countries of Europe—		
Prussia,	17,000,000	
France,	11,300,000	
Belgium,	12,000,000	
Austria,	4,000,000	
Saxony,	2,500,000	
Hesse, Bavaria, Hanover, Russia, Spain, Italy, &c.,	4,000,000	
		50,800,000
North America,		15,000,000
Other countries of the globe—		
India, China, Japan, Australia, Chile, &c.,	3,000,000	
		68,800,000
Total,		170,430,544

The average price of a ton of coal at the pit's-mouth is from 6s. to 8s., a sum which is more than doubled in France, and trebled in England at places remote from the scene of production. The grand total of 170,430,544 tons represents then, estimating the ton at 10*s*. 6*d*., a sum total of £89,476,035 or nearly £89,500,000 sterling. This is double the value of the precious metals, gold and silver, annually produced throughout the globe. The collieries take precedence, then, of the metalliferous mines, of those of California and Australia, as well as those of Mexico, Chile, and Peru. Decidedly the black diamond fetches its price, and it is doubly well named.

Fossil fuel plays besides, in the social life of the people of our day, a part of still greater importance than the gem, for industry lives by it alone. It has taken the place of wood, which has become dearer and dearer; and it is calculated that the whole of Europe, if covered with forest, would scarcely furnish every year, in cut wood and charcoal, a quantity equivalent to the annual consumption of coal.

Coal has also compensated for the deficiency of manual labour. The steam-horse has replaced to a great extent that useful slave the draught horse; and as it is never weary, is in action day and night, and takes no rest, all the living motive power of the globe would scarcely suffice nowadays to do the work which steam performs.

France, with a population of about 40,000,000, consumes some 20,000,000 tons of coal annually; while Great Britain, with a population of 30,000,000, absorbed in 1866 about 92,000,000 tons, after allowing for the 9,000,000 tons which were exported to foreign countries. In other words, the consumption of coal per inhabitant in Great Britain in 1866 was about three tons; while during the same period it did not amount to half a ton per head per annum in France. To arrive at a correct result, however, account must be taken of the quantity of wood and other vegetable combustibles burned in both countries, and to transpose the whole into an equivalent weight of carbon, or refer it to a standard quality of combustible.

The quantities of coal consumed by different countries can scarcely be compared in an absolute manner with each other, like the relative quantities of tobacco, sugar, coffee, and tea. The following table will give a sufficiently near approximation in round numbers for the four chief coal-producing and coal-consuming countries of Europe, for 1866 :—

PRODUCTION OF COAL IN EUROPE PER HEAD IN 1866.

Name of Country.	Total Quantity Raised.	Number of Inhabitants.	Production Per Head.
United Kingdom,	101,630,544 tons.	30,000,000	3˙39 tons.
Belgium,	12,000,000 "	5,000,000	2˙40 "
Prussia,	17,000,000 "	20,000,000	0˙85 "
France,	12,000,000 "	40,000,000	0˙30 "

CONSUMPTION OF COAL IN EUROPE PER HEAD IN 1866.

Name of Country.	Total Quantity Consumed.	Number of Inhabitants.	Consumption Per Head.
United Kingdom,	92,000,000 tons.	30,000,000	3 tons.
Belgium,	10,000,000 "	5,000,000	2 "
Prussia,	15,000,000 "	20,000,000	0˙75 "
France,	18,000,000 "	40,000,000	0˙45 "

See England! Coal does not merely furnish the indispensable aliment of the factories; it serves besides to freight ships. Should coal fail elsewhere, England alone could supply the world. She exports at this time more than nine million tons, the eleventh part of her entire produce. It is by means of their coal depots that the modern representatives of the old Phœnicians mark their maritime halting-places on the globe, and it is partly for the supply of their steamers that they thus transport coal from one hemisphere to the other. In the Mediterranean they are everywhere, especially at Gibraltar, Malta, and Alexandria; in the Red Sea, at Suez; in the Indian Ocean, at Aden, the Mauritius, Natal, Mozambique, and Zanzibar; then at Muscat, Bombay, Madras, Ceylon, Calcutta, Rangoon, Singapore, and the stations in the China and Japanese waters; in the Atlantic at Buenos-Ayres, Monte Video, Rio Janeiro, Bahia, Pernambuco; and at the Azores, Madeira, the Canary Islands, Ascension, Saint Helena, the Cape of Good Hope, on the coasts of Guinea and Congo. All these stations, all these anchorages, have stores of English coal; so has the entire archipelago of the Antilles, especially Cuba, Jamaica, Saint Thomas, and Colon-Aspinwall. Along the coasts of North America—at Quebec, Halifax, Boston, and New York—the British Colonies and the energetic Yankees compete with each other in supplying coal. In the Pacific it is Panama, Guayaquil, Callao, Arica, and Valparaiso which the coal-ships visit; and on the opposite parallel in the northern hemisphere, San Francisco, the Queen of the Great Ocean: finally, between the Indian Ocean and the Pacific it is the great island Australia, which of itself forms a section of the globe. The globe now belongs to those who can supply her with coal; and, as a celebrated statesman said in the House of Commons, all the nations which are without combustible minerals are the vassals of England.

Europe, even, cannot dispense with the services of the United Kingdom. If nature in the formation of coal-fields has favoured England at the expense of the other countries of the Old World, the greater number of the latter are, notwithstanding, possessed of Carboniferous deposits, which they work with great ardour; still, what is required to make the production equal to the consumption is derived from England. Here, again, we see the British coal-ships make their appearance, not in the Mediterranean only, where we have already followed in their wake, but all along the coasts of the Atlantic, of the English Channel, of the North Sea, and the Baltic. France, herself, is in this respect the tributary of Great Britain. The quantity of 12,000,000 tons of coal annually raised by the French is not sufficient to supply their requirements, and they take every year about 6,000,000 tons from abroad, or half the quantity they themselves produce, or a third of their own entire consumption, or two-thirds of the quantity exported from the United Kingdom. Great Britain, Belgium, and the Rhenish provinces supply the deficiency; Belgium contributing three-fifths, the other two about one-fifth each.

It is interesting to calculate the number of persons directly employed in producing the quantity of coal annually raised. This number would seem to amount to 300,000 (Smyth) in Great Britain. France, Belgium, and Prussia together employ 200,000, of whom 120,000 belong to Belgium and France, and 80,000 to Prussia. These last numbers give, on an average, one coal-getter for a yearly production of two hundred tons. On the supposition that the same proportion holds good for the numbers employed by the remaining collieries, a grand total of 700,000 men is arrived at, which very closely represents the number of the army of colliers. It amounts to precisely the number of combatants that first-class powers send into the field on momentous occasions; but how much more valuable is the army armed with the pick than that which carries the musket! The latter spreads ruin, fire, and blood in its path; the former leads directly towards progress. The first keeps men nearly in idleness, or in costly and unremunerative activity; the second comprises the most energetic of industries. Both make use of gunpowder; but while the one destroys, the other creates. Both are valiant, beyond doubt; but one exists only for war and destruction, the other is an army of peace.

If the production of coal is sensibly proportionate to the number of men employed, it is not so in proportion to the area occupied by the field which is worked. It is supposed that a very large number of the coal-fields may be barren, that is to say, may only consist of Carboniferous sandstones, shales, and limestones, either altogether destitute of coal, or nearly so, as is the case in Ireland, Russia, and in some parts of British America. On the other hand, a large quantity of coal may be concentrated beneath a comparatively small surface, where the coal-seams are very numerous, as is the case in Belgium; or are very thick and near the surface, as in England. The quality of a combustible is also an element in the calculation of the entire extraction, for there are some coals of indifferent quality

which are scarcely worked.   Lastly, certain deposits cannot be
advantageously worked in consequence of particular circumstances, such as distance from centres of consumption, high
price of carriage, &c.   It is, then, never possible to establish an
exact relation between the productions of various coal-fields
and the areas which they respectively occupy.   It would be
more reasonable, in all cases, to compare the total extraction
with the cubic contents of existing coal, which is arrived at by
multiplying the superficial area of the coal-field by the aggregate thickness of all the beds put together.   However this
may be, the Table below shows the estimated area of the most
productive coal-fields:—

ESTIMATED AREAS OF THE PRINCIPAL PRODUCTIVE COAL-FIELDS.

Locality of the Coal-fields.	Area in Square Miles of Coal Measures.	Relative Coal-bearing Proportion of the Whole Area.
North America (United States and British Colonies), . . .	180,000	80 per cent.
United Kingdom, ...	9,000	5·26 "
France, ... 1,800		
Prussia, ... 1,800		
Other German States, . . . 1,800	7,200	3·2 "
Belgium, ... 900		
Spain, ... 900		
Other countries (approximately),	28,800	12·8 "
Total estimated area of all the coal-fields of the globe, . .	225,000	

From the foregoing Table it appears that the area of the
coal-fields of North America alone is four times as great as
those of all the other countries of the globe.   As this immense
extent comprises, in most cases, coal-deposits which are extremely rich, and all but untouched, or only very slightly
worked, it is there, as we have already stated, that the great
reserve is to be found of coal for the future.   It should, however, be observed with regard to the 180,000 square miles
given in the table as the superficial area of these regions, that
the fourth part, or 45,000 square miles, belong to the coal-fields
of the British colonies which bound the southern side of the

s 2

Gulf of St. Lawrence, and are less rich in coal than those belonging to the United States (Map VIII.).

The superficial area of the coal-fields of the British Isles is only one-twentieth of that of North America; but it would be one-tenth if the estimate of 18,000 square miles, assigned by some authors, were adopted. In all probability the latter take into calculation the entire area of the Irish coal-fields, which, though of great extent, are very poor in coal, or only slightly productive; and perhaps in the English area they take into consideration not only the fields which are actually worked, but include, besides, the probable extension of the coal-fields beneath the Permian and other overlying deposits (Maps I. and IV. to VII.).

Even at the lowest estimate the amount of coal raised in the United Kingdom is greater than those of all the other countries in the world besides. France (Maps II. and III.), Prussia (Map X.), and the German States have separately, in the three cases, an equal extent of coal-bearing strata; but this, on the other hand, is but one-fifth of that generally assigned to England.

Lastly, the sum representing the area of the Belgian coal-basin is one-half that of France (Map II.); and the same holds good with regard to the Spanish coal-fields. These last, as has been elsewhere stated, are at the present time only worked to a small extent, furnishing scarcely 400,000 tons a year. This may at some future day prove the store from which Europe will derive its chief supplies, when coal becomes scarce through the exhaustion of other countries.

In order to render complete the Tables given at page 256, the production of the four great coal-producing countries of Europe in 1866 is compared with their corresponding areas:—

PRODUCTION OF COAL IN EUROPE PER SQUARE MILE OF COAL-FIELD IN 1866.

Name of Country.	Total quantity Raised.	Area of Coal-field.	Produce per. Square Mile.
United Kingdom,	101,630,544 tons.	9000 square miles	11,292
Belgium, . .	12,000,000 "	900 " "	13,333
Prussia, . .	17,000,000 "	1800 " "	9,444
France, . .	11,300,000 "	1800 " "	6,265

In the above Table the little country Belgium raises almost as much coal per square mile of coal-field as Great Britain, and it closely follows that country in the quantity which it produces and consumes per head of its population. It is partly to its canals, railways, and its well-kept roads, that Belgium is indebted for this fortunate position. Owing to the peculiar unity of its geographical surface, it is a country which possesses a greater extent of canals and railways than any other in the world, except England and some of the states of the North American Union.

On comparing the areas of the coal-fields with the entire superficial area of the chief coal-producing countries, it will be seen that for the United States the relation is one-twentieth or one-fourth, according as (in the first case) the whole extent of that immense empire is taken into consideration, or (in the second case), as is more reasonable, merely the extent of the coal-producing states.* For every twenty or for every four square miles of country, there is then on an average one square mile of Coal Measures in the United States. For Great Britain the proportion is $\frac{1}{19}$th, for Belgium $\frac{1}{18}$th, for Prussia $\frac{1}{90}$th, for Spain $\frac{1}{115}$th, and for France $\frac{1}{170}$th only. Italy is almost entirely destitute of true Coal Measures.

For the four great coal-producing countries, the total quantities may be summed up as follows :—

PROPORTION OF COAL-FIELD IN EACH COAL-PRODUCING COUNTRY IN EUROPE.

Name of Country.	Total Area.	Area of Coal-field.	Proportion of Coal-field.
United Kingdom, .	171,000 square miles.	9000 square miles.	$\frac{1}{19}$th.
Belgium, . .	16,560 " "	900 " "	$\frac{1}{18}$th.
Prussia, . .	162,000 " "	1800 " "	$\frac{1}{90}$th.
France, . .	306,000 " "	1800 " "	$\frac{1}{170}$th.

On comparing the annual lists of the coal raised in all the principal coal-producing countries, for the last fifty years, that

* The whole superficial area of the United States is 409,000 square leagues, or 3,681,000 square miles.

is to say, since the adoption of fossil fuel on a large scale by
manufacturers, a most curious economical law is established—
that the production of coal doubles itself in about every fifteen
years. Some countries which have been amongst the latest to
take up coal-mining, as Prussia or the United States, which
in 1822 had scarcely begun to raise coal, advance at a still
more rapid rate. Since 1830 Prussia has doubled her extraction every ten years; and the same for a time has been the
case in the United States every five or six years.

For subsequent years, some quantities, extracted from official
documents, are given below :—

INCREASE IN QUANTITY OF COAL RAISED IN THE CHIEF COAL-PRODUCING
COUNTRIES FROM 1830 TO 1865.

Years.	Name of Country, and Production in Millions of Tons.		
	England.	Belgium.	France.
1835	26,000,000 tons.	3,000,000	2,500,000
1850	49,000,000 "	6,000,000	4,500,000
1865	98,000,000 "	12,000,000	12,000,000

	Prussia.	United States (Anthracite of Pennsylvania only).
1831	1,500,000 tons.	"
1841	3,000,000 "	"
1851	6,000,000 "	4,000,000
1855	8,250,000 "	6,500,000
1865	17,000,000 "	10,000,000

As nothing tends to show that the progression everywhere
established should not continue, the question naturally arises
as to the time when the coal-fields will be exhausted, and as
to the nature of the fuel which will replace coal after its total
exhaustion—a double problem such as has never been presented
until now during the history of the world. Sir William Armstrong, in his Presidential address to the British Association at
Newcastle in September, 1863, in calling the attention of the
meeting to the rapidly increasing consumption of coal, argued
that in the course of two centuries all the available coal-seams
of the great northern coal-field of Durham and Northumberland would probably be worked out.

Sir Roderick Murchison, quoting this opinion to the British Association, when President of the Geological Section in 1865, could not refrain from bestowing a glance at the future, and seemed to ask himself with anxiety, what England would become if her collieries were exhausted ?   In 1866 he further pointed out to the British Association the areas in England over which coal was at present found, and those where it might reasonably be looked for in times to come.

Mr. Gladstone, struck by these opinions, spoke of them in the House of Commons, in order that the country, warned betimes, should not be some day taken by surprise.   Since then, Parliament has ordered an inquiry to be made by means of a Royal Commission, and Great Britain is now employed in taking stock, as it were, of her subterranean domains.

The duration of the coal-fields, which was formerly estimated by geologists at thousands of years, when the quantity raised was not a fourth of the present output, will, perhaps, not exceed a few hundred years.   It may be asserted with confidence that, in the countries of Europe which are the most largely worked, coal-mining will certainly be at an end in that time.

Mr. Edward Hull, of the Government Geological Survey, who, in the performance of his official duties, has had good opportunities of studying several of the most important British coal-fields, has (1857) estimated the total quantity of workable coal, situated at a less depth than 4000 feet from the surface, to be nearly 80,000,000,000 tons for the whole of Great Britain; and he has calculated that, at the present rate of consumption, all that quantity would be worked out in about eight centuries.   (" Coal-fields of Great Britain," 2nd edit., p. 245).

Mr. W. Stanley Jevons, whose book, " The Coal Question," excited considerable attention at the time of its publication, has been represented as saying that British coal would be exhausted within a century.   What Mr. Jevons really *did* say was very different from this. After examining the rate of increase during the last ten years, he assumes a constant rate of increase in the consumption of coal at 3½ per cent. per annum to continue until 1961, when the consumption at the assumed rate of increase would amount to two thousand six hundred and seven millions of tons (a manifest absurdity). He then says, *always hypothetically*, "*If* our consumption of coal continue to multiply for 110 years at the same rate as hitherto, the total amount of coal consumed in the interval will be *one hundred thousand millions of tons.*" We now turn to compare *this imaginary* consumption of coal with Mr. Hull's estimate of the available coal of Britain, viz., eighty-three thousand millions of tons within a depth of 4000 feet. Even though Mr. Hull's estimate be greatly under the true amount, we cannot but allow that "rather more than a century of our present progress would exhaust our mines to the depth of 4000 feet, or 1500 feet deeper than our present deepest mine." Then, after showing how rapidly the price of coal must increase with the increase of depth, he says, " I draw the conclusion that I think any one would draw, that *we cannot long maintain our present rate of increase of consumption; that we can never advance to the higher amounts of consumption supposed.*"

It is some consolation to find that Mr. W. W. Smyth, who, from his intimate personal knowledge of our coal-mining districts, is better qualified to express an opinion as to the probable duration of the British coal-fields than almost any other person, considers the assumption that the quantity of coal raised will go on doubling itself every twenty years, to be based on a fallacy; and that "although the numbers for certain years appear to fit such a conclusion, the increase to our production of from 2,000,000 to 3,000,000 of tons annually, serious as it undoubtedly is, will keep us within comparatively moderate figures for a long time to come, and at all events defer, as regards the country at large, the evil day for two or three centuries."*

The above estimate with regard to the duration of the coal-fields might be extended at the very most to twice or three times longer for countries like North America, whose immense deposits are nearly untouched; but on this point, as in all the other coal-possessing countries of Asia, Africa, &c., the coal can never be available, except in exceptional cases, for any but local consumption. Besides, coal, at least when it is to be used for great industrial operations, is not an article of sufficiently high price to bear the cost of long transport even by sea. At the present time, in the Indian Ocean, the Mauritius, and Isle of Bourbon, where there are depots of English coal, the steam-ships are the only consumers of the combustible mineral. For sugar-boiling and other native factories the price is too high. At Suez, in the Red Sea, the price has been known to be as much as £8 per ton.

May it be supposed that the consumption, in most European countries, will diminish some day, when all the network of railways being completed will allow some of the present metallurgical factories to be closed, and when a substitute for coal-gas for illuminating purposes shall have been discovered? In that case, it is still possible that the diminution of consumption which may be effected by these means may be, to some extent, counterbalanced by the increased quantities required for the greater number of locomotives and steamships.

There lies, then, in the eventual exhaustion of the coal-fields (an exhaustion which, from the data we possess, can be calculated with some approximation to truth), one of the gravest and most important of subjects—a question which, without being exactly threatening to the present generation,

* "On Coal and Coal Mining," by W. W. Smyth, p. 241.

is not the less deserving of attention henceforth, and demands the most serious consideration. Such is the opinion of England and Belgium, which are at this moment preparing a balance-sheet of their mineral wealth, and an estimate of the contents of their subterranean forests.

In all coal-producing countries the greatest anxiety is felt with regard to the means of raising coal from a depth of 3000 and 4000 feet;* and those comparatively thin seams of coal, and those of middling quality, which were neglected twenty or thirty years ago, are now considered worth working. Every possible precaution will henceforth be taken, in order to defer the evil day as long as possible. The utmost economy will be practised, and the most perfect and ingenious contrivances will be devised to reduce the future price to a minimum.

By the adoption of improvements, such as better methods of working, and a better study of the coal-fields, the ultimate exhaustion of the mineral fuel may be postponed, but cannot altogether be prevented. Some day or other this exhaustion *must* take place, even though the existing coal-basins were ten times more extensive and ten times more numerous than they are now supposed to be—even though fresh deposits were discovered in a hundred different places, and the coal could be economically worked at greater depths than 4000 feet, and a practical solution were arrived at of the numerous difficulties connected with temperature, ventilation, and with raising the water and coal from such enormous depths.

The steam-engine, as contrived by Watt (who was one of the greatest geniuses who have done honour to humanity), except as regards the improvement of details, which are being made every day, is the last expression of modern mechanics. A substitute may some day, possibly, be devised for this admir-able and wondrous engine, of which the combustible mineral

---

* In the province of Hainault, in Belgium, a pit has been sunk to the depth of 3411 feet at the Colliery *des Viviers*, at Gilly, near Charleroy. ("Coal and Coal Mining," by W. W. Smyth), p. 246.

is in truth the daily aliment, and for whose especial use coal
is worked.   Do not the researches which have been recently
undertaken by so many learned men on the *mechanical equivalent*
of heat, tend to demonstrate that the force which the combustible imparts to the engine is only the product of the solar
heat condensed in the carbon out of which the coal was formed
in past geological times?   Do not these same researches prove
that the three agencies of light, heat, and motion, are but three
modified conditions of one and the same physical force, and that,
consequently, the attempt to substitute some other substance
for coal for heating the furnaces of steam-engines, or to calculate on the discovery of a new economical motive power, would
be to seek to substitute one form of carbon for another; leading
us to turn in a vicious circle, or, at least, to revert to some other
form of carbonized matter, such as petroleum?   "What urges
that engine?" asked the great engineer, Robert Stephenson,
when he saw a train in motion on the Stockton and Darlington
Railway. His friend paused for a reply. "It is the sun,"
said Stephenson. "The solar heat fixed the carbon in the
plants which formed the coal, millions of years ago, and the
heat absorbed in doing that work is liberated now to raise the
steam." So nothing creates itself, nothing is lost in nature,
neither force any more than matter; and the locomotive engines,
as Stephenson has said, are only *the horses of the sun.*

It is certainly reasonable to look for a perfect calorific
machine in the production of steam that will effect the greatest
possible economy of coal, the largest proportion of which is now
often lost in the form of smoke. The saving thus effected would
be remarkable, since certainly not more than the tenth part
of the calorific or mechanical power of the coal is now actually
realized. In that sense, it may be said that our real economy
will some day consist in the smaller quantities of coal ordinarily
consumed. As to the adoption of a new generator of heat, or
motive power, the expedients which have been sometimes indi-

cated are scarcely more consoling or feasible than the suggestion that the Falls of Niagara should be turned to account to
supply motive power to the machinery of all the factories in
the world, which would have to be collected in its neighbourhood for the purpose ; water would then be employed to
compress air, and in that way obtain the most advantageous
and economical of motive agents—all which, though very well
in theory, is but little applicable in practice. Besides, to
establish factories in the neighbourhood of water power is not
only to go back to the past, but an arrangement that would
have the effect of rendering few establishments of the kind
possible at the present day. It is only in altogether exceptional cases, such as, for instance, tunnelling through the
Alps, that the use of compressed air becomes usefully and
economically applicable ; and it must be remembered that
an expenditure of mechanical force is required to compress
the air itself, to begin with.

Neither can the electro-magnetic engines, which were proposed some years ago, be made to replace steam-engines ; nor,
except in certain cases, can gas-engines working expansively,
about which so much has recently been said : for these last, in
generating a given force, consume a larger amount of fuel, very
often three or four times as much as the ordinary steam-engine
of the same nominal power. If they surpass it in some instances, especially for engines of small power (the Lenoir engine,
for instance), which is extremely doubtful, it is simply owing to
particular arrangements, and not from economy of fuel, which
they never realize. Still less must engines acting by means of
explosions be dreamed of, since from their very nature these
hardly admit of application, except for the propulsion of projectiles. Engines in which the generation of steam is attempted
to be effected by means of friction, are merely interesting
as curiosities ; the ingenious and skilfully-contrived engines
worked by combined vapours, especially that of M. du Tremblay,

have only furnished nearly negative results, as have also the hot-air machines—that of Ericson not excepted.*

Thus, in the present state of our knowledge, the steam-engine cannot be replaced by any other of a more simple or perfect kind. Whence, then, shall we borrow mechanical power and fuel when the coal is exhausted, or shall have become too costly, owing to its depth below the surface or the distance of the latest workings from the centres of consumption? At present the question seems to be incapable of solution, and science is nearly silent on the means by which the problem is to be solved. Some recommend planting fresh forests to replace coal at some future day, just in the same way that the last took the place of wood; but the world never goes backward. Besides, as we have stated, it has been proved that the entire surface of Europe, if covered with forests, would be insufficient at the present day to supply the requirements of industry, and would not furnish a quantity of wood and charcoal equivalent to the present annual consumption of coal. Then there is petroleum, which, perhaps, in many cases, might be used as a substitute for coal. Considerable quantities have been found in the United States and in Canada; both these supplies are, however, limited, and the duration of these oil-springs will be still less than that of coal; for even now the quantity is found to be diminishing.

If the petroleum obtained by the distillation of coal-shales is suggested, it must be remembered that the coal used in dis-

* Engines worked by combined vapours, or rather binary machines, are those in which the heat lost by the steam is made use of after it has acted on the piston of the cylinder, to convert into vapour a liquid more volatile than water, such as ether, chloroform, &c. For a given power as much as fifty per cent. of coal is saved in this way. M. du Tremblay, one of the ablest of French mechanicians, has attracted especial notice for the invention of these engines; but he has contended against almost insuperable difficulties, opposed by the volatile nature of the liquids employed, together with the difficulty of condensing them in summer temperatures.

In Ericson's engines the heated air acts by its expansive force on the piston of the cylinder in the same way as steam. In escaping it parts with all its heat to a very fine metal network, through which in its turn the cold air enters and becomes heated in passing through the network, and afterwards over a furnace. This system, as economical as it is ingenious, has not been able to save Ericson's machine from being condemned; the frequency of repairs, arising from the very nature of the invention, having caused its rejection in practice.

tilling the petroleum would have given the mechanical power
which belongs to the hydrocarbon obtained.

Instead of looking to wood and petroleum, the attempt
should rather be made to decompose water, and the limestones
which are so abundantly spread over the surface of the globe,
in an economical manner. Water is composed of the two
elements which are the most active sources of heat—viz.,
oxygen and hydrogen; and limestone contains carbonic acid,
and consequently carbon. An effort should also be made to
discover a new motive power in the application, if possible, of
compressed air; but to effect this object chemistry, physics,
and mechanics must be placed in requisition; the former two
for the purpose of more intimately solving the mystery of
elasticity and of the combination of substances, and mechanics
in order to devise a method of compressing air by some other
agency than that of fire or water power. How do we know
whether the wind, a natural source of motion so liberally
placed at our disposal by nature, may not contain some latent
application?

In our opinion there is also a true discovery to be made in
the utilization and condensation of the vast heat of the sun
which is now lost, or in other words we must learn *to bottle the
sunbeams.* This solution, which a man familiar with all the
speculations of science lately suggested to the author as a
pleasantry, we have in our turn adopted with entire conviction.
The English were the first to say that coal was "the sun in
a cellar." What becomes of all the heat which that incandescent planet pours on our soil during the long summer days, and
upon our cities and houses? Recourse may some day be had
to the burning mirrors of Archimedes, and the astonishing
experiments in combustion which were repeated by Buffon or
his pupils on the faith of the Greek geometrician may be
renewed; but here again the attempt appears to be scarcely
altogether applicable in practice. To make use of the sun-

beams for industrial purposes as a source of heat, by means of reflectors which concentrate and throw back its rays, is to suppose the daily, if not constant, presence of that planet; which reminds us of what is said about certain countries where it never rains, but where civilized life has scarcely made its appearance. Are not those plans still less favourable to the grand modern industry than the Falls of Niagara?

However that may be, it is doubtless in the sun that the combustible of the future is to be looked for. The most recent discoveries in physics with regard to heat justify our taking this view of the question, and it may be said to spring naturally from the curious experiments which have rendered illustrious the names of so many physicists in England, France, and Germany. In every instance it may be said that the exhaustion of the coal-fields will not involve the end of the world, at any rate of the civilized portion of it. As is the case between iron and the other metals, which are so indispensable for the progress of civilization, there exists a sort of *pre-established harmony* by which everything has been regulated in a much better manner than was supposed by the German philosopher. In that respect it is necessary to be a believer in final causes; for if iron and coal, which may be said to have been created from all eternity, have only been really sought for and worked in an active manner at our epoch, and if their speedy exhaustion may be almost predicted—especially as regards coal—which can never be burned a second time and is never used again like iron, it may also be taken for granted that when all the coal is exhausted something equivalent to it will be discovered, even though it should be in connection with the sun. Future investigators must direct their attention, then, to that orb, although it cannot be said at present in what particular direction their researches ought to be pursued. The germ of every invention, after lying dormant for ages, comes to light at its proper time; and just as the eolipyle of Hero of Alexandria waited for nearly two thousand years until Savery, Newcomen, and especially Watt, revived it in order to borrow from it the idea of the steam-engine—in the same way the mirrors of Archimedes seem destined to indicate to future inventors the mode in which they should seek for the new combustible of industry. To those who may express any doubts on the subject, based on the impossibility of making such use of the sun, we will reply : "Who would ever have supposed, on seeing the lid of a kettle raised by the force of steam, that the germ of a most mighty power was there ?"

The sun is then, beyond a doubt, the combustible of the future, and the torrid regions, which are now nearly desert, may perhaps some day witness a migration of civilized people in a mass in that direction, like those incursions of the barbarians into Europe in former times. Whether these predictions may appear paradoxical or not, the world will certainly not perish for want of coal, whatever may be the combustible which human ingenuity may devise to supply its place.

PART II.

METALLIFEROUS MINES.

METALLIFEROUS MINES.

CHAPTER I.

THE STAGES OF THE HUMAN RACE.

Primitive man.—*The Stone Period.*—Discovery of fire.—Prehistoric Times.—*The Copper Period.*—Tin.—Origin of Metallurgy.—Tyre and Sidon.—*The Bronze Period.*—Discovery of Lead, Silver, and Quicksilver.—The first smith.—Tubal-Cain or Vulcan.—*The Iron Period.*—The dream of the Alchemists.—Discovery of Cast-iron ; invention of cannon.—Marvellous progress.—Present phase of the Iron Age.—Steel.—Function of the precious metals.—The minor metals and the lesser planets.

EVERYBODY has thought, at some time or other, of the important part played by the metals in the life of nations. From the most precious to the basest, from the rarest to the commonest, they all play a special part ; and material progress is only effected by their means. The history of the metals constitutes the true history of inventions and labour.

Primitive man was compelled to fashion everything—his arms, utensils, and tools—out of bones or stone. Very often he chose chippings of flint which, at first, were left in their natural state, but were afterwards simply reduced roughly into shape, and finally were rubbed down to a smooth surface and polished. Although the earliest men were unacquainted with the use of pottery,* as their intelligence developed and their wants increased they made rude kinds of pottery out of clay, which at first were only baked in the heat of the sun.

* See papers by W. Boyd Dawkins, F.R.S., in *Quarterly Journal of Science*, vol. iii. p. 333 ; *Quarterly Journal of the Geological Society*, 1862 ; also, *Saturday Review* for November, 1866.

The cave-bear and hyæna, the hairy elephant, the large-horned stag or Irish elk, contended with him for food and shelter, and, with his clumsy weapons, it was with difficulty that he could defend himself against their attacks. As fire was, most likely, one of the earliest discoveries made by man, the hunter soon availed himself of it for cooking the products of the chase. The skins of these animals, sewed together with their tendons, served him for clothing; sometimes he carved their images on their bones, and to these rude attempts may be traced back, as it were, the origin of art.

This period, during which the human race was in its infancy, has been aptly characterized as the Stone Age, from the circumstance of man's ignorance of any metal at that time. Attempts have been made to calculate the length of time during which this state of things lasted, and the results arrived at are startling to the imagination. The great antiquity of the human race is now a recognized fact, and geology demonstrates that this period of early barbarism may have extended over tens of thousands of years. On the other hand, written history dates all civilized nations only from yesterday, and some of the children of the great human family—the Polynesians, for example—have not yet emerged from the Stone Age. Barbarism may, then, under some circumstances, be of almost unlimited duration.

The commencement of civilization only dates, in reality, from the discovery of fire and the metals. The Greek mythology, which is ever full of poetical imagery, tells us that Prometheus stole fire from heaven. It was, doubtless, made known to man in the first instance by the lightning, which may even have set fire to forests traversed by metallic veins: so that fire and the metals were discovered simultaneously, the latter being freed by the fire from the stony matter, or the matrix with which they were mixed. It is pretended by ancient authors that the metalliferous chain of the Pyrenees

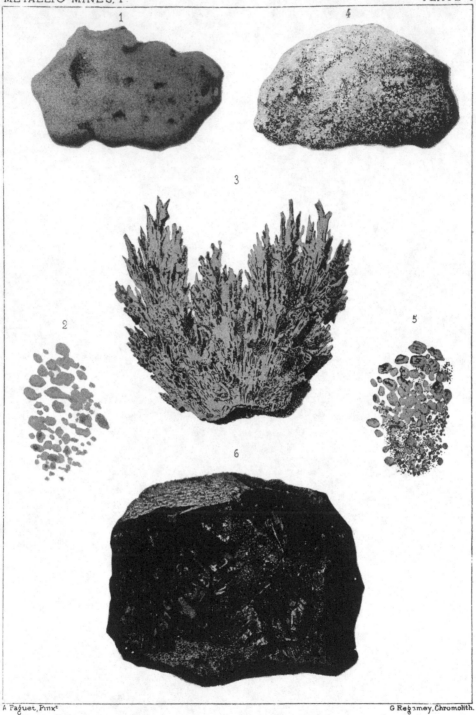

A Paguet, Pinx.ᵗ   G Regamey, Chromolith.

1. Nugget of Gold, *(California)*   4. Nugget of Platinum, *Choco, (New Grenada)*.

2. Scales of Gold, *(Australia)*   5. Scales of Platinum, *(Ural)*.

3. Capillary Silver, *Sonora, (Mexico)*. 6. Cinnabar or Native Vermilion. *(Spain)*.

CHAPMAN & HALL, London.   Imp Lemercier & Cⁱᵉ Paris.

(which signifies in Greek the "fiery mountains"), owes its name to a meteorological phenomenon of this nature; but this is very doubtful.

There are other explanations of the discovery of fire, for the field for conjecture is vast, and the earlier races of mankind have taught us nothing on this subject. Thus fire may even have been known from all time, through volcanic eruptions and lava-flows. The men who lived far away from volcanos might, perchance, have discovered it, if not through lightning, at any rate through the spontaneous combustion of wood after hot and dry summers, such as those of Algeria and California at the present day. It is also possible that the first spark given off on striking any hard body against a flint, of which the men of the Stone Period made such constant use, may have led the way to the invention of fire. Finally, the discovery may have been made in rubbing two pieces of dry wood against each other very rapidly, a mode which is still in universal use amongst all savages.

Whatever may have been the way in which this marvellous discovery was made, it is certain that it led (though, possibly, after a long interval) to that of the metals. Still, the discovery of gold may be as old as that of flint, in which case fable trenches on actual facts when it represents the golden age to be the earliest state of the human race. The precious metal is nearly everywhere spread over the surface of the earth, and the sands of running streams. The lustre and colour of gold attract the eye at once; but primitive man had no use for these brilliant scales disseminated in the sand (Plate I., fig. 2). He could neither fashion them, nor even join or solder them together; and when the woman, his wife, wished to add to the charms bestowed upon her by nature, it was by means of shells strung together by a thread that she made bracelets, necklaces, or ear-rings.

If the discovery of gold, at any rate in the first instance, exercised no influence on the civilization of the human race, it was otherwise with the discovery of the commoner metals. Once possessed of these last, at first of copper and tin, and then of iron, the human race suddenly made the most rapid progress, and history commenced from that time. Until then legend and tradition have only handed down to us fables—a shadowy mythology which the learned have not yet succeeded in altogether unravelling. These are the times called pre-historic, the golden age of the ancients, the Stone Period of actual science, dating back from the latest geological times.

The discovery of copper and tin have clearly preceded that of bronze, which is an alloy of those two metals. Copper might have been discovered either in the state of the natural —or native—metal, as it is called by mineralogists (Plate II., fig. 1), or in combination with substances, such as oxygen or carbonic acid, for which it has only a feeble affinity, and from which it may be disengaged by means of fire and fuel. The oxydulated ores of copper, the oxide and the carbonate (Plate II., figs. 2, 3, and 4), are of this kind, and become reduced to metallic copper when subjected to great heat with wood or charcoal. The metal was obtained by the earliest smelters in a state of tolerable purity, and with its characteristic beautiful red colour. It was easily fashioned into sheets or wire under the blows of a stone-hammer; but it was not hard enough to be of immediate use for very many purposes.

The ores which have been just mentioned form the upper part of the lodes—near their outcrop, as it is called—and is that which makes its appearance at the surface. The peculiar red colour of native or oxydulated copper, the pavonine or iridescent tints and metallic lustre of the sulphide, ought to strike the eye at once, no less than the blackish aspect of the oxide, and still more the blue and green tints of the carbonates of copper. The latter were doubtless easily recognized by the colours which distinguish them, and they acquired a certain value, since they have been employed in jewellery and painting from the earliest ages.

After the miners had explored the outcrops, they were naturally led to examine the cupriferous lodes in depth. Then, instead of the usual decomposed ore, altered by the action of the air and carbonic acid, and of comparatively low specific gravity, they met with an unaltered substance of a shining brass-yellow colour (Plate II., fig. 5), of high specific gravity, called copper pyrites, or yellow sulphide of copper, because it consists of a combination of copper and sulphur.*

A first step was made in working metalliferous mines ; but a greater step had to be made in the art of treating the ores— in metallurgy. Who first taught man to smelt copper pyrites, a mineral which even now is of so refractory a nature in our furnaces ? Who showed the smelter that he must first roast the ore before smelting it—that is to say, calcine it in contact with air, to get rid of the sulphur which it contains ? Or rather, was not the ore merely cast into the furnace without any preparation, and then purified by successive castings? What an immense series of indecisions and researches, in the prosecution of which the earliest chemists were formed, must have been necessary before the metal, alloyed with others rendering it not only harder but more brittle, yielded itself to the smelter with all the qualities requisite for ordinary purposes !

As it is seldom that the sulphur-ores of copper are not mixed with other ores, an impure copper was obtained, alloyed with lead, zinc, and especially with iron, from the very composition of the copper pyrites. Tin, not having been directly introduced into the alloy, would only be present by mere accident. In any case that was the first bronze, the first brass.

Who can now tell the immense interval between this discovery and that of true bronze, that alloy of copper and tin in

* The true composition of copper pyrites consists of nearly equal quantities of copper, sulphur, and iron, from thirty to thirty-three per cent. of each, being in fact a double sulphide of copper and iron.

calculated and studied proportions, of which the ancients made such great and remarkable use? In what country did the earliest miners make their appearance, who searched for and smelted tin? Was this metal first found in India, where such abundant supplies are still furnished by Banca and Malacca to all the markets of the present day? Did it come from Cornwall, or from her sister and neighbour Gallic Armorica—modern Brittany? These two countries were rich in ores of tin, which have never ceased being worked; and the ancient stanniferous gravels, which have been ransacked and turned over by the aboriginal miners, are even yet in existence.

The bronze-users appear suddenly and apparently simultaneously on the Asiatic border of Europe, and exterminated the stone-users. Probably tin was first discovered in India; but whether India, Gaul, or Britain were the first producers of the metal, there is no doubt that all those three countries contributed, the two last exclusively in the sequel, to supply the countries bordering on the Mediterranean with tin. The ore, easily recognized by its colour and heavy weight (Plate IV., fig. 5), was separated from the sand, gravel, and other foreign impurities with which it was mixed, by mere washing, and was then reduced by fire. As tin-stone is merely a combination of the metal with oxygen, the former was separated at once by smelting the ore with charcoal—a method which is still in use.

From those remote times, anterior to the Trojan war, and which form a sort of limit between fable and history, the Phœnicians, who were the first navigators of the Mediterranean, sailed through the Pillars of Hercules (the Straits of Gibraltar) and went to the Cassiterides,* or to Vectis,† to barter the pro-

---

* Various spots have been spoken of as the Cassiterides. M. Simonin says they were the islands at the mouth of the Loire, or the Sorlingues. Generally, however, the Scilly Islands have been supposed to be the Cassiterides. Unfortunately for this hypothesis no tin exists in the Scilly Islands. In all probability the Cassiterides signifies the western promontory of Cornwall, everywhere most abundant in tin, and from the sea appearing as a group of islands.

† The Isle of Wight has been named, but it is absurd to suppose that tin was carried from Cornwall to that island.

ducts of the East for the tin of the Kelts and Britons. Those bold merchants, true sons of the sea, went still further, and even proceeded to the Baltic Sea, in quest of amber; then, in face of a thousand dangers, they returned to their own country. The copper supplied in abundance by Asia Minor and the Isle of Cyprus,* was alloyed at Tyre and Sidon with the tin brought from Gaul and Britain, and thus bronze or brass was produced. The brass of Sidon is cited by Homer. Who first thought of making this alloy? Was it a Phœnician merchant, who had brought tin from the British miners, and probably copper, which the mines of Cornwall then produced, but not in abundance as now? Was he a smelter of Armorica? Strange to say, antiquity has not attempted to settle this point, and does not even allude to the circumstance. According to all ancient authors, whether Latin or Greek, bronze (æs, or χαλκος) is a simple metal obtained direct from the ore. Nevertheless, we know by chemical analysis that it is an alloy of copper and tin, in most cases in calculated proportions, varying with the object to be produced, whether arms, medals, or statues. Doubtless the smelters of those days had an interest in not divulging the rules on which they worked.

By whatever means it became known to the ancients, this alloy of copper and tin marks a period in the history of the human race as memorable as that which we have already noticed in the first part of this book, in connection with the extended use of coal. In order thoroughly to comprehend the immense phenomenon which was then called into action in the advancement of human progress, it is necessary to recapitulate all that has previously been stated.

The cutting and polishing of flints, the working of clay, the rude carvings on bones, have all marked, in the development of the human race, a first or dawning period of intellect and art, which, in common with all other archæologists, we have

---

* The Latin name for copper (*cuprum*) is derived from this island.

called the Stone Age. In this primitive period language was
invented, if man did not receive it at his birth. Then follows
the discovery of copper, or more generally of the working of
mines and of metallurgy, forming the second stage, which may
be called the Copper Age. The third stage is now marked by
the discovery of tin, and shortly afterwards of its alloy with
copper, to form bronze. The smelters must then learn how to
temper the new metal—that is to say, to harden it by im-
mersing it when hot in water, and afterwards to forge it or
to render it pliable by hammering it. The tools, which had
been nearly all made or contrived by the men of the preceding
ages—such as the hammer, the anvil, the chisel, and the lever
—were henceforward made of tempered and forged bronze, as
were also the saw, the wedge, knife, axe, fish-hook, and needle,
which had been previously made out of flint or bone. This
rapid and immense conquest forms the dawn of a new era—
the Bronze Period. The advance of civilization is thenceforth
rendered certain, and the fine arts have really sprung into
existence. Bronze may be used for all purposes; the plough-
share can be made of it, as well as the pick of the miner, the
hammer and compass of the architect, and the burin of the
engraver. A material eminently fusible, it may be made to
assume all forms, the most exalted as well as the most vulgar.
With it the art of moulding begins. It serves, in concurrence
with gold, to create money, to form a basis of commercial
values, and originates commerce, which had previously been
based on barter or exchange only. But defensive and offensive
weapons are also made of it—the heads of arrows, lances, and
javelins, swords, bucklers, cuirasses, and helmets; and the art of
war, which is as old as the world, advanced another step forwards by means of the new invention. At each of the stages
which we are about to describe the case is the same. "The
art of killing one another," as Montaigne says, "will advance
at the same time as all the other arts."

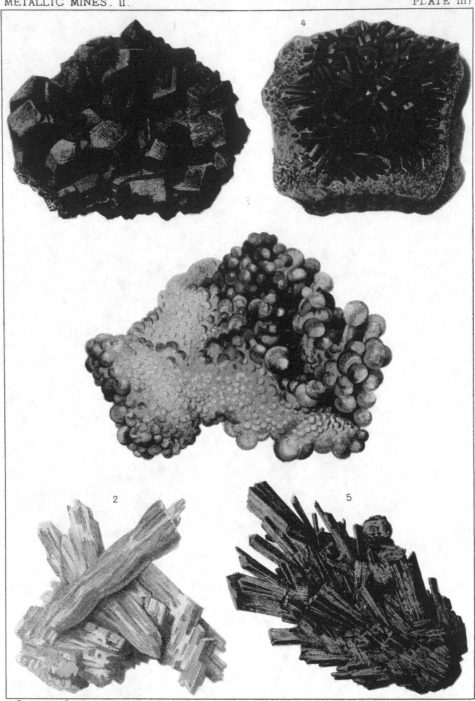

A Faguet pinx.   C. Regamey Chromolith

1. Galena (Sulphide of Lead) *Freyberg*.   3. Phosphate of Lead, (Pyromorphite) *Hofsgrund (Baden)*.

2. Cerusite (Carbonate of Lead) *Leadhills*.   4. Chromate of Lead, (Crocoisite) *Siberia*.

5. Sulphide of Antimony (Stibnite) *Felsöbanya (Hungary)*.

CHAPMAN & HALL, London.   Imp. Lemercier & C.ie, Paris.

" Arma antiqua manus, ungues, dentesque fuerunt,
  Et lapides, et item sylvarum fragmina rami,
  Et flammæ, atque ignes, postquam sunt cognita primùm :
  Posterius ferri vis est ærisque reperta."
                              —Lucretius, De Rerum Naturâ, lib. v.

" Arms of old were hands, nails, and teeth, and stones and boughs broken off from
the forests, and flame and fire, as soon as they had become known. Afterwards the
force of iron and copper was discovered."

The three metals, lead, silver, and quicksilver, were doubtless found simultaneously with bronze, and perhaps prior to it;
that is to say, at the same time as copper. Silver, a faithful
companion of lead-ores, must have been discovered with the
latter metal. The most common ore of lead is galena or
sulphide of lead (Plate III., fig. 1). As the Greek name
(γαληνη) denotes, it is brilliant, bluish, and frequently occurs in
crystals. The sulphur is given off by simple exposure to flame
in the state of sulphurous acid, and the lead melts. The
discovery naturally followed when men learned to recognize
all natural substances that were heavy, shining, and like metal,
and which bear a certain sort of family likeness, such as those
termed ores. Silver remained to be discovered—a metal of
rare occurrence in a native state (Plate I., fig. 3). The ore is
nearly always mixed with others, and is most frequently met
with in the state of a sulphide in galena; but the lead
possesses the curious property in smelting of carrying with
it the silver, and the result is not pure lead, but an alloy
of lead and silver. The lead when heated with access of air
becomes converted into the oxide, and forms litharge; while
the silver, which remains unoxydized, remains in the form of a
brilliant button, requiring a much greater heat to melt it.
Thus the brother metal to gold was found, and at the same
time cupellation,* an ingenious process, mentioned in the
Bible, by which silver is separated from its mineralogical alloy
lead, must have been discovered.

* The hearth of the furnace in which this operation is carried on in smelting-houses is in the form
of an enormous cup or cupola, whence is derived the name given to the process.

The common ore of quicksilver, called cinnabar or native vermilion, is also a sulphide of the metal (Plate I., fig. 6). This substance is of a beautiful red colour, sometimes crystallized, and with a brilliant lustre, but commonly amorphous and earthy. Like galena, it must have been discovered at an early period, for it possesses the property of parting with its sulphur when heated. The metal also vaporizes, but is condensed into mobile globules by contact with a cold substance : it is quicksilver or liquid silver (the *hydrargyros* of the Greeks and Romans). The ancients used it in its natural state as a solvent for gold, and in the state of vermilion as a pigment.

A long period may have elapsed since the discovery of bronze before that of iron was made, probably owing to the circumstance that bronze supplied all the requirements of the arts of peace, as well as of those of war ; while brass would naturally precede iron, because it is more easily worked :—

> " Et prior æris erat quàm ferri cognitus usus,
> Quo facilis magis est natura, et copia major."
> —Lucretius, De Rerum Naturâ, lib. v.

" And the use of copper was known before that of iron, as its nature is easier to work, and it is found in larger quantities."

Yet we find the Assyrians using iron to strengthen their bronze ornaments ; and we find Hesiod offering a prize of iron to the victor in some of the most ancient of games.

Nevertheless, miners were struck at an early period by the appearance of a somewhat heavy earthy rock of a red, yellow, or blackish colour, and soiling the fingers, which was abundantly diffused in certain countries, and forms the oxide-ores of iron (Plate VI., figs. 1, 2, 3). The discovery was completed on the day when the metal was extracted from its ore, and exceeded in value all that had been previously made. The first blacksmith has been celebrated in turn in the popular legends of all nations, as one of the greatest of inventors, by his discovery of smelting the ore of iron and casting the hard metal. In the

IRON

A. Faguet pinx.ᵗ   G. Reyamey Chromolith.

1. Pisolitic Iron Ore. *Berry (France)*.   3. Stalactitic Oxide of Iron. *Sirgen (Prussia)*.

2. Iridescent Oxide of Iron. *Cornwall*.   4. Specular Iron Ore. *Rio*.

5 Magnetic Oxide of Iron (Magnetite) *Siberia*

CHAPMAN & HALL, London   Imp.Lemercier &Cⁱᵉ Paris

Bible he is called Tubal-Cain, the "instructor of every artificer in brass and iron" (Genesis iv. 22); in Egypt, Ptha, the god of fire, whom the Greeks converted into Ηφαιστος, and the Romans into Vulcan, from whom volcanos derive their name : he is the Scandinavian Thor, and the British Wayland Smith. Then a new age begins, the Iron Age, the latest stage of humanity, since it is that of the present time under which such astonishing marvels are performed beneath our eyes.

The reduction of iron-ore, which has since passed through so many phases, and has always been a delicate process, must have been attended with much labour at the outset. In the stone hollow or receptacle forming the hearth in which the ore and the combustible were brought together, air must be blown with force in order to subdue the refractory rock of which the ore of iron consisted. The natural currents of air, penetrating freely through openings in the furnace, and of which recourse would certainly have been taken in the previous operations, were insufficient here. Very likely the first bellows were fashioned out of a leather bottle alternately filled and compressed with air, or the hollow trunk of a tree, in which a piston moved up and down as in the cylinder of a pump. These two comparatively feeble methods, remains of man's primitive industry, are still in use amongst the Malays and the African negroes ; so that while the Polynesians have remained stationary at the Stone Period, the Malays and negroes have scarcely reached the first cycle of the Iron Period, whose rude implements they have retained. It may be suggested that this was done by design, in order, as it were, to guide the civilized man of to-day in the obscure study of the commencement of man's early history.

When the early smelter found that the iron would not flow in the furnace, he took the ball of metal, beat it with repeated blows on an anvil, and drove out the slag, just as water is expressed from a sponge ; and then the metal, after having been replaced in the fire, was welded and forged.* Thenceforth susceptible of numerous applications, like bronze, it formed one of the family of common metals. In the art of war, owing to its property of forming steel, it partially superseded its rival, and swords, darts, lances, poignards, and all offensive weapons, were thenceforth made of iron. Shields, helmets, and cuirasses were likewise made of the new metal, or continued to be fashioned in bronze. The discovery of liquid or cast iron was not made for a long while; brass took the place of iron, for ages, for certain special purposes.

In this history of the metals it may not be out of place, in connection with the period of the Middle Ages, to relate the long dream of the alchemists, who dreamed of nothing less than to transmute one substance into another, to make gold out of lead, and to change the ignoble into noble metals. It is they who gave the names of the seven planets then known, or stars supposed to be such, to the seven metals which have been already mentioned, and made use of those cabalistic signs to denote them which have been retained by astronomers and geologists. Gold, silver, quicksilver, tin, copper, lead, and iron, were thus symbolized, respectively, by the Sun, Moon, Mercury, Jupiter, Venus, Saturn, and Mars;† and were represented by corresponding signs, as the circle, crescent, caduceus, thunderbolt, mirror, scythe, and lance with buckler. The first two were the noble metals, the eldest, and high members of the family; the four latter were ignoble or base metals, vulgar plebeians. Mercury was assigned an intermediate place between the two classes, as the marvellous solvent by whom the ignoble metals should some day be transmuted into noble metals. That was the grand secret to be discovered, the key

---

* The above is the process employed, not only by the Malays and negroes, but also by the smelters of the Pyrenees, Catalonia, Corsica, &c., from the time immemorial when they began to work in iron.

† The name of mercury, the pyrites called *Martial* by the older mineralogists, the chloride called *Salt of Saturn* in pharmacy, the precipitation known in Chemistry as the *Tree of Diana*—are all terms showing traces of the nomenclature of the alchemists.

to the Hermetic science. But to relate the story would lead us to exceed the limits imposed upon us, yet we must not leave the blowers and their retorts without remembering that chemistry is indebted to them for many discoveries, amongst others of some new metals, as of antimony,* for instance.

Neither must the discovery of cast-iron, or fusible carbide of iron, be passed over in silence, when treating of a period which was more fruitful in great inventions than is generally supposed. Cast-iron served at once for making shot and the guns, for gunpowder also was just invented. The English were the first to adopt the use of the new projectiles, and tried the effect of those terrible engines of destruction against the French, at the battle of Crécy, about the middle of the fourteenth century.

After the lapse of many ages, the high furnace, a giant structure built of brick and stone, in which cast-iron is smelted, has been long in use, and is everywhere heated with wood or charcoal. In spite of stringent laws regulating the management of forests, the moment is already seen to be at hand when the fuel will fail. Many attempts were made, but in vain, to use the combustible mineral for smelting purposes. Simon Sturtevant and Dud Dudley certainly used coal about the beginning of the seventeenth century in smelting iron, with a considerable amount of success. We find about 1657 that the smelting of iron with pit-coal ceased; and it was not until 1735 that an English iron-master, Abraham Darby, with the help of a Welsh shepherd boy, John Thomas, who was tormented by the genius of inventions, succeeded in applying coal, which he previously coked, to the melting of iron for castings. Stroke after stroke, the art of treating iron, or iron-founding, made the most rapid progress in England. Coking-

---

* The salts of antimony, especially the emetic, possess the properties of making people thin: whence the name given to the metal.

ovens for the purification or carbonization of the combustible fossil, reverberatory furnaces for the refinement of the cast-iron, and rolling-mills for rolling out the metal into sheets, were invented in succession. The great foundries sprung into existence, set in motion by steam-power, which Watt had just completed, and reduced definitively to practice. The English method of manufacturing iron was adopted on the Continent, and was followed by fresh improvements. To avoid the abstraction heat by the large quantity of cold air forced through the furnace, the air blown into it is now in almost all cases previously heated—still some iron is made by the cold blast in this country; and the waste gases (nitrogen and carbonic oxide), which under ordinary circumstances are allowed to escape from the furnace, are conducted to the steam-boilers and there burned, and so made to serve as fuel. France holds a high place in the profound changes which have been introduced into metallurgy, and contests with England the discovery of the tilt-hammer, which forges the metal mechanically. The steam-hammer of Nasmyth, however, belongs to Britain, and nearly every important improvement has been made by English iron-masters.

The present generation, playing its part in these transformations, has witnessed the invention of Bessemer for the manufacture of steel on a large scale. By this method steel of a superior kind can be obtained in considerable masses. This effects an entire change in the art of war, in the same way that international relations have been modified by means of the railway and the steam-engine. It has been shown elsewhere that coal first paved the way for those grand phenomena, which we now see completed by iron. Let us dwell on the part which is now played by iron and steel. Littoral fortifications, as well as ships, are clad with the former as with an impenetrable armour; ships of war find in it, besides, the element of their formidable ram; lastly, the new cannon are, as well as their shot, made of both iron and steel, their range exceeding all limits. But the beneficent arts of peace, likewise, are indebted for most magnificent developments to cast and wrought iron and steel. At the present day these three metals are in universal use; they have created new roads, replaced timber for shipbuilding, flooring, and roofing; and stone in bridge-building and the pillars for the support of edifices; for ornamental castings they have been substituted for bronze. No tool or weapon can be made without them. They constitute the parts of all machinery, and compose even the most delicate mechanisms. We enter upon a new phase of the Iron Age which may be styled the Steel Period—a phase which may prove glorious beyond all others, in helping to bring about—if not the abolition of war—at any rate that of the brutalizing labour of the workman or slave, which will soon be performed by machinery alone.

But if the common metals, especially iron, are so intimately bound up with the progress of civilization that the existence of a refined state of society cannot henceforth be imagined without their aid, the precious metals, on the other hand, play a part in this world which cannot escape us. In consequence of their exceptional qualities, their rarity, inalterability, and weight, they have not only become the sole representatives of value; but they are also those which, at all times, have helped in the colonization of different countries, by the fascination which they exercise on the minds of the multitude. Without mentioning here what has taken place in ancient times—would the two Americas at the commencement of modern times, or California and Australia in our days, have been so brilliantly settled, but for the existence of mines of gold and silver? The wonderful transformations which those two metals have produced in all those countries is well known, and it is sufficient for the moment to recall them to mind.

Whether precious or common, the seven metals of the ancients are still our own. The most indispensable of all,

U

iron, has only passed through phases from which it has emerged profoundly modified in the state of cast-iron and steel. On the other hand, it has been said—as if a gain should be counterbalanced by a loss—the art of tempering and forging bronze is no longer known; those who say this are most assuredly mistaken, as the bronze manufacturer, if required, could make cutting instruments equal to any produced by the Romans.

The metals discovered during the Middle Ages, or in our time—antimony, arsenic, zinc,* manganese, nickel, cobalt, bismuth, platinum; and quite recently, potassium, sodium, magnesium, palladium, aluminium, rubidium, cæsium, thallium, and indium—do not seem destined to play such important parts as their elders. Palladium is employed, in small quantities, by dentists as an alloy with gold and silver; and two other rare metals, osmium and iridium, are used for tipping gold pens, either in the state of "native alloy," or as artificially combined. Thallium has recently been used in glass, instead of lead; it is said to produce a glass of a very high refractory power, and, consequently, of great brilliancy. Sodium was made on a large scale at Newcastle, it being employed in the production of aluminium, which metal is now manufactured into sextants, opera glasses, and other articles, where lightness is important.

Like many other metals not mentioned here, which have been discovered in the laboratory—some of which have quite lately been detected in the sun, as well as in terrestrial substances—some of them only bear on operations, for the most part, of a purely chemical nature. Civilization might, in case of necessity, dispense with several of these secondary metals. In the world of mineralogy and metallurgy they are like the minor planets of the astronomical world, the learned

---

* Suspected by the alchemists, who gave to the oxide, white, light, and flocculent, the name of *lana philosophica* or *nil album*.

only are acquainted with them; the crowd does not see them, and is nearly ignorant of their names. Not so with the commoner metals, which are so essential to the wants and progress of the human race. The important part which they play in the life of the peoples is known : it remains to be seen how their history composes, to some extent, the very history of civilization.

CHAPTER II.

THE LABORATORY OF NATURE.

Sedimentary and eruptive rocks.—Metamorphism of rocks.—Formation of metallic veins.—Werner's theory; Humboldt's objections.—Linnæus' error.—The theories of Descartes, Leibnitz, and Buffon, accepted by M. Elie de Beaumont.—The sea of fire.—Volcanic emanations.—Mineral waters and metalliferous deposits.—Metallic veins : Laplace's hypothesis.—Characteristic deposits of the various terrestrial epochs —The Placers ; minerals contained in them.—Ferruginous springs.—Mineral truffles.—Poetry and metallic veins.

WHEN the formations composing the crust of the earth are examined, it is seen that some extend in continuous flat masses, divided into beds, which present every appearance of having been deposited under water; for they contain, amongst other fossils, petrified shells. Such are limestones, marls, clays, sands, sandstones, and coal (Plate XI., fig. 3).

Other formations, unlike the first, assume the form of abrupt masses, with jagged and irregular outline, while the rock has a vitreous, crystalline appearance, and contains no fossils. Forced from below upwards, it has pierced through the preceding deposits, which have been violently raised up by it. It has been subjected to the action of fire, or at all events of a very high temperature. To this new family belong the granites, porphyries, and volcanic rocks (Plate XI., fig. 3).

The first-mentioned formations are called, in geological nomenclature, *sedimentary;* the second are the *eruptive* rocks. The former are also sometimes called stratified, aqueous, and Neptunian; the latter, by way of contradistinction, massive or crystalline, igneous, and Plutonic.

The effect of eruptive rocks (or, more properly, of the forces connected with and resulting from internal igneous agency) in elevating the sedimentary deposits, has been to produce chains

Metal Mines

Plate XII.

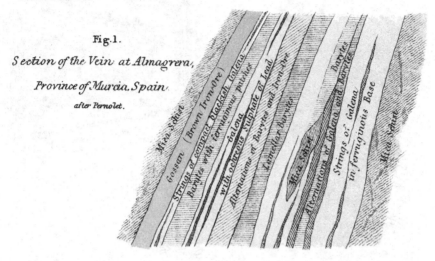

Fig. 1.

Section of the Vein at Almagrera,

Province of Murcia, Spain.

after Pernolet.

Fig. 2.

Section of the Vein at the "Virgen

del Carmen" Mine, Almagrera,

Province of Murcia, Spain.

after Pernolet.

Fig. 3.

Geological Section of the Artesica Mine,

between Carthagena & Cape Palos, Spain.

after Pernolet.

J. W. Lowry Sculp.t

Vincent Brooks, Day & Son, Lith.

Chapman & Hall, London.

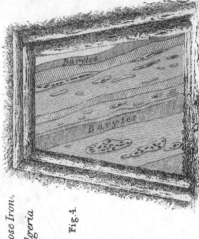

Plate XIII.

Sections of Metalliferous Veins.

Arranged by M. Simonin.

Metal Mines.

Inflexion of a Vein of Argentiferous Galena,
at Holzappel, Nassau.

After Burat.

Fig 1.

Veins of Blende & Galena at Holzappel, Nassau.

After Burat.

Foot Vein   Hanging Lode

"White-Rock" Vein

Foot-Wall Lode

Fig. 2.

Reticulated Veins of Cupriferous Spathose Iron,
near Ténès, Algeria.

After Burat.

Fig. 3.

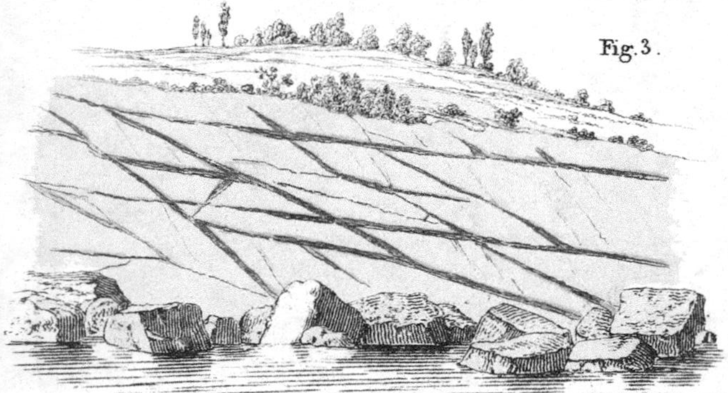

Veins of Grey Copper-Ore & Spathose Iron,
at the Mines of Mouzaïa, Algeria.

After Burat.

Barytes

Barytes

Fig. 4.

J.W. Lowry, feath.

of lofty mountains along certain meridians, and to impart to the crust of our globe its existing relief. The influence of these formations does not rest there: they have not only upheaved, but have besides fractured and burst through the stratified deposits, in which they have frequently produced, over a long distance, fissures, chasms, and large rents, such as are now produced during earthquakes or by the eruption of existing volcanos. Lastly, the vicinity or the contact of igneous rocks has not only changed the appearance and structure, but even the composition, of the sedimentary deposits, which, having become profoundly modified and altered, are called *metamorphic*. In that strange phenomenon, certain strata, as the clays, for instance, have assumed a new aspect; having become baked, and converted into a hard porcellanite, or changed by heat into bright coloured jasper, with a shining lustre. By the chemical re-arrangement of the elementary substances forming their constituents, and under pressure, a great many new minerals, such as garnet, hornblende, talc, and mica, have been produced or introduced into the original rocks; and in this way micaceous and talcose schists, gneiss, and marble have been formed, which serve as a passage between the eruptive rocks and the normal or unaltered sedimentary deposits.

It is across the foliations of schists, but more especially in the clefts produced by eruptive rocks, that the metalliferous substances which constitute metallic veins or lodes are chiefly deposited (Plate XII , fig. 1; Plate XIII., fig. 2).

Werner, a Saxon miner, who at the end of the last century and the beginning of the present, rendered the Mining School of Freiberg so renowned, imagined that all the substances contained in the lodes were derived from surface-waters. This hypothesis might be accepted as an explanation of the origin of the stony, barren matters which accompany all metalliferous deposits, and which are called by the French the gangue.*

* From the German *gang*, " vein " or " lode."

It also accounts for the ribboned and symmetrical structure of certain lodes (Plate XII., figs. 1, 2; Plate XIII., fig. 4; Plate XIV., fig. 3), peculiarities which are displayed by all the Saxon mineral veins.* Lastly, it explains in its way accidents, fractures of veins, what are called faults, throws, or heaves (Plate XIII., fig. 1; Plate XIV., figs. 1, 2; Plate XVI., fig. 2); but it is difficult to understand how, supposing Werner's theory to be true, the deposit itself of metalliferous substances took place, few traces of which are to be found at the surface. This objection, the gravest that can be urged against the principles of the master, did not escape the notice of his most illustrious pupils, Humboldt and Leopold von Buch, who are an honour to Germany, and the French engineer d'Aubuisson. Humboldt especially, in his memorable travels in America, had very soon perceived that everything in this world did not proceed from water only; that fire had also played a certain part in the formation of our globe, and that by a rational eclecticism an honest geologist should be at once Neptunist and Plutonist. We must not, however, forget that Werner has enunciated the principal laws bearing on the formation of metalliferous veins; that he has demonstrated, for example, that the deposits in question only consist of fissures filled up after their formation, which had not been suspected before his time. It must be remembered that from his time, not only miners but some learned persons believed with the ancients in a sort of growth or subterranean reproduction of mineral substances. The great Linnæus himself broached the axiom : *Mineralia crescunt*, or "minerals grow."

It remained, therefore, to modify the theory of the celebrated German on some points, and above all to explain in a more rational way the mode of formation of lodes. Before the time of Werner, the great philosophical naturalists, Descartes,

---

* Which gave rise to the saying on the part of Werner's opponents, that the professor pretended that *God had created the world on the model of Saxony.*

Leibnitz, and Buffon, allowed that metalliferous substances proceeded from the centre of the globe, and rising from below upwards became condensed on the passage. This natural idea has been adopted by nearly all modern geologists, with M. Elie de Beaumont at their head.

It has been argued that the Earth's crust floats like a raft upon the surface of a liquid sphere, forming a sea of fire. On this hypothesis, if the globe is supposed to be represented by an orange, the skin of the orange is taken to represent the crust of the Earth; the wrinkles and inequalities, the hills and valleys; the fleshy fruit, the great central sea of fire. Beneath our feet is the great laboratory of nature, whose furnaces are ever in activity. Why not, it is asked, allow that metalliferous emanations have proceeded thence, and been deposited in the fissures which contain the lodes, either by sublimation, in a state of vapour, that is, by the *dry way*, as in volcanic vents, or the chimneys of metallurgical furnaces? Another hypothesis supposes mineral lodes to be formed by chemical precipitation, or in the *wet way*, as in the analyses of our laboratories.

This second hypothesis answers nearly all objections, because it explains at the same time the formation of the ore as well as the gangue. Water raised to a high temperature, and even in a state of vapour, may then have played an important part in the formation of lodes; and what takes place under our eyes in the deposits of metallic oxides from mineral waters seems to favour this explanation. Such waters contain much earthy or stony matter, saline or metallic, which they often throw down in their course; like lodes, too, they are frequently in the vicinity of eruptive rocks and fill pre-existing fissures, so much so that M. Francois, a learned mining engineer, by the application of the principles of the geology of lodes to the search for mineral waters, has been able to add considerably to the list of thermal springs in France. Lastly, like lodes, these springs are met with in mountainous countries, generally near elevated and lofty spots, in such a manner that the maps of the mineral waters and of the metalliferous deposits of the same country sensibly correspond. The Alps, the Pyrenees, the Vosges, the mountains of Auvergne, the Cévennes, where all the French metalliferous mines are situated, are equally celebrated for their thermal stations. The names d'Allevard, La Molte, Luchon, taken at random in the Alps and the French Pyrenees, recall to mind both the names of metallic mines and the sites of mineral waters (Maps II. and III.).

The eruptive rocks which produced the fissures that have become subsequently filled with metallic deposits, are often pierced themselves by these latter substances, as though the appearance of the ores and of the igneous rock had been contemporaneous (Plate XVI., figs. 3, 4). In that case the rocks in question are called metalliferous, a term which is usually applied to them by geologists in a general way, to denote the part which they have played in the formation of mineral veins.

It has been stated that the outer shell or crust of the earth floats on a sea of fire. Laplace has demonstrated mathematically that our globe was at first only an incandescent mass, a true sun, which has gradually cooled down and become crusted over, and which now, by the insensible loss both of its atmosphere and its water, tends to pass into the same state as the moon. The terrestrial crust ought then to present, at the earliest period of its formation, its minimum thickness; whence it follows that the appearance of eruptive rocks and the fractures they produce should be much more frequent then than they have been since: and such is precisely the case. It is in the older schistose rocks, corresponding to the Primary period of sedimentary deposits, that the greatest number of metallic lodes are found. Gold, silver, platinum, quicksilver, and tin are sure to be found there, as are also antimony, arsenic, bismuth, nickel, and cobalt, and frequently, likewise, copper, zinc, lead, iron, and even manganese. Granite and

Metal Mines.

Plate XIV

*Intersection of Silver & Lead Veins
at Freiberg, Saxony.*

After Wessenbach & Burat

Fig. 1.

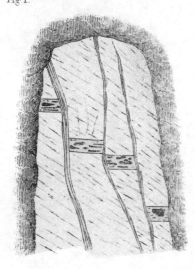

*Outcrop & Throw of a Copper Vein
at Weynähr near Obernhof, Nassau.*

After Burat.

Fig. 2.

*Strings of Argentiferous Galena,
and other Ores*

Fig. 3.

Section of the Copper Deposits of Chessy, Rhône.

After Élie de Beaumont.

Fig. 4.

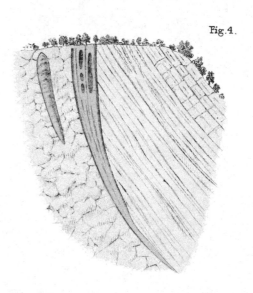

a. Ore (Argentiferous Galena, Copper & Iron Pyrites,
   Blende, &c.)

b. Gangue (Calcite, Quartz, Barytes, &c.)

J.W. Lowry sculp.

Vincent Brooks, Day & Son, Lith.

Chapman & Hall, London.

porphyry are the prevailing metalliferous rocks belonging to this period.

The Secondary period, comprised between the Permian and the latest Cretaceous deposits, contain much fewer lodes, and offers scarcely anything besides copper, lead, zinc, iron, manganese, and, accidentally, gold and silver. The deposits frequently extend upwards, as far even as the middle of the sedimentary formations, in regular beds or in masses, and in a network of interlacing veins (Plate XIII., fig. 3). Others are deposited at the junction with, or in the immediate neighbourhood of, eruptive rocks (Plate XIV., figs. 1, 2, 3) and form in some localities what are called columns or beads (Plate XV., fig. 4). Sometimes they have made their appearance with violence, all in a piece, behaving themselves like eruptive rocks, or else they are intimately mixed up with them (Plate XVI., figs. 3, 4). The appearance of fresh igneous outbursts, of green serpentinous rocks, is characteristic of this age of the earth.

During the Tertiary period the same phenomena were reproduced, only on a much smaller scale. The eruption of granitic, porphyritic, and serpentinous rocks is over, and that of purely volcanic rocks begins.

Lastly, great metalliferous emanations have entirely ceased since the Quaternary period, of which the present epoch beholds the development, and only deposits termed alluvial or of transport are formed now, amongst which are the gold-diggings.

The placers* are nearly always superficial deposits, or are situated at a slight depth. They occupy the beds of ancient water-courses or of old valleys, and are met with in the plains and even at a high level; but water-courses and existing valleys equally contain them. Sometimes they are associated with immense subterranean diluvial deposits of sand, rolled

* The name given by the Spaniards to the auriferous gravels of America. *Placer*, in Castilian, signifying "pleasure."

pebbles, and clay, marking the commencement of the Quaternary period. These clays, sands, and rounded pebbles,
which are often cemented together, and buried at considerable
depths (one hundred yards and more in California and Australia), are derived from the disintegration of rocks in place,
such as granite, quartz, schists, &c., which were already metalliferous. The placers generally derive their riches from the
spoils of these sources (Map XII., fig. 1). The valleys where
they occur are in fact almost always dependent upon the
mountains which are traversed by the lodes. Sometimes,
however, the placers seem also to have been traversed by
thermal waters, in which the metals have been dissolved
under the influence of alkaline bases, and have been afterwards deposited.

The ores met with in the placers are according to the
country—gold and platinum, either native (*i.e.*, in a metallic
state) in powder, or in grains, or in pepitas* or nuggets;
oxydulated iron-ore or magnetic oxide of iron in fine dust
attracted by the magnet; and lastly, oxide of tin in more or
less imperfect little crystals. The hydrated oxide of iron, or
Limonite, is also met with, but is not worked. In certain
alluvial formations it occupies distinct places, where it appears
in masses which are often of great extent, and occurs in the
form of grains or scales (Plate II., fig. 1). This ore is called
in the foundries alluvial iron-ore, or granular iron-ore, in
contradistinction to rock and mountain "mine" or ore contained in regular banks in all sedimentary deposits, and in the
lodes which traverse those formations, especially the oldest.

The influence of ferruginous springs affords an explanation
of the origin of nearly all the ores of iron, whatever may be
the nature of the rock in which they occur. In the Jurassic
system, which forms the central part of the Secondary rocks, and

---

* Pepita in Spanish signifies "pip-stone," or "little nut," because the gold of the placers presents such
an appearance.

Metal Mines..

Plate XV.

Veins of Copper at Terriccio, Tuscany.
After Burat.

Fig.1.

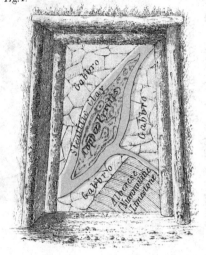

Veins of Copper at Terriccio, Tuscany.
After Coquand & Burat.

Fig.2.

Veins of Copper at Castellina, Tuscany.
After Cailloux.

Fig.3.

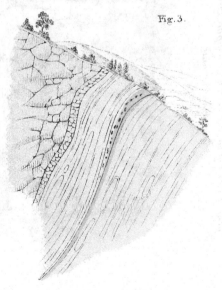

Veins of Copper at Monte Catini, Tuscany.
After Cailloux.

Fig.4.

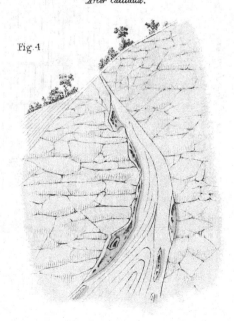

J.W.Lowry Sculp.

Vincent Brooks, Day & Son, Lith.

Sections of Metalliferous Deposits.

Plate XVI.

Metal Mines.

Common mode of occurrence of Bog Iron-ore
in fissures of pre-existing Rocks.
After Burat.

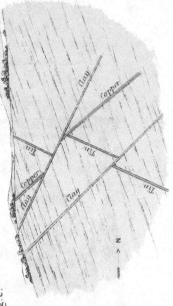

Fig. 1.

Copper- & Tin-Lodes at Huel Peever, Cornwall.
After Carne.

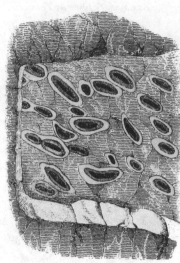

Fig. 2.

Parallel Veins of Copper Pyrites in compact Lievrite
at Temperino, Tuscany.
After Burat.

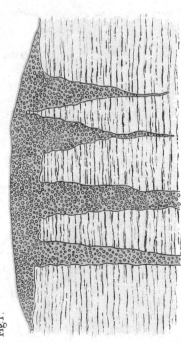

Fig. 3.

Concentric zones of Copper Pyrites & radiating Hornblende,
at Temperino, Tuscany.
After Burat.

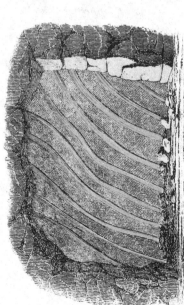

Fig. 4.

which is so called because its type is especially developed in the low chain of mountains sweeping round the north-west frontier of Switzerland, which is called the Jura, there are very extensive deposits of ferruginous ores. The prevailing ores of the Jurassic series, as the rocks of the Oolitic period are called on the Continent, are oolitic and pisolitic varieties, resembling masses of the roe of fishes, or peas cemented together, the waters of the rivers as well as those of lakes and seas having been at particular times, during the epoch in question, saturated with iron. In the formation of alluvial ores, although the ferruginous springs only filled basins of very limited areas, or shallow lakes, still the phenomenon remained the same; and the varieties which, from this mode of occurrence, have been named marsh-ore or bog iron-ore, lake-ore or meadow-ore, have originated, and are still formed, at the present time in a similar manner.

The formation of black oxide of manganese, commonly called Wad, and sometimes met with near the surface in irregularly-disseminated nodular concretions resembling masses of truffles, is due to the same causes which have tended to produce the above-mentioned ores of iron. These mineral tubercles have evidently been deposited from water impregnated with manganese, and which has percolated through the ground. It is one of the most curious of alluvial ores.

We have already described the various hypotheses which would account for the formation of metalliferous deposits in nature. From the very earliest geological times up to the present the work of speculation has gone on in a modified way, and it has continued in some degree to be the same until now. We think we see that, however varied and multiplied the accumulations may be, the law of their formation still continues to be marked by its simplicity, to the explanation of which, however, we have not yet found the key. The origin of these deposits presents nothing abnormal in the domain of physical facts, and no doubt they are compatible with a small number of laws, but none of those have been definitely established by science. Metallic veins are analogous to the great metallic reserves which societies have been accustomed to store up from the earliest historical times; except that the natural reserves, in spite of all that has been taken from them, are far from being exhausted: and in this respect especially the lines of the poet are literally verified :—

> "Le globe est un vaisseau frété pour l'avenir,
> Et richement chargé. . . . . . "

> "The globe's a richly freighted ship,
> Amply stored for future use."

CHAPTER III.

THE PRINCES OF THE MINERAL KINGDOM.

ORES.—Nuggets and grains of Gold and Platinum.—The Ores of Silver, Quicksilver, Copper, Tin, Lead, Zinc, Antimony, Arsenic, Cobalt, Nickel, and Bismuth ; Iron and Manganese.—Association of Ores.—Gangues ; their composition and uses.— Mystery to be explained.

THE ores contained in metallic veins are either native metals or the combinations of those metals with some simple or compound bodies, always present in small numbers, such as oxygen, carbonic acid, sulphur, &c. All these ores have a common family likeness, which is striking even at the first glance. They possess a well-marked metallic lustre, bright colours, a high specific gravity, crystallize in distinct forms, which are always the same for each species, the whole furnishing a sufficient number of characters to enable us to arrive at a knowledge of their nature and sometimes of their composition. With regard to their economic importance, they hold the first rank amongst mineral substances ; their beauty and brilliancy, likewise, give them the first place in cabinet collections, where they hold equal rank with precious stones—a title which they sometimes usurp. They are, it may be said, the princes of the mineral kingdom. Haüy, the founder of French mineralogy, included them in one single and brilliant family in his classification of inorganic bodies. The gangues in which they are contained and set are themselves often crystallized, and contribute by their comparatively subdued and modest tints to enhance the richness of tone of their more splendid associates. To enter here into a minute study of all these metalliferous substances—of these jewels of nature—would be out of place ;

but a few words may be said in reference to those which are
of the greatest use, or are most widely diffused.

The metals which are most commonly found in a native
state are those which are the most inalterable, such as gold,
silver, platinum, quicksilver, and copper. Gold and platinum
are scarcely ever met with except in that state, usually in
nuggets, and grains (Plate I., figs. 1, 2, 4, 5). Gold is too
widely diffused, and too much valued by everybody, to render
it necessary to dwell further upon it. Platinum is less familiarly
known. It was first discovered in the American placers by
the Spaniards, who threw it on one side as of no use; but it
was afterwards introduced into Europe about a century ago,
under a name having reference to its colour—*platina* (the
diminutive of *plata*, "silver," in Spanish) signifying "little
silver." It is the most inalterable of all the metals; it is also
the heaviest, being twice the weight of silver (bulk for bulk).
It is affected by no simple acid, but is attacked by nitro-muri-
atic acid, and is infusible except at a temperature exceeding
a white heat, which is 2000 degrees.

Platinum is found associated, but not alloyed, with gold in
the placers, and these two metals often occur in grains and
nuggets of various sizes, from a pin's head to a fist. There are
even auriferous nuggets, preserved in museums, weighing as
much as from 30 to 63 lbs., and which are consequently worth
from £1440 to £3024, calculating the value of gold at £4 per
ounce. What a windfall to meet with such "a find!"

But such cases are rare, and the fortunate finders, whose
good star has led to the discovery of such heavy and rich
ingots, become notorious for luck.

Silver, like gold and platinum, is also met with in a native
state. The fibrous and foliated varieties in filiform or capillary
and dendritic forms are very common in collections (Plate I.,
fig. 3). More frequently the ore is a combination of the metal
with sulphur, chlorine, iodine, bromine, &c.; silver having a

great affinity for all those bodies. The natural sulphides, chlorides, iodides, and bromides of silver are simply binary combinations, or are mixtures of each other. The sulphur-ore of silver is called *vitreous silver-ore* or *silver-glance;* when it contains antimony its colour is orchil-red, and it is called *red* or *ruby silver-ore.* The sulphide of silver is also mixed with galena or sulphide of lead (Plate III., fig. 1) in sufficiently large proportions to constitute a true ore of lead and silver. A very large quantity of the precious metal has always been extracted from this ore, especially in Europe. The name *horn-silver* or *corneous silver-ore* has been given to the chloride of silver on account of its appearance, and from the way in which it may be cut with a knife, like horn.

In America the backs or upper parts of silver-veins are often composed of earthy black or reddish pulverulent, ferruginous matters, which are known in Peru, Mexico, and Chile as *colorados, pacos, negros,* or *negrillos.* These earths generally contain immense accumulations of silver in the state of chlorides, sulphides, &c.

Quicksilver sometimes exists in a native state in small globules. Its real ore is the sulphide, *cinnabar* or *native vermilion,* of a beautiful red colour, and often crystallized (Plate I., fig. 6). Alloyed with silver, quicksilver constitutes a natural amalgam named *native amalgam,* and met with, occasionally, in very perfect crystals.

Native copper is a very widely diffused mineral in several localities, and occurs mostly at the outcrops of the lodes, in a massive form; but it also forms pockets in the lodes, and capillary, dendritic films, filling cavities and crevices in the rock, where it ramifies in crystals, or thin plates and sheets, and in delicate fibres (Plate II., fig. 1). The red oxides of copper, or *red copper-ore* (Plate II., fig. 2) and the black oxides, the blue and green carbonates—*azurite* and *malachite* (Plate II., figs. 3 and 4)—are associated with native copper. They give place, in depth, to the *yellow* or *variegated pyrites*—which are double sulphides of copper and iron (Plate II., figs. 5 and 6). The former has the colour of yellow copper or brass; the second presents the bright and variegated tints of a peacock's plume, whence the term *pavonazzo* given to this mineral by the Italians. The variegated variety is less common than the yellow; so also is the *copper-glance* or *vitreous sulphide of copper*, which is a simple combination of copper and sulphur, containing 79·79 per cent. of the metal. This latter sulphide, of a dull grey colour, is so soft as to be cut with a knife.

An ore of copper which must be now mentioned is *grey copper-ore* or *Tetrahedrite*, which the Germans call *Fahlerz*. Besides iron and copper, it also contains antimony, arsenic, and sometimes as much as thirty per cent. of silver. It is smelted for the copper and silver; but it is very difficult to treat, in consequence of its multiple composition, and gives much trouble to the smelters.

Tin, the faithful alloy of copper in bronze, is sometimes found associated with it in nature, as in *tin pyrites* or *bell-metal ore*, which contains 29·77 per cent. of copper and 27·44 of tin, and it alternates in a remarkable manner with copper-ore in the lodes of the Cornish mines. Tin is never met with in a native state, and scarcely exists except in the state of a crystallized oxide, *tin-stone*, of a chocolate-brown colour (Plate XV., f . 5). More r .rely, in the stanniferous sands, it is yellow, rose-red, and even colourless and translucent as crystal. The learned have given it the name *Cassiterite*, derived from the Cassiterides of the ancients, the islands of the Atlantic the precise positions of which modern geographers have not been able to fix, and whence, even before the time of Homer, the Phœnicians went to trade in tin.

Lead, like tin, is not found native, except in very exceptional cases; but the sulphates, the white or yellow carbonates, *native white-lead* (Plate III., fig. 2), are common in the upper

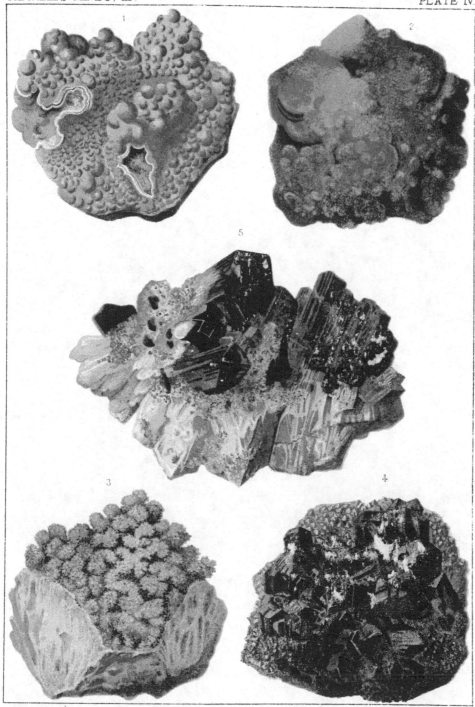

A.Faguet pinx.   Regamey Chromolith.

1 Silicate of Zinc (Smithsonite) *Cumberland*.   3 Carbonate of Zinc (Calamine) *Vieille-Montagne*.

2. Smithsonite *Stolberg*.   4. Sulphide of Zinc (Blende) *Kapnik (Hungary)*

5. Oxide of Tin (Tin-stone) *Morbihan (France)*.

CHAPMAN & HALL, London.   Imp. Lemercier & Cie, Paris

parts of lead-lodes; and in the cavities which line the walls of the lodes, the green phosphates, yellow aluminates, and red chromates (Plate III., figs. 3 and 4), the finer specimens of which are in great request by collectors.

The most widely diffused ore of lead is the sulphide, or *Galena* (Plate III., fig. 1), usually containing silver. It is of a bluish or steel gray colour, and very often crystallized. No very long time ago, the galena in large crystals or broadly laminated was supposed by miners to be always poor in silver; while the galena in small crystals or in fine grains was always considered rich in silver; but numerous facts have tended to show that those assumptions were often erroneous.

Galena, when mixed with the sulphides of antimony and copper, takes the name of *Bournonite*, in honour of the French chemist Bournon, by whom this species was first noticed. In the lead family this mineral occupies a place analogous to that filled by grey copper-ore amongst the copper group of ores. It is not held in much estimation by the smelters, although it frequently contains silver, and even gold.

Zinc is never found in the lodes as a pure metal; but the oxides, carbonates, and silicates of zinc, or *Calamines*—varieties of white zinc-ore which are very easily fused (Plate IV., figs. 1, 2, and 3)—are found, especially at the outcrops of the lodes, in considerable quantity and in enormous masses in the north of Spain and in Sardinia. The most common ore of zinc is the sulphide, or *Blende.** It occurs crystallized or in crystalline masses of a brownish colour, and sometimes it is of a honey or amber yellow (Plate IV., fig. 4).

Antimony,† like most of the metals, occurs in the upper parts of the deposits, in the state of oxide, and at a greater depth as a sulphide. The pure crystallized oxide is nearly

---

* From the German *blenden*, "to dazzle," from its bright or dazzling lustre.

† The chief use of antimony in industry is as a constituent of certain alloys, in order to make them hard. With this object in view it is alloyed with lead, to make type-metal, for casting into letters used for printing.

white or colourless, transparent, limpid, resembling the diamond; but it is very much softer, and does not refract light
like the incomparable gem in question. The compact varieties
of this oxide resemble limestone, and have a yellow or greyish
earthy appearance; but their great density soon proves them
to be a metallic substance. The sulphide of antimony (*Stibnite*)
is of a lead-grey to a steel-grey colour, and occurs in the form
of long prismatic or acicular crystalline groups, with a columnar structure (Plate III., fig. 5), and sometimes displaying
very curious assemblages of fibrous bunches or tufts.

Arsenic,* cobalt, nickel, and bismuth are met with native,
but most frequently as sulphides in certain lodes. *Mispickel*, a
mixture of sulphur, arsenic, and iron, with a tin-white colour
and a metallic lustre, gives off a characteristic odour of garlic
when struck with steel. The simple bisulphide of arsenic, or
Realgar, is of a beautiful red colour (Plate V., fig. 2); *Orpiment*,
or the ter-sulphide, is, on the contrary, of a citron-yellow
(Plate V., fig. 3). Cobalt and nickel are often combined with
arsenic. Arsenical cobalt, or *Smaltine*, is of a tin-white colour,
inclining to steel-grey; while the arseniate called *Cobalt Bloom*
is of a peach-blossom colour (Plate V., fig. 4). Arsenical
nickel, or *Copper Nickel*, is copper-red, inclining to Florentine-
bronze; the arseniate of nickel, or *Nickel Ochre*, is apple-green;
and the carbonate is named *Emerald-nickel*, from its beautiful
emerald-green colour (Plate V., fig. 5). Bismuth is most
commonly found in a pure or native state: it is silver-white
(Plate V., fig. 1). All the above metals form an interesting
family, and are applied to a variety of uses. †

---

* It is purposely attached to the family of metals, to which it is allied by all its physical properties,
and whose constant companion it is in many lodes. By chemists it is only considered to be a metalloid.

† Arsenic is used in medicine and in colour-making. Its poisonous qualities are well known in the
state of white acid or arsenious acid, which is familiarly called *arsenic* both in this country and in
France, where it is also sometimes known by the name of *mort-au-rats*, from the use which is made
of it in destroying rats and other vermin. Cobalt, nickel, and bismuth, especially the former two, are
alloyed with copper and antimony in making German-silver (Britannia-metal or *maille chort*). Nickel
is also used alone, or alloyed with copper, for making copper-coin. The pretty and well coined
sous-pieces current in France are struck in Switzerland and Belgium.

A. Faguet, pinx.ᵗ   Regamey. Chromolith.

1 Native Bismuth, *Cornwall.*   3. Yellow Sulphide of Arsenic (Orpiment) *Hungary.*

2 Red Sulphate of Arsenic, (Realgar) *Transylvania.*   4. Arseniate of Cobalt (Cobalt-Bloom) *Saxony.*

5 Hydrous carbonate of Nickel (Emerald-Nickel) *Pennsylvania.*

CHAPMAN & HALL, London.   Imp. Lemercier & Cⁱᵉ, Paris.

The ores of iron and manganese, which are nearly always earthy and often hydrated (that is, contain water in a state of chemical combination), are widely different from all the preceding. The family of iron-ores is numerous and interesting, whether the hydrated oxides (*Limonte* or *Brown Hæmatite*), compact or earthy, in a state of yellow wder of a rusty nt, occasionally pisolitic and sometimes ii ed (Plate VI., figs. 1 and 2); the anhydrous peroxides (*Red L natite**), which sometimes resemble the Brown Hæmatites in ou   appearance, and like them often occur in fibrous, columnar, mammillated or stalactitic, as well as in granular and compact forms (Plate VI., fig. 3); the micaceous, scaly, crystalline, or brilliant, hard and well crystallized varieties of Red Hæmatite (*Iron Glance*, *Specular iron*, or *Oligiste*),† also yielding a reddish powder, and frequently displaying red and changing tints (Plate VI., fig. 4); or, lastly, the magnetic peroxides, or *Magnetite*, either massive or in s.nall crystals of a steel-grey colour, and yielding a black powder (Plate VI., fig. 5). In case of doubt as to the name which should be given to a specimen of oxide of iron, a small piece of the doubtful mineral should be pulverized: a rust-coloured powder will be afforded in the case of Limonite, red from Hæmatite, black from Magnctite.

Carbonate of iron (the occurrence of which in the Coal Measures has been described) in a compact stony form, constitutes when pure and crystallized (Plate VII., fig. 1), *Sparry* or *Spathose iron-ore*‡ (*mine d'acier* or *mine douce*), an ore which is held in great estimation by smelters, who by no means regard with an equally favourable eye ordinary *Iron Pyrites*, or bi-sulphide of iron of a brass-yellow colour, the smallest particle of which, introduced into the furnace, renders the metal brittle;

---

* Hæmatite or bloodstone is said by Theophrastus, in his "Treatise on Stones," to be so called from its resemblance to clotted blood—Ἁιματίτης being derived from αἱμα, "blood."

† From the Greek word ὀλίγος, "rare." Haüy, when inventing this name, had forgotten the abundant and rich specular iron-ore of the Isle of Elba.

‡ Also called Siderite, from the Greek σίδηρος, "iron."

but the pyrites* is in request by collectors, on account of the superb groups of crystals it sometimes presents, or of its arborisations (Plate VII., figs. 2, 3); it also pleases the gold-finders, because it often contains gold; and it is largely consumed by the manufacturers of oil of vitriol (sulphuric acid), who extract the sulphur which it contains; and lastly, the variety called *Marcasite*, or White Iron Pyrites, when cut and polished, is used in jewellery.

The ores of manganese consist chiefly of oxides, of which the arborescent (dendrites) or crystallized varieties find a welcome place in mineralogical cabinets† (Plate VII., figs. 4, 5). The earthy varieties are employed in the manufacture of chemical products, and for some time past the iron-founders have also made use of them. Manganese produces a magnificent white cast-iron with broad scales, and imparts the qualities of steel to iron. *Wolfram*, a combination of oxygen and a very rare metal, tungsten, and containing iron and manganese, is also employed for this purpose, an addition up to $2\frac{1}{2}$ per cent. possessing the property of increasing the hardness and tenacity of cast-iron as well as steel.

The various ores which have been passed in review, from gold to manganese, are found associated together in the lodes, and seldom met with alone; moreover, they are intimately mixed with the stony materials called gangues, which also, like the ores themselves, often occur in beautiful crystals.

The gangues of lodes consist chiefly of Rock Crystal, or pure silica, and compact quartz having a similar composition; it is the usual receptacle of gold, tinstone, and, indeed, of most of the ores. Next follows Calcareous Spar, the purest variety of which is the well-known transparent Iceland Spar; carbonate and sulphate of baryta, detected by their high specific gravity;

* Still called in some words *Martial Pyrites*. The name pyrites comes from the Greek πῦρ, "fire," because pyrites strikes fire with flint. It was used for a long while instead of flint, in the locks of fire-arms.

† See Bristow's "Glossary of Mineralogy," p. 228.

A Faquet pinx.ᵗ

C Regamey Chromolith

1. Spathose Iron Ore (Carbonate of Iron; *Isère · France*).

3. Arborescent Iron Pyrites

2. Iron Pyrites (Bisulphate of Iron) *Elba*

4. Dendritic oxide of Manganese.

5. Oxide of Manganese, *Harz*.

CHAPMAN & HALL. London

Imp. Lemercier & Cⁱᵉ Paris

Fluor Spar, in cubical crystals of various tints of green, blue, violet, yellow, or colourless; and clay of different kinds, from hard and shining metamorphic schists to soft and plastic earth, of red, grey, or whitish tints. The granitic and serpentinous rocks likewise form the gangues of certain ores. Lastly, the gangues themselves may form distinct and independent lodes, of which sulphate of baryta and quartz furnish the most common instances. Quartz is used in the glass-works, and barytes for the adulteration of various articles, as white-lead and oxide of zinc, sugar, and starch, for which its heavy weight, whiteness, and low price render the latter particularly suitable.

But the gangues are often put to more worthy uses, as fluxes in the smelting of ores, the fusion of which they promote. Fluor spar is especially valuable as a flux for ores having a quartzose gangue; quartz being suited for those with a calcareous gangue. Even the very schists themselves which are traversed by the lodes, and which are refractory rocks, resisting the action of fire, can be utilized; for instance, for lining the interior of metallurgical furnaces. Is this curious assemblage of all these different substances, which seem to be collected at the same point for the purpose of mutually assisting each other some day, a prevision of Providence, or is it but the effect of mere chance? As for the existence of the lodes themselves, and of the ores which they contain, with their complex composition and ever-varying modes of occurrence, who can say whether those profound fissures, the working of which while it confers so many benefits on mankind, also causes much unhappiness, do not furnish an additional example of those numerous natural phenomena the supreme and ultimate object of which is for ever hid from us?

CHAPTER IV.

THE METALLIFEROUS WORLD.

EUROPE AND ASIA.

Great Britain.—Mines worked since the times of the Phœnicians.—Sweden and Norway.—Belgium.—The Rhine Provinces.—Prussia, the Harz, Mansfeld, Saxony. —The Germans, the Huns ; miners and smelters.—Gaul.—The iron-industry of France ; compared with England.—Spain.—Workings of the Tyrians and Carthaginians.—The pits of Hannibal.—The Romans, Moors, and Spaniards.—Portugal.—Italy ; Piedmont, Modena, Tuscany, Papal States, Calabria, Sicily, Isle of Elba, Corsica, Sardinia.—Balzac's speculation.—Calamine and argentiferous galena in Sardinia.—Smelting old slags in the Mendip Hills.— The Phœnicians, Etruscans, Romans, English, and Germans.—Policy of ancient Rome and Spain with regard to mining undertakings.—Ancient Greece.—Macedonia.—Turkish mining in the Taurus.—Judea, Arabia, Persia, India, Cochin China, Japan, Scythia.— The Altai mines of Siberia and Daouria.—Plumbago mines of Batougol.—The mineral wealth of the Ural.—The Demidoffs.

Now that we have become acquainted with the origin and aspect of metalliferous deposits, let us turn our attention to the countries in which they occur. Beginning with old Europe, let us first notice Great Britain, not less abounding in metalliferous wealth than in coal-mines. Lead-mines are met with in Ireland, Scotland, the Isle of Man, Cornwall, Wales, Devonshire, Somersetshire, and in most of the midland and northern counties of Yorkshire, Derbyshire, Cumberland, Westmoreland, Durham, and Northumberland, where the ore occurs in the Carboniferous Limestone itself. Mines of tin, copper, and zinc occur in Cornwall and Devonshire (Map IX.), the two metalliferous counties *par excellence*, whose workings have never been suspended, and are even carried on beneath the bed of the sea (fig. 102). The poets have styled Cornwall a cornucopia, or horn of abundance, a name which its shape renders not inappropriate. Derbyshire,

Cumberland, Cornwall, Wales, the Mendip Hills, the Isle of Man, and Ireland, also contain zinc. Iron-ores (oxides) are found in nearly all counties, especially in Lancaster, Cumberland, Devonshire, the Mendip Hills, the Forest of Dean, and many other districts, and as a carbonate, in most of the collieries, interstratified with the Coal Measures and near the limestone, which serves as a flux for the ore : the spathose ore of iron is also largely worked in the Brendon Hills of Somersetshire, and in Weardale, Durham. Arsenic, manganese, and other products, such as graphite—which was once largely worked in Cumberland, but is found there no longer—must not be omitted. Gold even is not wanting, having been discovered a few years ago in North Wales in veins of quartz similar to the quartz-reefs of Australia and California.

The antiquity of some of these mines is lost in the night of ages. The Cornish mines have been worked from the earliest historical times, and it is known that the navigators of Tyre and Sidon, and at a later period those of Carthage, went there in quest of copper and tin. The iron of Britain was equally celebrated, and was worked by the Romans in the Mendip Hills of Somersetshire and in the Forest of Dean in Gloucestershire. We are told by Cæsar that the Britons of his time coined it ; and the Romans, who were more skilled workmen than the Britons, knew how to forge well-tempered swords of iron, which for a long time enjoyed a sort of celebrity.

The metallurgical industry of Great Britain is far from having diminished since the time of the Phœnicians or of Cæsar, and has made continual progress.

The ores she does not raise in sufficient quantities to supply her own wants, which are chiefly those of copper, silver, and nickel, she imports from abroad. The cupriferous ores of Spain, Morocco, Gaboon, Congo, the Cape of Good Hope, Australia, Bolivia, Chile, California, Lake Superior, Cuba, and of the entire Mediterranean, are exported to and may almost be said

to be worked by England. They are no sooner drawn from
the bowels of the earth than they are on their way to Swansea or Liverpool, the centres of large smelting establishments.
Should they be of poor quality, they are separated by sorting
and washing, and then purified by means of a preliminary
fusion. In the part of this book which treats of coal, it is
shown that all these grand results are almost entirely due to
that substance (see pp. 50, 252).

In the same latitudes as Great Britain, Sweden and Norway are deserving of our attention, although in different ways.
There, in the midst of the granitic and schistose rocks which
form all the higher land of Sweden, are the copper-mines of
Fahlun, the silver-mines of Sala and Kongsberg, the iron-mines
of Dannemora, besides other mines of nickel and zinc. Fahlun,
Kongsberg, Sala, and Dannemora are equally celebrated,
and are visited by numerous travellers. At certain points, the
depth and immensity of the openings cause the beholder to
feel giddy; but the scenery in the neighbourhood of the mines
presents an appearance of rudeness and striking originality.
The iron-deposits of Sweden are especially celebrated. It
is there, again, that the miners toil for England, and the
renowned steel of Sheffield is indebted for its qualities to the
magnetic iron-ore of Scandinavia. The British Isles, which
are so abundantly provided with iron-ore, do not possess this
variety in any great quantity: the places where it chiefly occurs
being near Newton and Brent in Devon, and near St. Austell
and Penryn in Cornwall. The largest deposit of iron-ore in
Europe is probably that of Gellivara in Swedish Lapland, about
ninety miles from the head of the Gulf of Bothnia, where it
forms a bold hill rising out of swampy ground, made up of
a great number of parallel interlaminations of magnetic and
specular iron-ores, some of which are between one hundred
and two hundred feet in thickness.*

* "A Treatise on the Metallurgy of Iron," by H. Bauerman, p. 56.

On the other side of the North Sea is situated the industrious little kingdom of Belgium, a country essentially rich in mineral wealth.   Lead-mines are met with there at Vedrin between Namur and Charleroy, at Huy and Engis on the Meuse, and at Bleyberg on the borders of Rhenish Prussia; zinc-mines at Huy, Engis, Moresnet, and Corphalie—all, as well Vieille Montagne, situated in the province of Liége.   Lastly, iron-ores are present in considerable quantities.   Belgium is a country which, in proportion to her size, produces more iron than England.   A great part of her ore is exported, especially from Luxembourg and from the districts between the Sambre and Meuse; France taking more than 250,000 tons per annum.

The transition is a natural one from Belgium to Rhenish Prussia (Map X.).   Different metalliferous deposits are met with all along the valley of the Rhine, and on the flanks of the mountains which border its waters, or those of its affluents, the prolongation of the Vosges, the Taunus, Hundsrück, and Eifel ranges, and the mountains of Westphalia (Map X.). The most celebrated are the quicksilver-mines of Deux-Ponts; the copper, lead, and silver mines of Holzappel and Obernhof in Nassau* (Plate XIII., figs. 1, 2, Plate XIV., fig. 2), of the Eifel, Eschweiler, and Stolberg, near Aix-la-Chapelle; and lastly those of copper, iron, lead, and zinc of Ramsbeck in Westphalia in the Upper Ruhr.   The sands of the Rhine, also, are auriferous, and have given rise in former times to unprofitable workings, especially in the plains of Alsace.   Large deposits of Phosphorite have recently been discovered in Nassau, in the neighbourhood of Limburg and Staffel, and are worked for the manufacture of superphosphate of lime, which is now so much used in agriculture as an artificial manure.

Before leaving Prussia, it should be observed that zinc is also found abundantly in Silesia; as well as iron in various localities, amongst others in Siegen, one of the Rhenish provinces.

* Worked by the Romans, and worked uninterruptedly since the sixteenth century.

It is with iron from this ore that Krupp, the great cannon manufacturer, makes his celebrated steel, at Essen, on the border of the coal-field of the Ruhr. It is also with Siegen iron that the famous foils of Solingen are made, the excellent temper and elasticity of which are so highly appreciated by all fencing-masters and cavalry-officers.

Prussia and Germany go hand in hand, at least on the map: Germany in the metalliferous provinces of Saxony, Mansfeld, Brunswick, and Hanover; the Harz—the ancient Hercynian forest (Hercynia sylva) whence this district derives its name—being situated in the last mentioned state; the Harz, with its lofty mountains traversed by metallic veins, whose culminating peak, the Brocken, reminds us of the land of sorcery. Klausthal, Zellerfeld, and Goslar, centres of mines and foundries, are all situated in the Harz, whose other summits, Rammelsberg and Andreasberg, give their names to new metallurgical districts, some of which, like Rammelsberg, have been opened since the sixteenth century. Iron, copper, zinc, lead, silver, and arsenic, are worked in the Harz.

In the middle of Thuringia, or more correctly, in the middle of Prussian Saxony, is the argentiferous Kupfer-schiefer or copper-slate of Mansfeld, with impressions of fossil fishes belonging to the Permian period. A full account of these curious mines, which have been worked since the year 1200, has been given by Mr. Jervis in the *Journal of the Society of Arts* for 1861, vol. ix. pp. 592, 603, 616, 627.

In Saxony the famous mines of Freiberg (Plate XIV., fig. 1), Chemnitz, and Altenberg, are situated on the western flank of the chain of the Erzgebirge ("ore-mountains"). There, as also in the Harz and Mansfeld, the difficult art of subterranean engineering was revived in Europe since the eleventh century— the traditions of the art having been lost some time previously to the fall of the empire. The Germans, who had remained

* The name Harz is supposed to be derived from the old German word *Hart*, "a wooded height."

miners since the time of Tacitus, and on whom had devolved the mission of regenerating the old Roman world by invading it in the first instance, played their part to the end, and taught the sons of the Latins what they had forgotten—the practice of carrying on underground excavations and the mode of smelting ores. The Romans had, in the first instance, learned from the Etruscans the art of exploring the ground, and of extracting the metal from the gangue by the action of fire. The sons of Tyrrhenus, on emigrating to Italy, had carried with them the knowledge of this art from the East, whence it has also been communicated to the Egyptians, Greeks, Phœnicians, and Carthaginians—the East having derived it from India, the cradle of all human knowledge. All the arts and sciences diverged from this luminous source, either because primitive man discovered them himself, or that he was endued with a knowledge of them when first created. The secret when once learned is never wholly lost, for the vanquished may be said to transmit it to the conqueror—the nation which yields to that which takes its place; or else it is the conqueror who imparts it to the conquered people. So is it with literature and the fine arts, as well as with the pure and applied sciences: progress never ceases, and the human mind is eternally soaring onwards and upwards.

It was, then, from the Germans that southern Europe re-acquired a knowledge of the art of mining; but the art extended thence over the whole of Germany, as we have seen, especially in Mansfeld, the Harz, and the Erzgebirge. The old Austrian provinces of Bohemia, Hungary, the Tyrol, Istria, Styria, and Carinthia, were then famous for their mineral wealth, and it is thence that the impulse has been given since the eighteenth century. Styria and Carinthia, which were anciently called Norica, produced steel from the earliest Roman times: Istria possessed quicksilver, mines of which are still worked at Idria in Carniola: Bohemia and Hungary tempted

the Huns by their lodes of tin, copper, and gold, the search for
which was actively pursued by the descendants of the followers
of Attila. Lastly, the Tyrol became celebrated for its mines
and miners, and the Tyrolese of the present time form many
of the pupils in Spain, France, and Italy.

Gaul, the neighbour of Germany, learned like her to derive
advantage from her mineral wealth. According to Tacitus,
iron was the special product; but in Gaul the aboriginal tribes
not only mined for that metal, but for copper, gold, tin, lead,
and silver besides. Cæsar, in his Commentaries, makes frequent mention of the skill of the Gauls in excavating subterranean galleries for attack and defence, and he attributes this
skill to their knowledge of mining operations. Cæsar also
speaks of the copper and iron of Gaul, and he might, also,
have mentioned the deposits of gold and tin, the ancient
workings for which are still to be seen—at the foot of the
Cévennes and the Pyrenees for gold, on the plains of Morbihan
and Limousin for that metal and for tin.

All round the mass of basalt and granite forming the heart
of France—that great central plain from which all its great
rivers flow—the Gauls also worked lead and silver mines, the
ruins of which are still in existence. The metals, especially
tin, were carried to the mouths of the Loire, where the
Phœnicians, in their voyages towards Britain and the shores
of the Baltic, freighted their ships in passing.

Who can say with certainty whether the famous Cassiterides
are not as likely to have been l'Ile-Dieu, Noirmoutiers, and
Belle-Ile, islands at the mouth of the Loire, as the Scilly Isles
(called by the French Iles Sorlingues), lost in full ocean
opposite Cornwall, so fertile in wrecks? At a later period,
when the commerce of Tyre and Carthage passed into the
hands of the Greeks, especially the Phoceans of Marseilles, tin
was carried through Gaul to reach Marseilles, which had
then become the great metal-mart of the Mediterranean.

In the Middle Ages, at the commencement of modern times, mining operations were carried on in France with as much ardour as in our day. The barons and the religious orders of the feudal times undertook, at their own expense, great works, from which they derived enormous profits. At the foot of the Pyrenees, on the French side of the Alps, in the Vosges, Brittany, all round the central plain, especially in the mountains of the Lyonnais, of Forez, Vivarais, and in the Cévennes and the Monts Lozère, operations were begun which amply repaid the seekers. Jaques Cœur, the silversmith to King Charles VII. of France, found in this industry the origin of the immense fortune which excited so much jealousy, and ultimately lost him his master's esteem.

The names of some of these ancient workings are significant: as Sainte-Marie-aux-Mines, Plancher-aux-Mines, Croix-aux-Mines, in the Vosges, worked from the time of Dagobert and the goldsmith, Saint Eloi; these are the Coffres (from the patois *cobre*, "copper") of Aveyron. The places called Argentière, Aurière, and Ferrière, exist in many departments of France. Finally, the Pyrenees are deserving of the name of the French West Indies, given to them by writers of the Renaissance because of the abundance and the riches of their deposits. The working of the mines of America, which had the effect of lowering the prices of the metals, and drew miners from afar; the bad administration of the later kings, who gave up the ownership of the mines to incapable favourites; with a host of other reasons, such as wars and the revocation of the Edict of Nantes—led by degrees to the almost total abandonment of the French mines. For the last forty years they have been rising from their ruins, and returning insensibly, one after another, towards that period of prosperity which was characteristic of the past. The copper-mines of Chessy (Plate XIV., fig. 1) and Sainbel, near Lyons; the lead and silver mines of Poullaouen and Huelgoët in Finistère; of Pontgibaud,

in the Puy-de-Dôme; of Vialas, in Lozère; and of Argentière
in the High Alps—were the first to enter into this movement,
which has led to such glorious results as regards the national
industry of France.

Besides lead and silver, zinc, tin, copper, antimony, and gold
are extracted, but on a small scale. With regard to iron, the commonest metal, but the most necessary withal, France furnishes
abundant supplies of it, as may be seen on reference to Maps II.
and III. Besides the limonite-mines in Perigord, Berry (celebrated for its ores in the time of Cæsar), the Landes, Champagne;
the stratified deposits in La Comté, the Ardennes, and Brittany;
the iron-lodes in Isère and the Pyrenees; certain of the French
iron-ores have long enjoyed a great reputation, and amongst
others those of Ariège, Isère, and Perigord for their natural
steel, or their iron for converting into steel. Nowadays when
this indispensable metal is produced to the utmost possible
extent, exceptional qualities of ore are no longer relied upon,
and every stone is considered fit for smelting if it will yield
twenty-five per cent. of iron. But France does not possess
the same advantages as England in having the combustible,
the ore, and the flux occurring in the same strata—at the
very foot of the furnace. In having to bring all the first
materials together from a distance, France is at a disadvantage
compared with England; but the French foundries are well
fitted up and ably managed, and the iron-trade of that country
has made rapid advances during the last few years.

From France to Spain the transition is easy. The iron-
lodes which make their appearance on the northern flank of
the Pyrenees reappear on the south, and are continued in the
Cantabrian Sierra, which is a prolongation of the Pyrenees in
the Asturias, on the borders of the Gulf of Gascony. Catalonia
and Biscay are especially renowned for the iron-works which
have been carried on in those provinces from time immemorial.
The method of obtaining iron in the Catalan forge, which is

still in use in the Pyrenees, is as old as the world nearly, for it ought to date from the time of Tubal-Cain, the first worker in metal. The blast in this primitive process is neither produced by a pump nor a bellows, but by means of a *trompe*, or the hollowed trunk of a tree, in which a descending current of water produces a draught of air, which is thrown into the furnace.

The mountains of the Asturias, no less rich in iron than the Pyrenees, contain considerable quantities of zinc in the state of calamine and blende; which traverse the chain like the iron-lodes of the Pyrenees: in Old Castile, on the southern slope of the Cantabrian Mountains, these mines are worked to a depth of more than 1000 mètres (1094 yards). In that half-savage country, dotted with natural meadows like the flanks of the Alps, and like them visited early by snow, the jagged peaks bear high-sounding and proud titles, amongst others the *Pico d'Europa*, denoting, if not the culminating point in Europe, at all events that of these particular districts. Difficulty of access, distance from populous centres, and badness of roads, have retarded the working of mines on this side of the Asturias, as well as in most other parts of Spain; but on the other side, which faces the sea, mining is being most actively carried on. Ten years ago the industrial world was enthusiastic on the subject of Spanish calamines, and everybody talked of the zinc-mines of the Asturias. The ores, which are especially abundant at Santander, are sent to the Vieille and Nouvelle Montagne, which have found in them a supply which the deposits of Belgium and Rhenish Prussia were unable to furnish those works with.

There are lead and silver mines in another part of Spain —in the mountains of the Sierra Nevada, which traverse the provinces of Grenada and Murcia, and descend to the sea in the direction of Adra, Almeria, and Carthagena. These mines, mentioned by Pliny, were already celebrated in the

time of the Phœnicians and Carthaginians, and the pits of
Hannibal, who derived enormous riches from them, are still
shown. After the Carthaginians the Romans worked these
mines with ardour, and after the Romans the Arabs. The
surface of the ground is riddled with old pits, and dotted with
heaps of cinders and the ruins of furnaces, signs of these
ancient workings, some of which are more than thirty centuries
old. Spain ultimately conquered the Moors, and soon afterwards discovering America, discontinued these works. King
Ferdinand and his successors caused the Iberian workings to
be closed, in order to promote emigration to the distant
countries which had then lately been discovered by Columbus,
Cortez, and Pizarro, and to devote all their energies to the
mines of the two Americas. Political economy, in those days,
experienced strange aberrations. It is only forty years ago
since these very mines in the Sierra Nevada have been taken
up by enterprising persons, and from the very outset, as in
former times, they have yielded unexpected results—furnishing
now one-third of the lead used in the world, and a notable
proportion of silver. Although attention has been drawn to
these localities almost entirely in connection with the above
metals, the mines of zinc, iron, and manganese, especially
those of Cabezo-de-la-Mina, the rich ores of which supply
the foundries of Marseilles, should also be mentioned.

Parallel with the Sierra Nevada runs the Sierra Morena,
which traverses the northern part of Andalusia and the province of La Mancha, celebrated as the birthplace of Don
Quixote. Almaden, or "the mine," as it was called by the
Arabs, and Almaden del Azogue,* as it is called by the
Spaniards, the most celebrated quicksilver-mine known, is
situated on the northern flank, and has been in continual work
for the last three thousand years. Under the dominion of
ancient Rome the works were regulated by a special edict.

---

* In the Castilian language *azogue* signifies "quicksilver."

On the other flank of the Sierra Morena are the celebrated copper-mines of Huelva, near Rio Tinto, which was so called from the circumstance of one of its head-streams being stained of a green colour by the vitriolic salts produced in the ancient workings by the decomposition of the copper pyrites. The old openings, and the mounds of rubbish and cinders met with in these places, date from the time of the Phœnicians. Near at hand is Cadiz, the Gadir or Gades of the Tyrians, where the Phœnicians formed depots of the quicksilver, copper, iron, silver, and lead derived from the adjacent mines, as well as the tin brought from the Cassiterides and the north of Iberia. Gold was likewise produced in abundance by Betica, the present Andalusia, as well as by other districts, especially Gallicia and the Estremaduras (both Spanish and Portuguese), where the old workings for gold and tin are still visible. In consequence of increased facilities of transport, attention has lately been directed afresh to the extensive deposits of phosphorite which were long known to exist in Spanish Estremadura at Logrosan, as well as near Caceres and Montanchez; and large quantities of the mineral have been brought to this country for conversion into superphosphate of lime by the manufacturers of artificial manure.

Spain was in ancient times the great purveyor of the precious metals to the Carthaginians, as well as to the Romans, to which countries she stood in the same relation that America afterwards did to the modern people of the Middle Ages, and that Australia and California do to the nations of Europe at the present day. Lusitania, now called Portugal, contributed with Spain to supply the countries bordering on the Mediterranean with metals. Ancient mines are met with in that country, some of which—and amongst others copper, tin, iron, lead, antimony, and gold—might be reopened with good results.

The deposits on the French side of the Alps traverse the

enormous mass of schists and granites separating that country from the Italian peninsula, precisely as those of the French Pyrenees reappear in the mines of Iberia; the lead and silver lodes of the Col-de-Tende being, doubtless, a prolongation of those of Argentière in the High Alps, or those of Pezey in Savoy. Piedmont, at the foot of the Alps, as its name indicates, is a country rich in metals. The Val Anzasca is renowned for its auriferous pyrites: the Val d'Aosta for its copper pyrites. Some of the mines on the Piedmontese slopes reappear in Switzerland—in the Valais, for instance—where argentiferous lead is obtained. The mountains of serpentine bordering on the Gulf of Genoa are rich in ores of copper, which at the present moment are not sufficiently esteemed: and the mountains of Modena, where the statuary marbles of Carrara are quarried, contain iron, lead, silver; and copper mines, which extend to the most inaccessible summits of their peaks.

The Apuan Alps, adjacent to the Modenese chain, and forming the northern frontier of Tuscany, are in turns traversed by veins of quicksilver, magnetic iron-ore, and argentiferous copper and lead ores. Silver, in particular, was worked by the ancients. The Etruscan city of Luna, the ruins of which exist on the neighbouring sea-shore, had for its emblem a crescent, the symbol of silver, dedicated to Diana. In the Middle Ages, the mines of this district were the subject of violent disputes between the lords of the locality and the neighbouring republic of Lucca, which finally gained possession of the mines, and coined the silver which she got from them into money. Under the Medici most of these mines were opened afresh, and some, amongst others Bottino (bottino, "mine-pit"), are still in full work.

In central and southern Tuscany, as in the north, there are numerous metalliferous districts, amongst which the copper-mines of Terricio and Castellina in the centre, and Monte

Catini near Volterra, deserve special notice (Plate XV.). The latter mine has been known ever since the times of the Etruscans, and the present proprietors derive enormous sums from it every year. Further off are the mines of Campiglia (Plate XVI., figs. 3 and 4), also opened by the Tuscans, and which furnished the greater part of the bronze used by the ancient Etrurians: then the various mines of the republic of Massa-Maritima, where iron, lead, copper, silver, alum, and sulphur were worked in the Middle Ages with such success, that this little state, since the twelfth century, had the honour to be the first country in Europe provided with a regular code of mining laws. The country was called Massa-Metallorum, or Massa-of-the-Mines, to distinguish it from its namesake in the north, Massa-Carrara. Ancient pits, and the ruins of old foundries, may be counted by the hundred in travelling through this district, which is covered with marshes and ravaged by fever; and it is also in the same districts, between Massa and Monte Catini, that the famous *soffioni*, or vapour-vents, occur, which were first utilized with success by M. Lardarel in the extraction of the boracic acid from the lagoons.

In Sienna and Grosseto silver and copper mines again make their appearance; and with those metals occur, in southern Tuscany, quicksilver at Selvena, Pian-Castagnajo, and Castellazzara, and antimony at Montauto and Pereta. All the above districts are situated on the west flank of the Apennines, or rather on a littoral chain which is merely the continuation of that of the Gulf of Genoa, and which the Pisan geologists, Messrs. Savi and Meneghini, have correctly named the metalliferous chain.

Countries may adjoin without bearing much resemblance to each other; an instance of which is afforded by Tuscany and the Papal States, the former being rich in metallic mines, while the latter are so poor. The Calabrias are more favoured in this respect, and possess iron-lodes and ancient silver-mines.

Perhaps a portion of the metal from which the Syracusan medals were struck, which with those of Marseilles are the most beautiful Greek coins known, may have been procured from Calabrian sources. There were also formerly ancient metal-mines in Sicily, but now nothing is seen except its sulphur-mines: and there, where brigands are allowed to carry on their trade, and roads are wanting, industry cannot enjoy any great amount of prosperity.

Opposite Italy there are three islands which must not be forgotten: Corsica, French politically and at heart, but Italian in language and geological structure; Sardinia, a prolongation of Corsica under the sea; and the Isle of Elba, the smallest of the three islands, and though a mere speck by the side of the other two, the most celebrated, on account of its important iron-mines, which have been worked uninterruptedly for the last three thousand years. It is still as it was in the time of Virgil, and as it had been long before, an inexhaustible source of the metal:—

> "Insula inexhaustis Chalybum generosa metallis!"—Æneid, book x.
>
> "An isle renowed for steel, and unexhausted mines."

In Corsica and Sardinia, along the chain of mountains which forms the axis of those two islands and extends to their contours, there are numerous metallic mines. At Cape Corso there is antimony at Ersa; towards Bastia, lead and iron; in the direction of Corte, iron and copper; and silver towards Calvi. In Sardinia, from north to south, and chiefly in the province of Iglesias, next to that of Cagliari, there are also numerous veins of iron, copper, manganese, and antimony, but more especially argentiferous lead-ore. The foundries of Marseilles derive an inexhaustible supply from Correboy, Ingurtusu, Gennamare, Monteponi, and Montevecchio, which is traversed by an enormous lode situated between the granite and the ancient schists, and extending over a length of three

leagues. Industrious Massaliotes or Genoese resmelt the old
slags which are met with here and there in the island, forming gigantic mounds, and have obtained very large sums of
money from them. Less fortunate than these improvised
smelters, Balzac, the French novelist, visited Sardinia with the
same object in view. On board ship he took the captain into
his confidence, and the latter, who was a cunning Genoese,
took up the affair, but without success; and Balzac returned
empty-handed, undeceived once more in those dreams of
fortune which he vainly pursued all his life. The merchants
and others connected with industrial pursuits who, ten years
ago, took up the notion which a poet and a sailor failed to
carry into practice twenty years previously, were the only
persons fitted to bring it to a successful issue.

An enormous deposit of calamine (carbonate and silicate of
zinc) has recently been discovered in Sardinia. One firm at
Swansea has contracted to purchase 4000 tons of the ore a
month, to yield not less than sixty per cent. of metal.

The slags of argentiferous lead which are found in so many
places in Sardinia, especially in the southern part of the island,
at Villa-cidro and Domus-Novas, near some ancient foundries,
are due to the Pisans and Genoese, who held possession of
the island in the Middle Ages. The Phœnicians, Carthaginians,
and Romans ruled there previously, and also worked the
mines; and more skilled in the smelting of ores than their
modern successors, who have been for the most part merely
sailors, the work has been better executed.

Only a small quantity has ever been obtained in Spain
from the ancient residue of the smelting operations—whether
Phœnician or Carthaginian—and M. Simonin has ascertained
by actual analysis of many of the ancient Tuscan heaps
of slags that they contained only a very small quantity of
metal—copper, lead, or silver. In this country some profit
has been made of late years by resmelting the old heaps of

slags left by the Romans on their lead-mines in the Mendip Hills of Somersetshire. It is a remarkable fact that the ancients, who were altogether unacquainted with chemistry, have sometimes treated ores with as much success as the moderns ; and it must not be forgotten that there were people essentially miners and smelters, like the Phœnicians and Etruscans, who were also sailors, merchants, and colonizers. Nowadays, the English and Germans, at the same time voyagers and colonists, also take the lead in the arts of mining and metallurgy. The subterranean cultivation of the ground is no less indispensable than that of the surface in founding colonies, of which California and Australia furnish contemporaneous instances ; and it may be said that the most colonizing races are at the same time endued with greatest aptitude for agricultural and mining pursuits. The English have the advantage of being a maritime and mercantile nation, which, in conjunction with a taste for long and adventurous travels, and self-reliance consequent on the enjoyment of full liberty of action, in that form of it which is commonly denoted *self-government*, is the secret of their brilliant success in their remotest settlements.

In ancient times the Romans so fully understood all the prestige which mines would exercise on the minds of emigrants, and all the advantages they offered both to a rapid and durable development of colonies, that a law was passed by the Senate prohibiting metalliferous mining undertakings being carried on in the Italian peninsula. Pliny, in his "Natural History" (lib. iii. xxiii. xxvii.), makes frequent mention of this law when alluding to the mineral riches of Italy. Although he does not give us the reasons which induced the Senate to make the enactment in question, he leads us to infer that their intention in all probability was to encourage, by means of distant enterprises, a vast emigration to Spain, Sardinia, Greece, and Asia Minor, all of which were then recently conquered provinces; and, moved by the conservative spirit by which ancient Rome was actuated, the Senate desired, also, to preserve intact for future use the subterranean riches of their own peninsula; and, finally, to encourage Italian agriculture, which might be injured at certain points by the working of mines at home. We have seen how Spain, at a subsequent period, impelled by the colonization of distant and recently conquered countries, had, on the discovery of America, also prohibited the working of her own native mines; but what was good in the old Roman times was little applicable at the period of the Renaissance, and the example of Spain, ruined at home by her colonies, and doubly ruined by their loss, proves the truth of this beyond a doubt.

Greece, not less than Italy and Iberia, was anciently the scene of prosperous mineral undertakings. Argentiferous slags of the time of Pericles have been discovered at the gates of Athens, and have laid the foundation of similar industrial undertakings to those of Sardinia. The disciples of Plato and Aristotle were more skilled in philosophy and rhetoric than in metallurgy, and these cinder-heaps in Greece, like those of Sardinia, contain large quantities of silver. The ores in question were got from the mines of the country, and it was with this silver, as with the bronze of Chalcis, Corinth, and Cyprus, that most of the objects of art of ancient Hellas, such as medals, coins, cups, vases, and statues, have been made.

Like Greece, Macedonia was for a long time renowned for its mines. The natural magnet or lodestone (Μαγνης) took its name from the province of Magnesia, where it was found. This variety of iron-ore first led to the discovery of magnetism, the mysterious properties of which must have caused considerable astonishment in the minds of Thales and his disciples, and was only entirely revealed to the learned of our own time. The Middle Ages, it is true, founded the compass on the natural magnet; and that was all.

The Macedonian silver-mines were no less celebrated than her iron-mines. When Paulus Emilius conquered the country, the Romans honoured him with a double triumph—for the annexation of a new province to the empire, and for the richness of its mines. Perhaps the silver-mines taken possession of by Paulus Emilius were the same which at a later period were worked at Sidero-Kapsa during the Middle Ages.

The ill-governed Greece and Macedon of the present day no longer produce any metals, and it is the same with the islands at the extremity of the Mediterranean, like Cyprus, the isle of copper, where the site of its ancient lodes is no longer known. The same may be said of all the other once celebrated mines of Asia Minor. The Pactolus which watered these countries and flowed through the states of King Midas, is now dried up, or at all events no longer runs with gold.

The Turks, by whom some of the mines in Anatolia are worked in the same slovenly fashion as their collieries (page 114), obtain copper from them more particularly, especially at Tokat in the Taurus. The metal is carried to Trebizond, and thence it is exported to Marseilles, where, being too impure for direct use, it is purified and refined. Judea and Arabia were likewise the theatre of vast workings, traces of which are seen in the enormous mound of ancient copper-slags at the foot of Mount Sinai. Nowadays the desert extends all around before the incursions of the Bedouins, and the whole country has reverted to an almost savage state; the natives are ignorant of the art of mining and smelting the ores, and are even ignorant of the sites of the places where most of the undertakings formerly so flourishing were carried on. They may be said to have lost the very trace of the veins. In some parts of Syria iron as well as copper is still worked, and the steel of Damascus is well known. In ancient times the Assyrians and Babylonians were acquainted with the method of forging this metal, and a certain number of iron

Fig. 103.—Interior of the Graphite Mine of Batongol, in the Saiansk Mountains, Eastern Siberia.

objects of art, half destroyed by rust, have been disinterred amongst the ruins of Nineveh.

Persia, India, Cochin China, and Japan have likewise always cast-iron, copper, and other metals. Indian steel is renowned; and the brilliant and solid colours which ornament the old porcelain of the countries of the far East are nearly all metallic. As alloys, the gongs and sonorous tam-tams and the bronzes of China and Japan, are known. Those countries also make use of mercury and zinc, which serves for money in Cochin China, and, finally, gold is worked there; but silver-mines are very scarce.

Let us close this rapid review with the land of the Scythians. It is said that at the foot of the Altaï (or "golden mountain;" *alta* signifying "gold" in Mongolian) are the most ancient deposits known. The mines of Siberia and Daouria are in this locality, towards the river Amoor, and from them are extracted copper, iron, lead, silver, and gold. The Kirghiz, the Kalmucks, the Tartars, and the descendants of the ancient Scythians, carry on the mining operations which were commenced by their ancestors; and it is to them that the Czar, retaining a barbarous custom, sends unfortunate political offenders. There, also, on the west of Lake Baikal and the town of Irkutsk, towards the frontiers of China, at the foot of Mount Sayan, are the plumbago-mines of Batougol, which were discovered and worked by a Frenchman, M. Alibert, whose beautiful specimens everybody has seen at the different Industrial Exhibitions, and whose crayons, adopted by all the artists, have thrown into the shade those of the famous Mengin.

On the west of Siberia are the Ural Mountains, metalliferous on both flanks, which contain hundreds of mines: they are the Peru of Russia. Rich in gold, iron, copper, and platinum, they also yield precious stones in large numbers. Around Ekaterinburg, towards the middle of the chain, some

of the workings are clustered. At Nischne-Tagilsk and
Kuschwinsk in the Ural the Russians work a magnetic iron-
ore of such a high quality, that scarcely sixty years ago it
was the only one equally esteemed in Europe with the iron-
ores of Sweden ; they also extract thence platinum, associated
with gold in the alluvial sands and gravels. In the Ural
Mountains are also found the famous malachites, which in
their natural state are smelted for the copper they contain,
and of which the finest specimens have been so successfully
introduced into the decorative arts of jewellery and architecture. A hundred years ago some blacksmiths, of the name
of Demidoff, discovered these deposits, and the Czar, by way
of reward, granted them a concession of the mines.   It is well
known that these men afterwards became millionaires. If
the working of metalliferous mines be often a lottery, it must
be borne in mind that enormous prizes sometimes fall to the
lot of fortunate speculators.

CHAPTER V.

THE METALLIFEROUS WORLD.

AMERICA, POLYNESIA, AFRICA.

Veins and Mountains.—Mines of the Andes.—Chile.—Bolivia.—Peru and Cerro de Pasco.—Travels of Copper and Silver.—States of Colombia.—Central America.—Mexico.—The Apaches and Comanches.—Sonora and Raousset Boulbon.—The Region of Diggings.—California, its great natural wealth; gold-mines; quick-silver-mines of New Almaden, and their annual production.—Silver-mines of Nevada; the Washoe Region; the Comstock Lode.—Fissure-veins and Mother-veins.—Annual production of bullion in the Pacific States and Territories.—Mineral resources of the Rocky Mountains.—Copper-mines of Lake Superior.—North and South Carolina.—Canada and Greenland.—Nova Scotia; profits of gold-mining.—Cuba.—French Guiana.—Australia.—Tin-mines of Banca and Malacca.—Borneo.—Africa; the Cape, Congo, Senegal, Algeria, Tunis, Egypt, Abyssinia, Mozambique, Madagascar.

IN our course through the Old World we may easily perceive that the metallic veins occur chiefly in elevated districts, and are nearly always to be found on the sides of high mountains. In France deposits are generally met with in the Pyrenees, the Cévennes, and the Alps; in Spain, in the mountains most difficult of access; in Italy, on the other side the Apennines and the littoral chain; in England, in the denuded rocks of Cornwall; in Germany, in the Harz Mountains of Saxony, Hungary, and the Tyrol; in Russia, in the Ural chain and its spurs. Here lie hidden the natural minerals, in conformity with the theory now almost universally admitted, that wherever there are eruptive rocks, and upheavals of mountain-chains, there also are to be found metallic deposits. Nature seems resolved to make man pay the price of her favours, by generally directing the course of the veins towards rugged and desert places.

In America these principles are developed on a very large
scale; for instance, around the line of the Andes, from Terra-
del-Fuego to Behring's Straits, and to a height of about
18,000 feet, run mineral veins. The whole chain is metalliferous, the composition of the veins varying according to the
spots where they occur; and the deposits themselves varying
with the position of the central focus whence they radiate,
and the period of their formation. But everywhere the presence of metals has given rise to mining operations, and almost
always to a very rapid colonization.

The accompanying Map (XI.), showing a general view of the
main geological features of the metalliferous districts of Chile
and the south of Bolivia, is constructed from data collected
by Mr. D. Forbes during his travels in that part of South
America (from 1857 to 1863); the geology of the north-east
part of the Desert of Atacama not visited by him is based on
the statements of Philippi ("Viage al Desierto da Atacama,"
1860).

The metallic deposits of Chile are of two very different
geological ages, and are respectively associated with the
appearance of two distinct eruptive rocks, viz.:—

1. The old granites, which break through the slates and
metamorphic schists of the coast.

2. The new diorites, traversing all the subsequent strata,
including those of the Cretaceo-oolitic period, but not the Tertiary beds.

The geological age of the first of the eruptions cannot as
yet be determined from want of fossil evidence, but from comparison with similar outbursts further north in the Andean
chain, it would appear probably to be Upper Silurian or Devonian; copper, iron, and zinc ores are found in connection with
these rocks, but as yet gold is the only metal which has been
explored on the large scale in Chile in these formations.

The gold is found in the granite itself (composed of quartz,

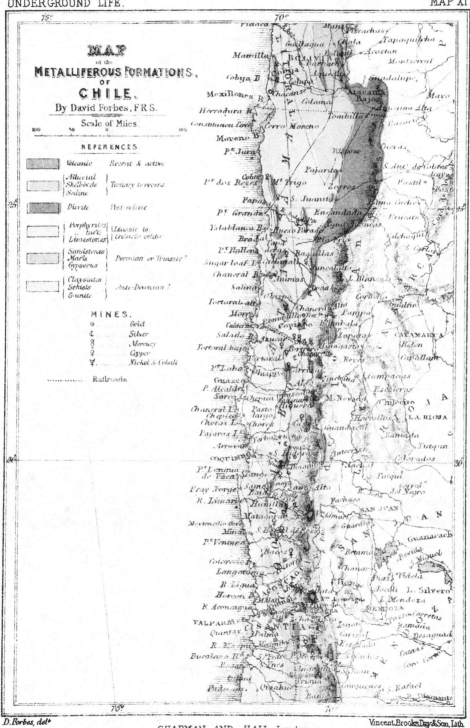

MAP
of the
METALLIFEROUS FORMATIONS,
OF
CHILE,
By David Forbes, F.R.S.

Scale of Miles.

REFERENCES.

	Volcanic	Recent & active
	Alluvial Shellbeds Saline	Tertiary to recent
	Diorite	Post-eocene
	Porphyries tuffs Limestones	Liassic to Cretaceo-colite
	Sandstones Marls Gypseous	Permian or Triassic"
	Claysiates Schists Granite	Ante-Devonian ?

MINES.

	Gold
	Silver
	Mercury
	Copper
	Nickel & Cobalt

Railroads

D. Forbes, delt.

CHAPMAN AND HALL, London.

Vincent Brooks Day & Son, Lith.

orthoclase felspar, and mica), which is auriferous, as well as in quartz veins, both in the granite traversing the schists in its vicinity.

Where, as in the neighbourhood of Valparaiso, this granite is found decomposed *in situ* to great depths (100 feet and more), the streams cutting through it are found to be rich in gold, as at Quilpue, Passa-Honda, Casa Blanca, Catapilco, &c.; at the last mentioned place the quartz veins in the granite are also worked for gold, whilst other quartz veins in the neighbouring rocks are reported to be very rich, in Araucania, Valdivia, &c.

The great mineral wealth of Chile, however, has its source in the metallic lodes of a far later period, which traverse the Liassic and Cretaceo-oolitic strata, which, being composed in great part of porphyrite tuffs, breccias, and interstratified porphyrites, have been called the metalliferous porphyries of Chile—a name which conveys an erroneous impression, since these rocks, when in their normal state, are never metalliferous, and only become so at those points (as seen on Map) where they have been disturbed by the eruptions of diorites (composed of hornblende with felspar).

The diorites themselves frequently, as at Carrisal, Tres Puntas, Ojancos, Tambillos, &c., contain the metallic lodes, but more frequently the lodes are found traversing the stratified rocks in their immediate vicinity.   Thus we have—

The gold veins of Chanaral, Mineral del Inca, Copiapo, Andacollo, El Toro, Tiltil, &c.

The silver lodes of Tres Puntas, Chanarcillo, Bandurias, San Antonio, Arqueros, Catemu, San Felipe, San Lorenzo, San Pedro Nolasco, &c.

The copper lodes of Tocopilla, Gatica, Cobija, El Cobre, Paposo, Taltal, Pan de Azucar, Chanaral, Puquios, Copiapo, Ojancos, Cerro Blanco, Carrisal, Andacollo, Tamaya, Catemu, El Volcan, San Jose, &c.

The nickel and cobalt lodes of Chanaral, Pan de Azucar, Tres Puntas, Buitre, San Simon, &c.

Other metals, such as iron, lead, zinc, antimony, arsenic, and vanadium, are also found in these lodes, often in great abun-

dance, but as yet have not become an object of practical exploration, principally from the want of means of cheap transport
and fuel. It is curious, however, that with all this abundance
of different metallic ores, the metal tin has not yet been discovered in any part of Chile.

In Chile the gold, copper, and silver have drawn the population towards districts that were formerly wild and uninhabited, and Copiapo, Huasco, and Coquimbo, on the Pacific,
owe their birth to the mines, which have also contributed to
their commercial development. Silver created the two first
of these towns, and copper the last mentioned. Even into
the very heart of the desert of Atacama, a plain of burning
sands—the Sahara of these countries, which separates Chile
from Bolivia—have the gold-seekers advanced. These hardy
cateadores * now look back with regret to the legendary times
when the diggings produced nuggets so large that the gold
was weighed by the Roman steel-yard.

On the table-lands, where the miners often wander, and
where many deposits of ore still remain partially unknown, the
greater part of the workings have for many years yielded the
richest results, and there have besides been extracted, and
are being still extracted, to the great delight of the learned, a
whole series of new species in the family of minerals.

There are famous mines in Bolivia, on the highest summits
of the Andes; and near Chuquisaca,† or La Plata, is Potosi,
which has become proverbial, and which, from the period of
conquest to that of emancipation—that is to say, for two centuries—has furnished in massive silver more than six milliards
of francs to Spain (£240,000,000). In Bolivia the depths of
the Andes contain throughout the white metal, whence the
name of Plata, or silver, which the Spaniards gave to this

* From the Spanish *catear*, "to seek."

† From the terms *choque saca*, in the language of the native Indians meaning "bridge of gold,"
in consequence of the immense treasures which were carried across the river at this point on the way
to Cuzco, the town of the Incas.

Fig. 104.—Auriferous Placers of the Tipuani Valley in Bolivia   After a drawing in water-colours by Deherrypon.

province and its capital, and which they retained until 1825, when they gave to them, after the great General Bolivar, the heroic liberator, the name which they now bear. The neighbouring republic (the Argentine Confederation), of which the capital is Buenos Ayres, and which, like Bolivia, is separated from the viceroyalty of Peru, also contains silver-mines, which we shall revert to during our circumnavigation of the two Americas.

The greater part of the mines of Bolivia, which were worked in old times with such industry, are now abandoned, and the same fact is to be deplored with regard to Peru, Central America, and Mexico. The depth of the works, the influx of subterranean water, and the frequency of domestic conflicts, have combined to cause many of the mines of Spanish America to be abandoned. This evil has existed from the beginning of the present century, that is, from the period of the commencement of the wars of Independence. The impoverishment of the strata on descending to great depths has sometimes been assigned as the cause of these successive abandonments, but such has very seldom really been the case.

Bolivia not only contains silver, but also gold, which exists in numerous veins, or *veneros*, in the valley of Tipuani, one of the tributary streams of the Amazon (fig. 104). Tin is also found in abundance in the alluvial sands; and copper, in grains, nodules, and plates, in the red sandstone, at Corocoro, near La Paz. The mineral produced from this mine, Barilla, as it is called, is well known to the smelters, and the Bolivian metal is appreciated in all the markets. It has sometimes been found mingled with native silver.

Next in order is Peru, the cherished land of the Incas, opening up its treasures; in the south is Huantajayo, with its ancient silver-mines; and at no great distance, Iquique, where the workings of the natural saltpetre beds have afforded a livelihood to the inhabitants of the country after the metal

mines were abandoned.   In the centre is Huancavelica, where
rich mines of mercury have long been worked.   Quicksilver is
indispensable in the treatment of the precious metals,* and
nature appears, in more than one instance, to have designedly
placed mines of mercury side by side with those of gold and
silver.

In the latitude of Lima, the city of kings, and the ancient
capital of the children of the sun, stands Cerro de Pasco, the
country of silver-mines, on an elevated table-land of the Andes,
at a height of above 14,000 feet above the level of the sea.
These mines have been not less actively worked than those of
Mexico and Bolivia, and continue to be worked at the present
time.   But the civil wars in Peru have been as much fraught
with evil as those in other Spanish Republics, and Cerro de
Pasco is far from presenting the animation of the past; some
of its mines are inundated, and only the refuse (*los desmontes*)
are now worked.   This refuse, merely spotted here and there
with metallic particles, which the ancients did not deign to
collect, but threw aside with other matter altogether sterile, is
now carefully sorted.   The English, who are always to be
found where there is a field for energy and enterprise, went to
these mines with their apparatus and machines, and have succeeded in restoring many of the workings to their former
vigour and splendour.

The journey from Lima to Pasco is long and painful; the
traveller must cross the Andes on foot or mounted on a mule, or
perhaps be carried by Indians, and thus reach heights where
the rarefaction of the air renders breathing very difficult.†

---

* The metallurgical treatment to which ores of gold and silver are subjected almost always consists in the formation of an amalgam, that is, a solid combination of mercury with the precious metals.
The separation of the mercury is subsequently effected by distillation in a retort; the precious metal
remaining in a cake at the bottom of the apparatus.

† The particular inconvenience which is then experienced is known, throughout Spanish America,
by the name of *soroche*. The Indians attribute it to the emanations from veins of antimony across
the rocks of the Andes, but it is not necessary to search so far for the cause of suffering; it is the same
physiological phenomenon as that to which one is subject in a balloon-ascent, on rising to a height of
from 15,000 to 18,000 feet.

Storms are frequent and violent. The journey lasts for several days; platforms of ice are traversed; the traveller sleeps in the *tambos*, a sort of common shelter or caravanserai, which date from the time of the Incas. At Cerro de Pasco, near the equator (about 11° south latitude), woollen clothing is indispensable, the mean annual temperature being from 6° to 8° Cent. (43° to 46½° Fahr.); the thermometer rises at times to 12° Cent. (53½° Fahr.). As on the highest summits of the Alps, the vegetation is stunted; the orange and date trees, coffee plants, bananas, indeed all the luxuriant vegetation of the tropics, covered with leaves, flowers, and fruits, which adorn the gardens around Pisco and Lima, have quite disappeared before Cerro; and a few scanty grasses are the only natural plants of these elevated spots.* On the horizon appear the last peaks of the Andes, covered with glaciers, and clothed in their mantle of eternal snow (fig. 105).

Descending to the shores of the Pacific, where the climate becomes more genial, the Chincha Isles may be seen, rich in guano, but not inexhaustible. The marine birds—cormorants, pelicans, and penguins—which gorge themselves with fish, still contribute to the formation of this precious manure. Nevertheless Peru, which draws from the working of these remarkable deposits the greater part of her wealth in silver, will do well to consider other means of revenue for the future. Meanwhile it is always guano, and at Iquique saltpetre, that merchant ships go in quest of in these waters. The English steamers, which carry on a regular traffic between the last ports of Chile and Panama, load in passing several bags of sugar at Pisco; and at Callao, the port of Lima, recently bombarded by Spain, ingots of silver, *la plata fina*, which is its Castilian name. They are stowed carefully away, one by one, in the hold of the vessel, in presence of the curious

* The town contains 15,000 inhabitants, almost all miners and smelters. They use for fuel the dung of the llama (the camel of Western America), clumps of knotted turf, or coal extracted from neighbouring mines. The silver-foundries consume, almost exclusively the fossil combustible—coal.

passengers. Some of the ingots are inscribed with the name
of the place they come from, the name of a factory, or simply
initials.   The colour is a greyish-white, dull, and dirty ; the
texture granular and porous ; but they cannot be mistaken,
and the extreme care which is bestowed on the ingots, and
the precision with which the purser of the vessel enters
them in his register, would be sufficient, in case of need, to
testify to their value.   When the lading is finished; the
voyagers with their precious metal continue their route
towards the Isthmus, and set sail for England, the great
centre of commerce for the colonial world.

Bolivia and Chile, like Peru, send their precious metals to
Southampton, Liverpool, and London, by the rapid route of
Panama.   Copper goes by way of Cape Horn.   Being less
valuable than silver, it is precluded from the more expensive
modes of transit, and therefore travels by sailing vessels instead
of by steamer ; and as only the smelters expect it, and not
financiers or bankers, it is of no great importance whether it
arrive a little sooner or later —none but the smelters are affected
by it.   But if silver fail, on the contrary, a crisis will often
be the result ; and if, on the other hand, it arrive in time, it
produces a corresponding animation in commercial affairs.

From Callao to Panama the route approaches the equator
and the beautiful river Guayaquil, until New Granada
(recently united to the Republic of Ecuador with that of
Venezuela) is reached.   The name of Colombia, which was
borne by the Confederation of the three States, besides the
advantage of uniting them in brotherly union, also possessed
that of recalling the name of the great discoverer of America.
There are no remarkable mines in the United States of
Colombia, except the platinum-mine at Choco, on the Pacific,
near the port of Buenaventura (New Granada).   There was a
time when the washings (*lavaderos*) of this country yielded gold
also in abundance ; and in the province of Panama auriferous

Fig. 105.—Silver-mines of Cerro de Pasco, Peru.

had been done in the sixteenth century, and which had varied
had been a ground of contention... for...in at two...
In 1858 a...

mines, formerly worked by the Indians, may everywhere be met with. These Indians even wrought the metal as gold-smiths and jewellers. At the present time these mines are in the hands of a few bold adventurers; the deposits are chiefly spread around the Chiriqui, on the borders of Panama and the state of Costa Rica (Central America); and in the latter region the States of Nicaragua and Honduras present interesting mines of gold and silver, which once attracted the Spaniards from their conquests, and which have recently arrested the attention of powerful European companies.

In the mountains of Mexico, running from south-east to north-west, metals lie hidden; the States of Guanaxuato, Guad-alaxara, and Zacatecas being the richest in silver of the whole Empire, whose great capitals were created by means of that metal. The celebrated mines of Valenciana, Real-del-Monte, Pachuca, Real-del-Oro, and many others, have furnished thousands of millions. At Guanaxuato the mother-vein (*veta-madre*) occurs; at Zacatecas, the great-vein (*veta-grande*)—two of the largest metallic veins in the world, which can be traced over an extent of four leagues. They are like immense walls of quartz, from 22 to 60 feet thick at the surface, and from 120 to 240 feet deep, bounded by limestone, porphyry, and schists, and form the matrix of numerous minerals. Mexico, in spite of its continual troubles, of all the Spanish republics has always produced the greatest quantity of precious metals; mercury, tin, iron, lead, copper, graphite, and coal abound no less in this vast country, whose favourable position and the riches of whose soil induced Cortez to bestow on it the name of New Spain.

The Valenciana mine, on the Veta-Madre of Guanaxuato, was opened in 1760 on a part of the vein on which some work had been done in the sixteenth century, and which afterwards had been neglected as unpromising for almost two centuries. In 1768 a rich "bonanza" was struck at a depth of 240 feet,

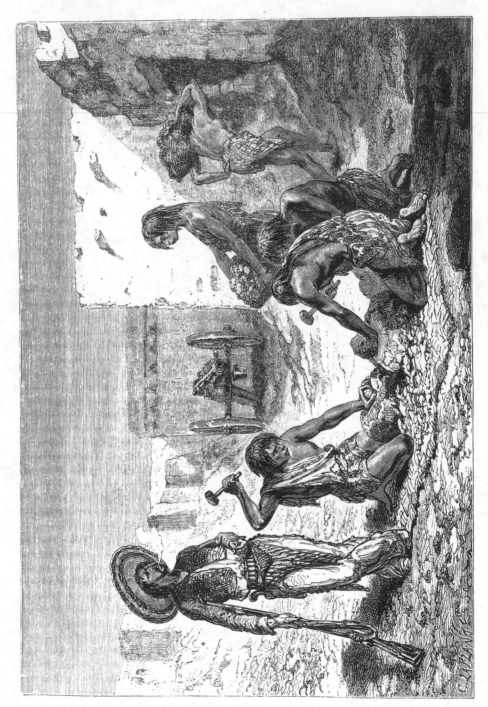

Fig. 106.—Apache Prisoners condemned to the Mines of Chihuahua, Mexico.

from which a million and a half of dollars were annually extracted. Still from 1788 to 1810 the yearly produce averaged 1,383,195 dollars. A town of 7000 inhabitants was built near the mine, which furnished employment for 3100 people. The rich ore was found only to extend to the depth of 1200 feet, below which it was then too poor for extracting. In 1825 the Anglo-Mexican Company, at a considerable outlay, freed the mine of water, with which it had been filled for fifteen years; but the ores yielded no profit in consequence of the cost of draining the mine. It is now owned by the United Mexican Company, and for a number of years has yielded immensely by the large amounts of poor ores which fill the vein.*

On the northern frontier of the Mexican State is the Province of Chihuahua, which is also rich in silver, but where the greatest disorder prevails. The Apaches and Comanches, both savage tribes, at enmity with each other, mingle their devastations with those of banditti and guerillas; and under such conditions as these industry can prosper but little, although the miners constantly return to the working of the mines, which are so rich as to yield large profits. The placers themselves abound; and the Indians, in default of lead, grind the nuggets on a stone to make them round, and convert them into bullets and small shot, with which they charge their carbines. Indians and Mexicans are continually at war with each other, and the Mexicans, who are almost always victorious, condemn the prisoners to the mines, or at least to break up the ores (fig. 106).

Between Chihuahua and the Gulf of California is situated Sonora, where are found gold, silver, and mercury, but which is also infested in the interior with the cruel Apaches, who scalp the unfortunate miners without mercy. Sonora is the country which drew the attention of the brave Raousset

* " The Comstock Lode," p 64.

Boulbon, in his adventurous expeditions, when he endeavoured to oppose a French barrier to the Anglo-Saxon invasion of the Yankees. He had heard of the existence of mines of solid silver, and that they were so abundant that the Indians made balls of this metal, as well as of the auriferous nuggets; that mercury was to be found everywhere in lakes; and that the streams ran over pebbles of gold. What attractions for future colonists! He, who thought only of the conquest and civilization of the country, advanced boldly and confidently; but after military successes and adventures of a most romantic description, he was taken sword in hand. His idea, which an exalted personage has since attempted to carry out, was not then approved in Mexico; he was condemned to be shot, and died like a hero.

To the north of Sonora extend California, Oregon, the territory of Washington, and British Columbia. All these countries form one vast domain of diggings, and are the chosen regions of gold-seekers. Since the time of Cortez, who discovered Lower California, and whose ships ploughed the waters of the Gulf of California, the Spaniards have given the name *Eldorado*, or the golden country, to these remote shores of the Pacific. The name originated with the Aztecs, as if they foresaw what these countries would become in time. In 1848, when the United States had obtained, by a stroke of the pen, from the republic of Mexico—which was then incapable of colonizing its more distant provinces—Upper California, New Mexico, and Texas, gold was suddenly found abundantly in California, on an affluent of the Sacramento. It is well known what followed, and how the whole world, generously invited to profit by this great discovery, sent to the land of gold a crowd of hardy emigrants.

The commencement of colonization was most tumultuous and difficult, for it was not always honest workers who hurried thither. Escaped convicts from Australia, lawless squatters from the American states of the Far West, *rowdies* from New York, and, in fact, the scum of the whole earth. spread alarm everywhere. The revolver and knife generally settled all disputes, and it was found necessary to establish committees of vigilance and Lynch law as a safeguard against assassins, robbers, and incendiaries; by this means, the bandits, struck with terror, took flight, and the real hard labour of the mines became a wholesome discipline for all. In this manner the goldfields were quickly transformed into model countries, and the cultivation of the fields and forests was developed under the happiest auspices, and simultaneously with the underground operations.

All along the Pacific coast of California are harbours, frequented by merchant-ships from all parts of the world, and even by whaling-vessels; the coast-line is almost straight. Towards the middle of this line the Bay of San Francisco opens on the view. So vast is it that all the fleets in the world might anchor there with ease; it communicates with the sea by a narrow entrance-passage, called poetically the Golden Gate. The littoral provinces, besides fishing and commerce, pursue agriculture, while the southern provinces produce olives equal to those of Andalusia, and wines similar to those of Champagne and Bordeaux. In the north wheat and maize are cultivated, and the land everywhere repays with interest the labours of the colonists, the virgin soil yielding fruit a hundredfold. Garden-produce is also most abundant and varied, the vegetables grown in the United States attaining here an extraordinary size; and, lastly, the forests contain treasures of their own, especially the beautiful red fir (*Pinus Lambertinus*), so much esteemed in the whole Pacific, which attains a height of 200 feet, with a circumference of 57 feet, and yields a pure amber-coloured turpentine.

A chain of mountains called the Coast-Range, taking a direction parallel to the coast, runs through the farming-provinces;

the mining-districts are traversed by the Sacramento and the
San Joaquin, two rivers which present the singular phenomenon
of having opposite courses and the same mouth. The Sacramento flows from the north, and the San Joaquin, which ultimately falls into it, from the south. These rivers, with their
affluents, rising in the sides of the Sierra Nevada, which bounds
the northern portion of the state of California on the east,
might truly be named Pactolus, without any figure of speech.
It was on American River, one of the affluents of the Sacramento, that gold was discovered in 1848 (Map XII.). The
first nugget was found at Coloma, at the saw-mill of Captain
Sutter, a Swiss colonist, formerly captain of a Swiss regiment
in the service of Charles X. He had emigrated to the United
States from France in 1830, and finally settled in California,
on the banks of the Sacramento, where the town of that name
now stands.

Gold-mining in this country of Eldorado has resulted in a
colonization so perfect, that more than one great European
state may find somewhat to envy in its condition. The
precious mineral is everywhere distributed in the valleys of the
Sacramento and the San Joaquin, and the veins of gold-quartz
are disseminated through all the spurs of the Sierra. The
gold which has been extracted from these deposits, circulating
through the country, has facilitated the cultivation of land on
a large scale, and led to the opening of roads, railways, canals,
and telegraphs. Canal-making, for irrigating the country and
supplying the diggings with water, has been carried out to an
unprecedented extent. Towns have risen out of the ground,
as it were, by enchantment; what were at first merely miners'
camps are now opulent cities; and thus California exists, far
from the ken of Europe, which only sees in this remote region
the country so terribly agitated by the first rush of immigration.

Favoured beyond measure with respect to the variety of
its subterranean riches, as well as of its agricultural produce,

MAP
OF
CALIFORNIA AND NEVADA.

*Shewing the position of the
Principal Mineral deposits.*

EXPLANATION.

Coal.   Gold.

Copper. Silver.

Mercury.

0   25   50   100   150
Scale of Miles.

California does not yield gold alone; when the miners, quitting the diggings which were worked out, began to organize their investigations, they were not long in discovering coal, mercury, and copper. The last named metal exists in such abundant quantities from north to south of the state that the mines, opened in 1862, already afford very profitable results. In the vicinity of the richest mines in the state of California, a town has been founded which has received the appropriate name of Copperopolis, or the city of copper. Coal, the existence of which was dimly suspected in 1861, has also furnished its contingent; and, lastly, mercury at New Idria and New Almaden, especially the latter, has equally astonished the colonists. Whilst, as a rule, quicksilver-ore is poor, and contains at most from ten to twelve per cent. of mercury, here the average contents are higher; the price of the metal has fallen from twelve to six francs the kilogramme, and several European mines, especially those of Tuscany, have been altogether abandoned. The mercury of New Almaden has driven that of Almaden in Spain from all the American markets. On the settlement by the supreme court of Washington of the famous lawsuit which had been carried on for many years, mining operations were soon resumed, and some rival mines—those of New Idria, Enriquetta, &c.—were opened simultaneously with those of New Almaden, without any of these enterprises militating against the others. This is, at the present time, the most important source of mercury in the world. The time is long gone by when the Indians, opening narrow galleries in the schistose rocks of San José, penetrated into the veins of cinnabar, in order to extract therefrom a little red earth to paint their faces with. It was they who, even before the discovery of gold, conducted the first Mexican colonists to these precious deposits.

The quantity of quicksilver produced in 1866 at New Almaden, up to the month of October, amounted to 30,029 flasks, of 76½ lbs. of mercury to the flask, or 2,297,070 lbs. The quantity of ore mined and reduced in 1865 was 31,948,400 lbs., or about 16,000 tons, and the general average of all the ore reduced, allowing 3 per cent. of dirt (*tierras*), was 12·43 per cent. The average per centage of 1864 was 16·40 per cent., and for the ten years preceding was 22·20 per cent.[*]

Separated from California by the snowy chain of the Sierra, the State of Nevada has been celebrated, since 1860, for its silver-mining. In November, 1859, the news of the discovery of silver-mines near Lake Washoe was confirmed at San Francisco; and in June, 1860, the mines of Washoe, the central western portion of the State, had already sent such rich results to Europe, that the French Ministers of Finance and Commerce despatched a mining engineer to Nevada to make a close inspection of these wonderful mines. It seemed as if the world were about to be inundated with silver, as it had been by gold ten years previously; and what would those economists now say who had only recently counselled that the value of gold and silver coin, should be lowered in order to restore the balance between the standard metals? Whilst the French engineer visited Nevada and prepared his report, the miners of Washoe continued working their veins of metal. At the present time the mines on the eastern slope of the Sierra Nevada annually produce about 16,000,000 dollars of silver, chiefly from the Comstock lode; the total yield of gold from the quartz mines of California probably not exceeding 8,000,000 or 9,000,000 dollars per annum.

The Comstock lode, in the State of Nevada in California, may be ranked amongst the richest and most productive metalliferous deposits ever encountered in the history of mining enterprise; its productive capacity, as now being developed, surpassing, if the mass of its ores do not in richness equal, those of the most famous mines of Mexico and Peru.[†]   Its total produce has been, from 1862 to 1865 inclusive, about 48,000,000

Report upon the Mineral Resources of the United States. 1867.   † Ibid.

MAP XIII

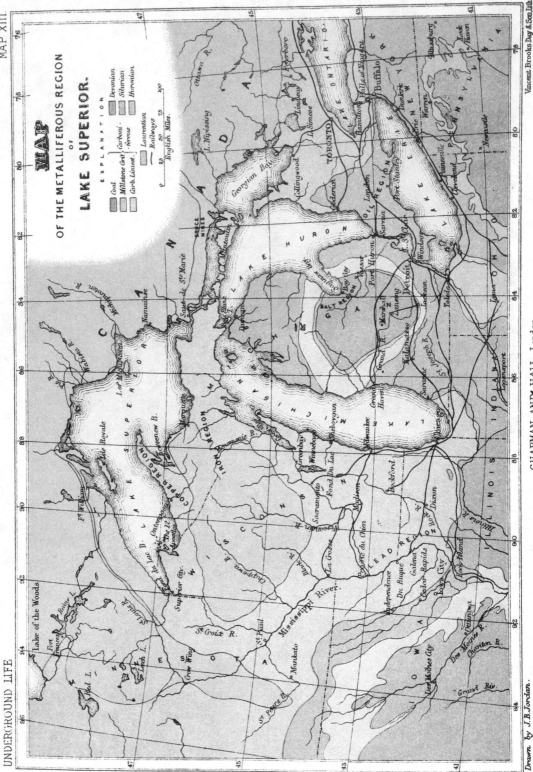

MAP

OF THE METALLIFEROUS REGION

OF

LAKE SUPERIOR.

EXPLANATION

	Devonian
Coal	Carboni-ferous	Silurian
Millstone Grit		
Carb. Limest.		Huronian.

Laurentian
Railways

0   25   50   75   100
English Miles.

CHAPMAN AND HALL. London.

Drawn by J.B.Jordan.   From Map in Sir W. Logan's Geological Report on Canada. Compiled from various sources.

Vincent Brooks Day & Son lith.

dollars, of which about 32,750,000 dollars were silver and 15,250,000 gold : the production for the year 1866 was little less than 15,675,000 dollars.

In 1854 the total amount of silver produced was, according to Professor Whitney, 47,443,200 dollars, of which sum 7,864,000 came from European, and 39,451,200 from American mines. The produce of silver of the Comstock vein in the last three years was about twenty-three per cent. of the entire amount furnished by all the silver-mines in the world. It exceeds the aggregate produce of all European countries, and equals that of the entire western coast of South America, which in 1854 was 11,099,200 dollars ; the only country to which it is inferior being Mexico.

"This extraordinary productiveness has made the Washoe region more famous for its mineral wealth than many places where silver-ores have been mined and extracted for centuries. It has attracted a numerous civilized population to a country which before was sparsely inhabited by wandering Indian tribes, and which, by its desert character, seemed to debar for ever human industry and arts ; it has built cities in this desert, and roads across high mountain ranges, and at the present time accelerates the connection by steam of the Atlantic and Pacific coasts of America ; it has created a new branch of mining in the Pacific States, which, through the discovery of the Comstock vein, has assumed gigantic proportions all over the great basin ; it has given successful employment to large amounts of capital, and rescued the trade of California from imminent decline at a time when the placer-mines were rapidly decreasing in importance."[*]

"The principal ores of the Comstock lode are—Stephanite, vitreous silver-ore, native silver, and very rich galena ; in small quantities occur—pyrargyrite, or ruby silver, horn-silver, and polybasite.[†] Besides these are found—native gold,

---

* "The Comstock Lode," by F. Baron Richthofen, Dr. Phil., p. 8.   † Ibid, p. 28.

iron pyrites, copper pyrites, zincblende, carbonate of lead, and pyromorphite (both the two last named being very scarce)."

The veins of Freiberg in Saxony, Kongsberg in Norway, Chanarcillo in Chile, Pasco in Peru, Catorce in Mexico, and Austin in Nevada, consist of a number of small fissures, which exhibit in depth nearly the same characters and richness as near the surface. Another class of silver-veins, which are prominent on account of their magnitude and unity, exhibit wherever they occur one great mother-vein, or *veta-madre*, surrounded in most instances by other smaller and less important veins. To this class belong the Comstock lode of California, the veins of Schemnitz and Felsobanya in Hungary, the Veta-Madre of Guanaxuato, and the Veta-Grande of Zacatecas, which from 1548 to 1832 yielded about 666,000,000 dollars; while the veins of Potosi in Peru, and the Biscayna of Real-del-Monte in Mexico, have to be referred more to this than to the former class. These great mother-veins, though few in number, furnish by far the greater portion of the silver produced throughout the world. (The Comstock Lode, p. 66-67.)

In enumerating mercury, copper, gold, and coal, the mineral riches of California are far from being exhausted; iron, lead, and silver have also been found, and their turn for being regularly worked will come. Meanwhile platinum, found in the alluvial sands, and with it precious stones, such as rubies and diamonds, which did not at first attract attention (there being few things so little striking as gems and metals in their native state), are now sought after everywhere. Agatised wood, opal, petroleum, borax, and sulphur, have also been discovered in large quantities. The petroleum of California corresponds to the rock oil of the States of the Atlantic; the sulphur and borax rival the analogous products of Sicily and Tuscany. It would seem as if no mineral substance were wanting in the gold country; onyx-marble (oriental alabaster) has been found there which equals that of Algeria.

Mr. Swain, superintendent of the branch mint at San Francisco, a gentleman possessing both the means and the disposition to inform himself on the subject, estimates the product of gold and silver for Oregon, California, Nevada, and Washington Territory, as follows:—

	Dollars.
In 1861,	43,391,000
In 1862,	49,370,000
In 1863,	52,500,000
In 1864,	63,450,000
In 1865,	90,000,000

Well-informed parties estimate the product for 1866 at 106,000,000 dollars, an estimate which appears exaggerated to some persons, while others pronounce it far below the actual yield.

Assuming the estimate of the product of bullion, as above given, to be approximately correct, it will be seen that the mines of the Pacific slope produce annually upwards of 100,000,000 dollars of the precious metals—a quantity more than four times as great as the total product of the world less than thirty years ago. The gold and silver product of the Pacific States and Territories for the year 1866 is therefore nearly double in amount all the combined bullion in the national treasury of the United States, and all the banks in the country, the total amount of which is estimated at 69,700,000 dollars.*

After having stated such astounding results of mineral industry as had never before been known, let us consider the Rocky Mountains, to complete the inventory of the subterranean riches of the United States between the basin of the Mississippi and the Atlantic. In these localities iron-ore, associated with coal, has contributed to the development of the greater part of the States, especially those of Pennsylvania, Ohio, Kentucky, Tennessee, Virginia, and Michigan. Pittsburg, in

* Report upon the Mineral Resources of the United States.

the State of Pennsylvania, is the great town for coal and iron, and copper-foundries, being often called the Birmingham of the United States.   In the State of Michigan, on the borders of Lake Superior, is Marquette (Map VIII.), where a magnetic iron-ore has been found during the last six years, analogous to that of Sweden and the Isle of Elba, and of which it already furnishes annually three hundred thousand tons—treble the produce of the Isle of Elba.   There are in that place moun-tains full of immense deposits of iron.

The state of New York owes equally a part of its fortune to iron.   Its capital has been proudly christened the Imperial City.   It is the most populous town of the United States, and contains, with its annex Brooklyn, nearly a million and a half inhabitants; it is the first American port in the Atlantic, but is also the great mart for all the United States for iron, without which article there could be no machines, and con-sequently no great industrial countries.   In other countries of the United States, around the heights of the Mississippi, the Missouri, Illinois, Iowa, and Wisconsin, lead is found.   Galena, the centre of important mines and enormous foundries, owes its name to the mineral which it exports—galena, or sulphuret of lead.   These deposits were comprised in the concessions of the famous bank of Law.   Equally rich as inexhaustible, twenty years ago they caused even the Spanish exports them-selves to tremble for a moment.

The copper discovered in different parts of the immense republic, but particularly in the neighbourhood of the river south of Lake Superior (Map VIII.), also deserves mention. From time immemorial these beds had been worked by the Indians, and were discovered, as early as the sixteenth century, by the Jesuit missionaries.   Sixteen years ago, when the Americans returned to them again with activity, the very qualities of the native metal, malleability and elasticity, were the cause of serious difficulties in its being worked: for the copper-ore of Lake Superior almost always occurs in masses, sometimes of enormous size, mingled with white veins of pure silver. These masses of native metal have to be sawed or cut with a cold chisel, which is a slow and laborious task, in consequence of their extreme toughness. It is stated by Dr. Percy, on the authority of Mr. Petherick (the well-known mining engineer), that in 1854 not fewer than forty men were engaged at Minnesota during twelve months in cutting up a single mass of native copper weighing about 500 tons. (Percy's " Metallurgy," vol. i. p. 309.)

The value of copper has since then fallen one quarter, and it has never again attained its former price. Pittsburgh has concentrated in vast works the fusion of the greater part of the minerals of Lake Superior; an example which has been followed by Boston. In the neighbourhood of the capital of Massachusetts foundries are erected, which are likewise supplied by the minerals of Chile and Canada; and a new Swansea, a thousand leagues from the former, is formed on the Atlantic.

Shall I now speak of other minerals also worked out in the United States—those of zinc, nickel, and gold? Before the discovery of the Californian deposits, some parts of the United States were renowned for the exportation of the precious metal. The two Carolinas were then foremost in this industrial work, but have since fallen to the lowest rank; who, among the gold-seekers of the world, will think of employing them as long as the mines of California and Australia are unexhausted? The zinc and nickel mines are, however, more flourishing, and might even be made capable of supplying the whole of Europe with those metals.

In Nova Scotia, between Cape Sable and Canseau, there is a gold-bearing country over 250 miles long and fully 25 miles wide, in every part of which as productive mines may be found as any now worked.

According to the report of the chief commissioner of mines

for the province for the year 1867, the yield of gold in the year ending December 31, 1862, was 6737 ounces; in 1866 it was 24,162 ounces; and for the twelve months ending September 30, 1867, it was 27,583 ounces. The crushing-mills employed at the latter date were thirty-five, of which twenty-seven were worked by steam-power and eight by water; the quantity of quartz crushed was 30,673 tons, and yielding on an average 17 dwts. 23 grains of gold per ton, the maximum yield being 26 ounces, 13 dwts., 8 grains per ton; the total yield of gold during the twelve months was 27,583 ounces, 6 dwts., and 9 grains, of which 49 ounces were obtained from alluvial mines. The average remuneration in the gold-mines for each man, counting 313 days in the year, and the gold to be worth 18½ dollars per ounce, was 2 dollars and 45 cents per day, or 765 dollars per year, a result probably without a parallel in any country.

Canada, brother of the United States, at least as regards its political constitution and its contiguity, has been, equally with its powerful neighbours, gifted by nature with mineral riches. Copper and gold are especially ranked among its exports, elevation of latitude not acting as a cause for complete mineral sterility. Were we not so much terrified by the degree of cold at the poles, we should find metals as far north as Greenland, where ores of copper, lead, and recently of aluminium, have been dug up by courageous workers.

Descending the Atlantic, that sea which, like a great river, separates the Old World from the New, and bearing towards the American shore—the traveller will arrive at Cuba, the largest of the Antilles. This island is not only celebrated for its Havannah tobacco, grown in the district of *la Vuelta Abajo*, and considered amongst the mildest and finest-flavoured made anywhere; and for its fine sugar and coffee plantations, and rum-distilleries; it is also famous for its copper-mines. In the neighbourhood of Santiago (which is the capital of the Island, while Havannah is the principal trading-port) are some green rocks, and it is in these that the copper-ore abounds. The mines, discovered many years ago, immediately attracted the attention of England; they are very rich, yield abundantly, and resemble, in the geological conditions of their formation, the mines of Tuscany and California.

French Guiana may be called the country of mines. Apronaque, with its metallic sands, restored the French colony (ruined in 1848 by the emancipation of the slaves), and to a great extent reclaimed the convicts, in the same way that the mines of California reclaimed some of the Australian convicts.

It is very difficult to assign a definite frontier to Guiana, to separate it from Brazil, as the boundary has been in litigation for ages. The provinces and localities bear appropriate names, in favour with the miners; such as *Minas-Geraes*, the universal mines, where precious stones are found; *Ouro-preto*, the town of black gold; then there are the chains of Diamantina and Esmeraldas. But if Brazil is the country of diamonds, topazes, and gold, it has also the advantage of possessing coal and iron. The first-named minerals could not give to this vast empire all the development of which it is capable; the last-mentioned, whose importance is fully felt and acknowledged in the present day, does not exist in sufficient quantity to achieve that result. The only localities on the eastern coast of South America, between the river Amazon and the river Plate, where the existence of coal has actually been determined, are the two southern Brazilian provinces of S. Pedro do Rio Grande do Sul and Santa Catherine, and the neighbouring republic of Bando Oriental or Uruguay. The friends of the Latin races in South America ought now to be well pleased. Hitherto they have only had to applaud, in all this immense continent, the marked but intermittent efforts of two noble states, Brazil and Chile; it is well known what the others are.

2 A

In the provinces which have yet to be considered, Paraguay, Uruguay, and the Argentine Republic, the same spectacle is presented as in Peru, Bolivia, Columbia, Central America, and Mexico; namely, that of countries favoured by Heaven but ruined by man, where disorder reigns, and perpetual dissensions take place.   In Paraguay, the overweening power of the Jesuits, and the strange teachings of Doctor Francia and General Lopez, are subversive of progress; while Uruguay, by its savage rivalry with La Plata, only tears itself to pieces. There is no longer any industry there; the Argentine Republic alone shows traces of underground labours—the name it now bears is that of La Plata, given to it by the Spaniards in consequence of the discovery of silver.   At the foot of the Andes, towards Mendoza, and separated from Chile by the American Alps, may certainly be found some ancient mines, some of which were re-opened a few years ago.   Those of Uspallata are the most celebrated.   In some places the slags are supposed to date back to the times of the Incas, and it is said that some of the veins worked by them have never been regained by the Spaniards.

The children of the Sun had extended the sceptre of their rule even to these distant regions, which formed the limits of their magnificent empire.   When Pizarro conquered Peru he found the country civilized, as Cortez had already found to be the case with Mexico.   The Peruvians and Aztecs were civilized races, practising many branches of industry; that of working in the precious metals being familiar to them.   They understood the art of smelting silver-ore by combining it with lead, and of purifying the precious metal.   They heated their furnaces by means of natural currents of air.   They kindled fires on the summit of the Andes, and the magic spectacle presented by these furnaces when at work by night, struck the conquerors with astonishment the first time they witnessed it. The final process of refining the silver was performed at home

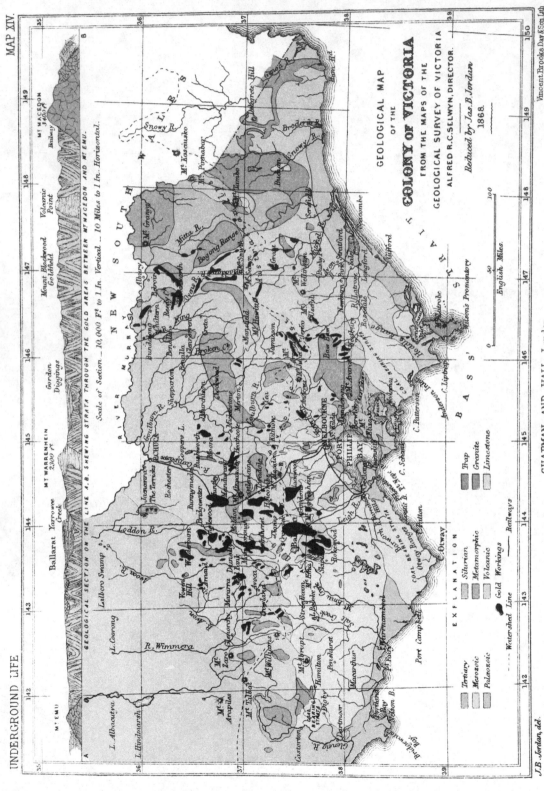

MAP XIV.

GEOLOGICAL MAP
OF THE
COLONY OF VICTORIA
FROM THE MAPS OF THE
GEOLOGICAL SURVEY OF VICTORIA
ALFRED R.C. SELWYN, DIRECTOR.

Reduced by Jas. B. Jordan
1868.

EXPLANATION

	Silurian
	Metamorphic
	Volcanic
	Gold Workings

- - - - Watershed Line
——— Railways

	Tertiary
	Mesozoic
	Palaeozoic

	Trap
	Granite
	Limestone

English Miles.

Vincent Brooks Day & Son Lith.

J.B. Jordan, del.

CHAPMAN AND HALL, London

in a large crucible, a number of persons grouping themselves around, each furnished with an air-cane, with which they blew the air into the furnace to oxidize the lead.

In Peru the people of the country were not only miners and smelters, but they also cultivated the soil. A well-irrigated country, with open roads everywhere, extended on either side of the Andes; and in these favoured regions, which possess light, heat, and water—all that is needed to make the earth productive—the sun rewards the labourer with interest; but still the intelligent worker spared no pains. The fertilizing properties of the guano, only lately discovered by ourselves, had not escaped the observation of the Incas, and the production* and extraction of this singular manure were regulated by the most stringent laws.

Let us now conduct the reader through the Straits of Magellan towards Australia. Here are the Blue Mountains and the Australian Alps, which run from Melbourne to Sydney; and this is another gold-country—New South Wales, the province of Victoria (Map XIV.). Bathurst and the river Macquarie, Ballarat and Bendigo, at the foot of Mount Alexander, are spots well loved by the miner. Since the discovery of gold in 1851, the Australian mines of Victoria have shown themselves as inexhaustible as their elders of California, and produce annually nearly thirteen millions sterling. The analogy between the two countries may be carried further, for, like California, Australia is not only favoured in the production of gold, but also in that of other mineral substances, of which copper is one. Even before this metal was found in the infant State of the Pacific, Australia displayed its malachite riches. The working of this mineral preceded that of gold; and every one has seen, in London and Paris, at the international exhibitions which the first great capital inaugurated

---

* It was forbidden, on pain of death, to kill the marine birds, which were constantly at work, after surfeiting on fish, in the formation of our modern guano.

in 1851, the splendid specimens of Australian copper. The Burra-Burra mine in South Australia, not far from Adelaide and the river Murray, vies with the most celebrated, and its produce is highly esteemed by the smelters of Swansea.

Lead, silver, iron, and coal are not wanting in the Australian soil, which, as a last point of resemblance to California, has experienced the development of its agricultural resources, thanks to the gold-workings; and now, should the nuggets fail, the colonization is accomplished. Wool, wood, and numerous other native products, would suffice to maintain life in the littoral provinces of this great continent, the interior of which alone remains to be cleared. Towns like Melbourne, Sydney, Victoria, and Adelaide—sisters of the Californian cities of San Francisco, Stockton, Sacramento, Marysville, and some older than these—can now, for the most part, compete with the largest commercial capitals in the world. The queens of the Pacific are on a par with those of the Atlantic, and the discovery of gold, in conjunction with the working of the mines, has effected these wonders.

Having given some account of the mineral riches of Australia, those of Van Diemen's Land, New Zealand, New Guinea, and the Sunda Isles next come under notice. The tin-mines, or rather tin-streams (washings) of Banca and Billaton are as remarkable and as productive as any in the world. From Malacca is obtained the Straits' tin, as it is called in trade. These colonies have made the fortune of persevering and industrious Holland;* and the soft metal has contributed to it in a great measure, the ore (tin-stone) occurring as stream-tin in valleys, and on plains, which are generally along the sides of the valleys. The rich island of Banca gives annually a clear profit of from three hundred and fifty to four hundred thousand pounds sterling to the country.†   Then Borneo, and

---

* The port of Amsterdam at this moment regulates the price of tin in all the markets of the world.

† "Banca, and its Tin Stream-works," by P. Van Diest, translated from the Dutch by Dr. C. Le Neve Foster, F.G.S.

the Spanish Philippines, with their auriferous deposits, might well allure patient workers, such as the Chinese coolies, before California and Australia had attracted all the seekers.

As yet no mention has been made of Africa. If it be the country of disinherited races, descendants of Ham, it must, at least, be acknowledged that nature has endowed it as lavishly as other countries with veins of rich ores. The geographer Maltebrun has observed with truth, that if the sea rose to a higher level round Africa, or if that continent were to sink somewhat beneath the sea, the water would preserve the same contours as at present. We should still have the Africa of our maps, only reduced in size, geometrically, so to speak. What does this remark of our sagacious author imply? That a continuous chain of mountains surrounds Africa. There is then a probability, according to the now received theories of geology, of meeting with ores everywhere not far from the African coast. And if the centre of this great continent is not yet much known, in spite of bold and enterprising travellers who have sacrificed their lives in traversing it, to explore those savage countries, miners might at least, by keeping near the coasts, be sure of finding their trouble amply rewarded.

The Cape, Congo, Gaboon, and Morocco are well known to copper-miners. At Congo, the provinces of Angola and Benguela, those gems in the crown of Portugal, contain large deposits of malachite. These deposits appear to be connected with those of the Cape and Morocco, and thus would be marked out an immense metalliferous line extending from the south to the north of Africa, along the western coast.

In Senegal the iron and gold mines have attracted by pre-ference first the natives, and then the colonists. And here it might be predicted that the most precious of all the metals would one day become a fruitful agent of immigration and progress for this sleepy and indolent colony.

In Algeria, in spite of the presence of lead, silver, copper, mercury, antimony, and iron, if not gold also, colonization is scarcely better developed than in Senegal. How is this to be accounted for? Is it because of the obstacles which the administration offer in authorizing the working of the mines? But the necessary permission is for the most part already obtained. Is it then because of incapacity on the part of the French nation for such works? But the French have proved, on their native soil, and in California to the number of more than 15,000, that they are equal to the difficulties of underground labour. Must the cause then be sought for in the sword, and a government more military than civil? The French mines in Africa are not in a flourishing condition; the workings of argentiferous lead at Gar-Rouban,* near the frontier of Morocco, and at La Calle, at the other extremity of Algeria, near the boundary of Tunis, are surrounded by blockades and small forts, to defend the miners from the incursions of Arabs and Kabyles. This incessant fear of the enemy is highly unfavourable to the development of mining-operations.

The Algerian copper is obtained from Tenez and from the too famous Mouzaïa (Plate XI., figs. 3 and 4); England receives from the last-named locality argentiferous copper-ores, mixed with arsenic, which are most profitably treated in some of the smelting works of Swansea. Near Bôna there are some magnificent iron-mines, those of Mokta el Haddid,† which bid fair to compete with those of the Isle of Elba. This deposit is one immense vein, which may be worked in the open air, explaining the name given by the Arabs to these mines. Lastly, in the province of Constantine, veins of mercury and antimony are to be met with, but they have only recently been worked with success. Oxide of antimony

---

* Gar, in Arabic, signifies "cavern," Rouban is the name of a tribe; Gar-Rouban is therefore, properly speaking, the "cavern of the Roubans." In fact, the traces of ancient excavations, dating back to the time of the Moors, may be seen on this spot, hollowed out of the ground like many other caverns.

† Literally, "the cutting or quarry of iron."

has been found there in considerable quantities, forming the upper part of the deposits. It was compact or crystallized; and fifteen years ago M. Simonin saw this ore, in both these two conditions, used in the furnaces of some metallurgical works situated near Marseilles. The manufacturer who worked the mines in Africa, and the Provençal foundry at the same time, made a rapid fortune. The Algerian mines reduced the price of antimony one-half, as those of California lowered the price of mercury, and those of Lake Superior the price of copper. A certain number of small workings, particularly in France at the foot of the Cévennes, where the peasants from time immemorial had melted the antimony themselves, came to a complete standstill, and the metal has never risen to its former value. Before 1848 it was worth about £6 per cwt., and since then it has not commanded more than between £2 and £3.

Tunis, Egypt, Abyssinia, and the western coast of Africa are not deficient in mineral treasures, the former being rich in iron, copper, lead, silver, sulphur, &c.; but Oriental customs have nothing in common with modern industry. In Egypt copper-mines are to be met with all along the chains of mountains which skirt the Upper Nile, which were worked in the time of the Pharaohs, and afterwards abandoned. How many idols and amulets have been wrought out of the bronze from these mines! The Egyptians also excavated their gold-mines, and the fact is mentioned by Herodotus;* this gold was used to fabricate the elegant ornaments which adorned the mummies, and of which they have been despoiled to enrich the museums of Europe. Abyssinia also has veins of copper, gold, iron, and silver mines. But it has already been remarked, in speaking of coal, how little King Theodore understood of the working of mines.

* Herodotus, in a famous passage which has puzzled all commentators, speaks also of some mines in India, which, according to him, were worked by *ants*.

In Mozambique there exist mines of copper, as at the Cape, and over the whole western coast of Africa; and there are also mines of tin and iron worked by the natives,* and even auriferous mines—at Sofala, for instance. Many of the older archæologists saw in this town the Ophir of the Bible, where the queen of Sheba sent her ships for gold and precious stones; but Ophir has been assigned to so many different localities that it is doubtful whether antiquarians will ever arrive at a satisfactory conclusion on the subject.

In view of the coast of Mozambique is Madagascar, where there are iron-mines which have been worked for centuries. The Hovas, the present possessors of the island, make from this ore an excellent steel, out of which they forge knives, tools, and arms, particularly the points of their lances or arrows.† The ore is generally magnetic, which accounts for its value as an iron for conversion into steel. At Tamatave, along the coast, on a bank of sand, constantly washed by the tides, the ferruginous oxide glitters in black grains, mingled with spangles of mica and quartz. That is a true iron-placer. The veins from which this iron-sand is derived cross the granitic mountains of the interior.

There may be seen also at Madagascar, lead, silver, copper, gold, graphite, petroleum, &c. In 1863, forming part of the scientific mission which the violent death of Radama II. was so soon to disperse, M. Simonin obtained some ingots of lead and copper melted by the natives. Rasoaherina Manjaka, ‡ placed on the throne by the conservative party, in the place

---

* The natives of Africa work the ore by methods which recall to mind the primitive processes still in use in Italy and Spain. In a very low furnace-hearth, often on the ground and quite open, they place the ore and the fuel. The bellows is a leathern bottle, a conical cylinder of leather, or the trunk of a tree hollowed out in the form of a pump. In each system, the bellows is double, and the movement alternating. The leathern bottle is characteristic of the pure African races, and is also to be met with in the Indies.

† The Hovas, of Malay origin, employ the pump, also peculiar to the coloured races, to blow their furnaces. The Malays have emigrated to the great African island, from a period of which tradition preserves no record.

‡ In the language of Madagascar, *Ra* signifies "nobility," *soa* means "beautiful," *herina*, "strong," *manjaka*, "king, queen, ruler."

of her murdered husband, again put in force the old laws of
the empire, which had been abrogated by Radama II.   The
first article of the Malayan code is formal : " Whosoever shall
discover, excavate, or make known gold or silver mines, will
be put to death."   Could the Hovas have guessed, when
thus transforming their code politic into a code of mines
borrowed from foreigners, that it was chiefly to the existence
of these two metals, gold and silver, that America, and more
recently California, Australia, and Nevada, owe their brilliant
and rapid colonization ?   However this may be, when the
propitious moment arrives, the first nugget will appear in the
great African island ; the news will spread through the
world instantaneously, and bold pioneers will run thither from
the four quarters of the globe, to colonize the gold-fields.
Should fate decree in that day that the whites of the great
country (the name given by the Madegasy to the French),
shall be the first to present themselves, indefeasible rights
will be again put in force, and the Island of Madagascar will
at length receive the name of Oriental France, which was
given to it by Colbert.

Our course is ended, and we may apply to the earth,
everywhere rich in metals, the salutation the poet addressed
to Italy :—

          " Salve, magna parens frugum."

Nature has enriched the earth with veins of metal from pole
to pole, and not a single country has been left unendowed ;
it might even be said that scarcely half the mineral wealth of
the globe is yet known.   But is there any apparent law
governing the distribution of these hidden treasures ?   As
regards coal, it appears to have been produced designedly in
temperate climates, where civilization sheltered itself at the
time of the discovery of coal-mines.   With metals—especially
precious metals, a contrary law appears to exist—the accumulation of these being chiefly in tropical countries.   It is not,

as the early geologists sometimes affirmed, that the motion of the earth's rotation at the beginning, when all was in chaotic fluidity, threw off from the centre to the surface the heaviest substances towards the points animated with the greatest velocity. It is known that other laws than that of centrifugal force must have presided at the birth of metallic deposits. Gold and silver were, by the design of the Creator, placed along a sort of metalliferous equator, as if to attract irresistibly civilized man to the colonization of those countries, which otherwise he would not attempt. We will leave this hypothesis to the meditation of philosophers, and of all those who delight to dive into the great secrets of nature.

CHAPTER VI.

EUREKA.

Boring.—Chance and the Savants.—Discovery of the silver-mines of Peru by a Shepherd.—Discovery of the gold-fields of the Sacramento valley in California, by Marshall.—Captain Sutter.—Isaac Humphrey, &c.—Discovery of Gold in North Carolina by Reid.—Discovery of Gold in Australia.—Remarkable forecast of Sir Roderick Murchison as to the probable occurrence of Gold in Australia.—The Rev. Mr. Clarke.—Chile; Discovery of the silver-mines of Chanarcillo by Godoy.—The Brothers Bolados.—The Irishmen of Allison-Ranch, California.—Mining operations in Tuscany.—M. Porte and the mine of Monte-Catini.

WE shall now proceed to describe the manner of discovery of metallic mines. Boring, which is of the greatest use in seeking for coal, is of little or no service in the search for metals, on account of the disposition of the veins or lodes. It has been shown how the veins are formed, and that they approach the vertical in nearly all cases. The borer who seeks them under the soil generally loses his labour, there being no continuous strata, or beds spread out horizontally, as in coal, but only disseminations, in fissures or vertical cracks. The boring implement would thus probably pass near the veins of metal without drawing forth the smallest particle, and unless it happened to light on a large mass, or a completely normal vein, which is rarely the case, the bore inserted in the ground would be to no purpose.

The countries in the midst of which metalliferous formations are distributed, repudiate altogether the idea of conducting their researches by means of boring. It has already been said that an eruptive rock is almost always associated in some way with metallic veins, which are often found placed in contact with and even running through it. Now the eruptive deposits are composed of substances not only crystalline, but also very hard; and the borer would weary himself to no purpose in attempting to penetrate granite, porphyry, and compact schists. Thus, unless it is a question of bedded deposits, as is often the case with iron-ore, the wonderful implement which finds beneath the surface artesian wells, coal, salt, petroleum, and even the gases which contain boracic acid, ought to be rigidly proscribed in the search for mines of metal.

To what, then, is the discovery of these mines due, when strongly marked, significant outcrops, extending to the surface of the soil, long failed to awaken attention? To chance partly, and partly to science. Science, yet in its infancy, has at present made comparatively few discoveries; chance has generally brought to view the metallic veins, and it thus often happens that those find who do not seek. A shepherd, a poor labourer, and sometimes even children, of which examples will be given, are chosen by nature to reveal to the world the treasures she has hidden beneath the earth. It was not Columbus, nor Cortez, nor Pizarro, who discovered the greater part of the American mines, but an Indian hunter; and the most famous silver-mines in Peru were discovered in the same way. One day a shepherd, leading his flock to feed on the slopes of the Andes, lighted some brambles to prepare his frugal repast, when a pebble, heated by the flame, attracted his notice by shining like silver. He found the stone massive and heavy, and finally carried it to the mint at Lima, where it was tested, and proved to be silver. As the Spanish laws in America, to encourage the discovery of mines, make them the property of the discoverer, the shepherd worked his mine, and soon became a millionaire. This story is no mere invention, but a fact which led to the discovery of the mines of Cerro de Pasco in 1630, in the manner related.

Almost all mining districts would furnish similar incidents. The rich gold-fields of the Sacramento valley in California, for instance, were found by a Mormon labourer named Marshall, who was then employed at a saw-mill, erected by Captain Sutter* on the south fork of the American River at a place now called Coloma (fig. 107).

It was on the 19th day of January, 1848, that James W. Marshall, while engaged in digging a race for a saw-mill at

Fig. 107.—The saw-mill of Coloma (California) as it was at the time when the first nugget was discovered.

Coloma, about thirty-five miles eastward from Sutter's Fort, found some pieces of yellow metal, which he and the half dozen men working with him at the mill supposed to be gold. He felt confident that he had made a discovery of great importance, but he knew nothing of either chemistry or gold-mining, so he

* Captain John A. Sutter, a man who had seen many vicissitudes and adventures in Europe and the wilds of America, arrived in California from the Sandwich Islands in 1839. In 1841 he obtained a grant of land and built a fort, which soon became the refuge and rallying point for Americans and Europeans coming into the country. The pioneers of that day all bear testimony to the generosity of Captain Sutter at a time when his fort was the capital, and he the government for the American colony in the valley of the Sacramento.

could not prove the nature of the metal or tell how to obtain it in paying quantities. Every morning he went down to the race to look for the bits of the metal; but the other men at the mill thought Marshall was very wild in his ideas, and they continued their labour in building the mill, and in sowing wheat and planting vegetables. The swift current of the mill-race washed away a considerable body of earthy matter, leaving the coarse particles of gold behind; so Marshall's collection of specimens continued to accumulate, and his associates began to think there might be something in his gold-mine after all. About the middle of February, a Mr. Bennet, one of the party employed at the mill, went to San Francisco for the purpose of learning whether this metal was precious, and there he was introduced to Isaac Humphrey, who had washed for gold in Georgia. The experienced miner saw at a glance that he had the true stuff before him, and after a few inquiries he was satisfied that the diggings must be rich. He made immediate preparation to go to the mill, and tried to persuade some of his friends to go with him; but they thought it would be only a waste of time and money, so he went with Bennet for his sole companion.

He arrived at Coloma on the 7th of March, and found the work at the mill going on as if no gold existed in the neighbourhood. The next day he took a pan and spade, and washed some of the dirt from the bottom of the mill-race in places where Marshall had found his specimens, and in a few hours Humphrey declared that these mines were far richer than any in Georgia. He now made a rocker and went to work washing gold industriously, and every day yielded to him an ounce or two of metal. The men at the mill made rockers for themselves, and all were soon busy in search of the yellow metal. Everything else was abandoned; the rumour of the discovery spread slowly. In the middle of March Pearson B. Reading, the owner of a large ranch at the head of the Sacramento valley, happened to visit Sutter's Fort, and hearing of the mining at Coloma, he went thither to see it. He said that if similarity of formation could be taken as a proof, there must be gold-mines near his ranch, so after observing the method of washing, he posted off, and in a few weeks he was at work on the bars of Clear Creek, nearly two hundred miles north-westward from Coloma. A few days after Reading had left, John Bidwell, now representative of the northern district of the State in the lower house of Congress, came to Coloma, and the result of his visit was that in less than a month he had a party of Indians from his ranch washing gold on the bars of Feather River, twenty-five miles north-westward from Coloma. Thus the mines were open at far-distant points.

The first printed notice of the discovery of gold was given in the California newspaper published in San Francisco on the 15th of March. On the 29th of May the same paper, announcing that its publication would be suspended, says:—" The whole country, from San Francisco to Los Angeles, and from the sea-shore to the base of the Sierra Nevada, resound with the sordid cry of *gold! gold! gold!* while the field is left half planted, the house half built, and everything neglected but the manufacture of picks and shovels, and the means of transportation to the spot where one man obtained one hundred and twenty-eight dollars' worth of the real stuff in one day's washing ; and the average for all concerned is twenty dollars per diem."

The towns and farms were deserted, or left to the care of women and children, while rancheros, wood-choppers, mechanics, vaqueros, and soldiers and sailors who deserted or obtained leave of absence, devoted all their energies to washing the auriferous gravel of the Sacramento basin. Never satisfied, however much they might be making, they were continually looking for new placers, which might yield them twice or thrice as much as they had made before. Thus the area of their labours gradually extended, and at the end of 1848 miners were at work in every large stream on the western slope of the Sierra Nevada, from the Feather to the Tuolumne River, a distance of one hundred and fifty miles, and also at Reading's diggings, in the north-western corner of the Sacramento valley.

The first rumours of the gold discovery were received in the Atlantic states, and in foreign countries, with incredulity and ridicule; but soon the receipts of the precious metal in large quantities, and the enthusiastic letters of officers in the army and of men in good repute, changed the current of feeling, and an almost unparalleled excitement ensued. The spring of 1849 witnessed the beginning of the most extensive immigration that the world has ever seen. All the adventurous young Americans east of the Rocky Mountains wanted to go to the new Eldorado, where, as they imagined, everybody was rich, and gold could be dug by the shovelful from the bed of every stream.

Adventurers poured into California from all quarters of the globe: first from Mexico, Chile, and Peru; then from the Sandwich Islands, China, and New Holland; lastly from the United States and Europe. During the six months between the first of July, 1849, and the first of January, 1850, it is estimated that 90,000 persons arrived in California from the east, by sea or across the plains, and that one-fifth of them perished by disease during the six months following their arrival; such were the hardships they had to endure, and the privations to which they were subjected.

Before 1850 the population of California had risen from 15,000, as it was in 1847, to 100,000, and the average increase annually for five or six years was 50,000.

As the number of mines increased, so did the gold production, and the extent and variety of the gold-fields.

In 1849 the placers of Trinity and Mariposa were opened,

and in the following years those of Hamath and Scott's valleys. During the last sixteen years no rich and extensive gold-fields have been discovered, though many little placers have been found, and some very valuable deposits, previously unknown, have been brought to light in districts which had been worked previous to 1851.*

We have remarked that nature sometimes makes use of children in revealing the richest mines ; and M. Marcou, who has made some highly interesting geological researches in North America, relates the discovery of the mines of North Carolina. He obtained the following details from the son of the discoverer :—

Towards the close of the last century, during the war of Independence, some German troops entered the British service, under the title of the Foreign Legion. This regiment was quartered at Charleston (South Carolina). A certain number of desertions took place from its ranks, and one of the soldiers, named Reid, succeeded in reaching the neighbouring state, North Carolina. There, in the county of Cabarrus, on the confines of civilization, he took possession of a piece of land as the first occupier, or squatter. An impromptu colonist like the veterans of Augustus, the old soldier cleared the ground. He had built a hut with the trunks of trees ; but he could hardly find means of subsistence, the country being almost a desert.

Things remained thus until many years after, in 1799—when this man was married and, like many colonists, had a numerous progeny—three of his children, while playing on the banks of a stream, perceived a yellow stone. Their father, when they showed him what they had found, did not trouble himself about the matter ; all stones being mere pebbles to him, he concluded this mineral to be of no value. But as the

---

* " Reports upon the Mineral Resources of the United States," by Special Commissioners J. Ross Browne and James W. Taylor, 1867. Pp. 14-16.

stone weighed fifteen pounds, he placed it on the floor near the door of the hut, which it served to keep open or closed— these good people being so poor that they had no latch to their door.   However, one day Reid showed this stone to an inhabitant of Concord, a neighbouring village, since become a town, and this man, seeing a rounded pebble, yellow, certainly, but without the least brilliancy, declared that it was a metal unknown to him.   The stone was returned to its place in the hut, and was shown to friends as a curious rock specimen.

Three years after Reid, on going to the market of Fayette-ville, took with him this mysterious piece of metal to show it to a goldsmith. " You ought to find out what it is," his neighbours had said to him, " perhaps you have some treasure there."   The goldsmith immediately declared it to be gold, and asked him to let him test it.   Reid left the nugget with him, and returning a short time after, he was asked what he would take for his pebble.   The inexperienced colonist, who could not have been aware that his pebble was massive gold, imagined he was naming a large price in demanding three dollars and a half, about 15s., which the goldsmith readily paid him. At the weight of fifteen pounds the nugget was worth £875 sterling.

Thus, it took four years to ascertain that the yellow stones of the streams of Carolina were gold.   The mountain at the foot of which the first nugget was found was so rich, that it was afterwards called by the Americans the Bull of Gold Mines.

In Australia pieces of gold had been found by mere chance by more than one person, but the real discovery of the precious metal is greatly owing to geology.

Soon after his return from the auriferous tracts of the Ural Mountains, Sir Roderick Murchison, on examining the specimens collected by Count Strezelecki in East Australia, was struck by the great similarity of the rocks of the two countries, and in his Presidential address to the Geographical Society in 1844, he alluded to the possibly auriferous character of the great eastern chain of Australia.   Impressed with this resemblance, and convinced in his own mind that gold would sooner or later be found in the British colony, Sir Roderick, at a meeting of the Royal Geological Society of Cornwall* held at Penzance in 1846, urged upon the unemployed Cornish miners, many of whom were then emigrating to the Australian copper-mines, the importance of turning to account their practical experience in tin-streaming in examining the alluvial deposits of that country for gold and encouraged them to search for gold in the same way that they streamed for tin in the gravel of their own district.

It is just to mention that the attention of the people of New South Wales was, in 1847, also directed by the Rev. W. B. Clarke, an active and enterprising geologist living in Sydney, to the auriferous character of the rocks of that country, and that he indicated, as Sir Roderick Murchison had previously done, the striking resemblance between the rocks of the Ural Mountains and those of the meridional range of Australia, which he had by anticipation termed the " Cordillera." †   It also appears that Mr. Clarke obtained gold in the Colony in 1841, and that he exhibited a nugget from the basin of the River Macquarie to the members of the government at Sydney in that year ; also that traces of gold had been found by Count Strezelecki as early as 1839.

Some time after this announcement by Sir Roderick Murchison, a shepherd of Scottish origin, Macgregor by name, came from the Blue Mountains to sell some grains and nuggets of gold to the goldsmiths of Sydney. He refused to reveal whence he procured it, and carried on from place to place his innocent and profitable trade.   At a somewhat later period

* Trans. Royal Geol. Soc. Cornwall 1846. p. 324.

† "Siluria," by Sir Roderick Impey Murchison, Bart., p. 460, 4th edition.

still, a man of the name of Smith, at that time engaged in
some iron-works at Berrima, was led by some of Sir Roderick
Murchison's observations which had been republished in the
Australian newspapers, to search for gold in the year 1849.
One day he went to the governor of the colony with a nugget
in his hand, " See what I have found in the country," said he ;
" give me five hundred pounds, and I will show you the place."
The governor wished him to lower his terms, but the miner
was firm, and they separated without settling the matter.
The great value of the gold-mines of Australia received its
first real practical solution from Mr. Hargreàves, an old Californian miner, who coming with the advantage of his experience
in that country, re-made the discovery in 1851, and got the
reward from government. " The first discovery was made on
the banks of the Summer Hill Creek and the Lewis Pond
River, small streams which run from the northern flank of the
Conobalas down to the Macquarie. The gold was found in
the sand and gravel accumulated, especially on the inside of
the bends of the brook and at the junction of the two water-
courses, where the stream of each would be often checked by
the other. It was coarse gold, showing its parent site to be
at no great distance, and probably in the quartz veins of the
metamorphic rocks of the Conobalas. Mr. Stutchbury, the
government geologist, reported on the truth of the discovery,
and shortly afterwards found gold in several other localities,
especially on the banks of the Turon, some distance north-east
of the Conobalas" * (Map XIV.). This was on the 9th May,
1851, and a few days after there were upwards of a thousand
miners on the spot ; at the end of some months, however,
things assumed a very different aspect. Fresh mines, and
fresh beds of auriferous quartz, became known one after the
other in New South Wales, and soon in the province of
Victoria ; and, as in California, the whole world, eager to

* " Lecture on Gold," by J. B. Jukes.

secure a part of these treasures, hastened to the spot. The names of some of the camps of Victoria, which became *par excellence* the auriferous province of Australia, crossed the seas; and the European journals immediately electrified all their readers by their calculations of the enormous riches to be extracted from the gold-fields, whose fertility appeared to be inexhaustible.

Whoever may be the discoverers of a mine, they seldom profit by their discovery. By a fatal law of nature, the man who finds a vein of metal does not generally enrich himself thereby; he makes the fortune of others, not his own. In 1859 Marshall was forgotten in California, and he had become poorer than ever.

The richest veins of silver in the favoured country of Chile are those of Chañarcillo, discovered in 1831. Godoy, a mountaineer, hunted guanacos in the Andes. These ruminants, closely allied to the llama, alpaca, and vicuña, in South America fill the place of camels and dromedaries, and in the silver-mines of Peru they are employed in carrying the ore. The fleece of the American camel furnishes the material for the excellent fabric manufactured in England by the name of alpaca.

One day as Godoy was hunting guanacos, being fatigued, he seated himself under the shelter of an enormous rock, and was struck by the colour and brightness of a projecting part. He chipped the stone with a knife, and finding that he could cut it like cheese (to use his own expression) he took a specimen of it to Copiapo. The mineralogists of the country, being expert in the science, at once perceived it to be chloride of silver. That substance, which is known by the name of horn silver, because it has the texture of horn, is designated by the Chilian miners, in their peculiar vocabulary, as *plata-plomo*, or silver-lead.

It has already been mentioned that by the Spanish laws the discoverer of a mine became its possessor. Godoy offered half his mine to Don Miguel Gallo, one of the oldest miners in the province of Copiapo, and the arrangement entered into between them was that which usually takes place under similar circumstances. Gallo agreed to furnish the necessary funds for working the mines, and the two parties were to share the profits. The mine was solemnly baptized under the name of *Descubridora*, or "that which is discovered." By a lucky chance which often occurs on opening a vein of ore, the miners came at once to masses of silver, and from the very first large profits were made. Godoy, like almost all those who discover mines, would not wait for the conclusion of the works; but allured by the hope of finding still richer veins, he sold his share for the price of 14,000 piastres (about £2800 sterling), wandered for some years in the Andes, dissipated his money, found no more mines, and died without a penny.

However, the news of this brilliant discovery had attracted to Chañarcillo numbers of miners from all parts of Chile. It always happens thus, from one end of the world to the other; if an unusually rich mine is discovered, all the miners, yielding to an irresistible temptation, rush from their working-places to the new spot in a fever of excitement. The Spaniards have a very expressive term for this ardent pursuit of mineral wealth: they call it *el furor minero* ("the miner's frenzy").*

Amongst the numerous Chilian miners drawn towards Chañarcillo, those upon whom blind Fortune first lavished her gifts were two brothers named Bolados. They owned a poor *rancho* (small farm) in the valley of Copiapo, and a drove of asses, which, in default of llamas, they employed to carry wood to the smelting-houses, where they smelted the silver-ore, and in that manner they gained a miserable livelihood. They had scarcely arrived at the mines when they discovered in a crevice, opened by some earthquake, perhaps even during the

† In 1858 California was nearly depopulated by the rush to the mines of Fraser River, in British Columbia, and in 1860, to the silver-mines of Utah. Some of the first deserters of the Californian diggings returned, looking very sheepish; they afterwards learned to manage matters better.

actual formation of the metallic vein—an enormous block of silver-ore. This was near the mine of Descubridora, and the place still bears the name of *manto de los Bolados*, or "vein of the brothers Bolados." According to M. Domeyko, more than sixty quintals of silver were extracted merely from the crust of this mass, and the block which remained, and which was obliged to be cut with a chisel, weighed more than thirty-three quintals, and was composed of a mixture of native silver and chloride.

The cutting, transport, and fusion of this mass of ore were so easy, that the Bolados, although entirely devoid of practical knowledge and capital, succeeded, in less than two years, in extracting silver to the amount of more than 700,000 piastres (about £140,000 sterling). Dazzled by their prosperity, they only thought of enjoying their gains, and whilst they squandered their money at Copiapo, then only a small village, in gambling and dissipation, their mine was suddenly exhausted.* They had not contemplated a reverse of this kind, and became poorer than before their discovery, not possessing even their asses!

These events are more like tales of glowing fiction than facts of every-day occurrence, and yet similar histories might be collected from one end of America to the other, in British Columbia, California, Mexico, Nicaragua, Peru, Bolivia, and in many European countries likewise.

Although prospectors of lodes for the most part know so little how to profit by their good fortune, many cases might be cited where their discoveries have greatly enriched them. Every one talks in California of the famous mine of Allison-Ranch, at Grass Valley, in the province of Nevada (Map XII.), which was discovered in 1852 by some poor Irishmen who were at work in a neighbouring mine. Notwithstanding the specimens which had been collected displayed the metal

* It is not rare in Chile for the outcrops only of mines to be rich, or as it is called, *en benefice*.

to the naked eye, the finders were not satisfied with this discovery. At that period the working of quartz-reefs was regarded as little productive, and the prudent disciples of St. Patrick carefully concealed the back of the lode with earth, to elude all competition. Under such circumstances no precautions could be too great.

At the close of the year 1855 these miners were working again at the same mine, when they again directed their attention to their vein of quartz. Numerous and rich mines had recently been found in the vicinity, and they determined to explore their own. They had only advanced a few feet into the rock when the richness of the quartz became manifest. Soon exceeding all ordinary proportions, the lode became worth as much as three hundred and fifty dollars per ton.

In October, 1856, the Irishmen erected a mill to crush their ore. Fortune plays strange tricks sometimes; this quartz was not only the richest, but also the most easily pulverized of all the numerous surrounding mines. The Irishmen's mill, at the end of 1859, with only eight miserable stamps, was crushing about twenty tons of ore per diem, whilst other mills only accomplished from eight to ten. The produce of the ore was then from 205 dollars (more than £40) per ton.* Thus every pound of stone was worth nearly sixpence.

The fortunate Irishmen, so unlettered that they could not even sign their names, knew not what to do with so much gold. They had built a chapel to thank God for his favours; caused handsome villas to be erected near their old mine; placed their workmen in exceptional positions; and at length, to amuse themselves, they went by turns every week to San Francisco with their ingots of gold. Retaining their simplicity with an income as large as that of many princes in Europe,

---

* The average richness of quartzose ores worked in California, in 1859, was only from ten to twenty dollars (or from £2 to £4) per ton.

they had delegated the superintendence of the works to an overseer, and passed a peaceful existence, more favoured than in their native Erin, which they could scarcely regret. The mine continues to prosper, and is one of the richest and most productive in the state. The average thickness of the lode is about eighteen inches, and the rock yields from thirty to one hundred and fifty dollars per ton. According to the best information obtainable by the State Geological Survey, 14,858 tons were reduced between March, 1857, and December, 1861, and the average yield was fifty dollars per ton, or 942,900 dollars in all. Since the summer of 1862 the mine has paid better than before. The lowest workings are nearly five hundred feet deep, and the lode at that depth is three feet wide, with rock that averages one hundred dollars to the ton. The owners refuse to furnish any statement of their receipts or expenditure, but the men employed in the mill say the yield is 40,000 dollars per month, or 400,000 dollars for ten months' work in a year; and of this sum two-thirds or more is clear profit. The claim has been worked for a length of about 1400 feet.* (State Report.)

In the silver-mines of the state of Nevada, bordering on that of California (Map XII.), two Irishmen were not less fortunate than their countrymen of Allison-Ranch. Having migrated thither from some mine where they were doubtless discontented, these two miners went to Carson City to improve their fortunes. Their names were Gould and Curry, and their mine and company were called after them. In 1860 they were laboriously occupied in working the lode, and fate did not appear to regard them more favourably than before. Monsieur Laur, a mining engineer sent to America by the French government, advised the two miners to relinquish their

---

* In most districts of Nevada, and in many of California, the miners' regulations allow one hundred feet of lode to every discoverer or first occupant, and one hundred feet to each of his associates. Under statute the miner may claim for each person in his company two hundred feet on the lode, by two hundred feet wide.

work, appearances being against success; they had stumbled
on a reef of quartz, which might be at most auriferous.   The
Irishmen, however, persevered; but their finances were exhausted, and they wanted more powder and implements.
They went from time to time to San Francisco to procure
provisions, and a grocer there supplied all their requirements,
as in that country tradesmen deal in many different kinds of
wares, thus making fortunes much more quickly than where
they confine themselves to one line of business; and it may
truly be said of them that they live in the land of gold.
These two miners then, Gould and Curry, were not worth a
penny, but it did not signify, the merchant trusted them, and
gave them unlimited credit, even to champagne; for mining
labour induces thirst, and the miner of Nevada, as well as of
California, is by no means insensible to the pleasures of the
bottle.   They think nothing there of giving for wine five
dollars (about twenty shillings) per bottle, and even more.
Instead of paying in money, the miners made over to the
grocer a foot, then a yard of the lode, until at last the latter
became proprietor of two-thirds of the mine.

 One day Gould and Curry placed their hands on a block of
ore, as the brothers Bolados had done in Chile.   Only a small
part of the mine remained to them; the lucky merchant held
nearly all the shares, and yet the two miners secured millions
for their own portion.   More fortunate than the Chilians whose
history has been related, their prosperity was followed by no
reverses.   When they remember the advice of the engineer
to abandon the mine, they must entertain but a mean opinion
of geologists, who, however, do not live beneath the earth,
and may therefore be sometimes mistaken.   Including the
product of the year 1857, the Gould & Curry Company have
taken from their mine a grand total of 14,000,000 dollars, of
which 6,500,000 dollars have been spent in general disburse-
ments and improvements; a little over 3,000,000 dollars for

work done by custom mills; the balance, something over 4,000,000 dollars, having been paid to the shareholders.

One more anecdote, which has become legendary in Tuscany, will close these histories about the discovery of mines, and which will confirm, while epitomising, all the facts which have been narrated.

Tuscany, at two periods of its history, under the Etruscans and during the Middle Ages, has been the theatre of the most prosperous mining operations. Towards the dawn of modern times, different reasons combined to cause these enterprises to be completely abandoned, especially the fierce conflicts of the republics amongst themselves, the incursions and ravages of the *condottieri*, epidemics, the black pestilence, and lastly, the fall in the price of money, of which the discovery of America was the cause.

The Medicis having subjected the whole of Tuscany to their house, and after them the princes of Lorraine, had vainly endeavoured, on many occasions, to re-open a part of the works so unhappily suspended. Affairs were in this condition when, in 1830, a Frenchman, M. Porte, who had accompanied the Princess Eliza to Italy when she was named Grand Duchess of Etruria under the first Empire, and who afterwards remained in the country, resolved in his turn to revive all these old workings. What he spent in time, intelligence, and skill to attain his ends those who knew him can tell, for he has left many lasting mementos in Tuscany. One mine especially engaged his attention, that of Caporciano or Monte Catini, in the valley of Cecina; but from 1830 to 1837 the result of his researches was very unsatisfactory, so much so, that his resources being almost exhausted, and his confidence shaken, just when he might have found the lode he sold the mine, and turned his attention to others, which held out better hopes. These were that of Accesa, near Massa, where immense works had been carried on in the Middle Ages; and that of Rocca

Tederighi, in the Siennois, which had also been partially excavated by the ancients.

M. Porte had scarcely disposed of Monte Catini when a block of massive ore was found which paid all expenses in one year, and left more than £4000 net profit. Years rolled on, and the profits increased, soon becoming tenfold. For fifteen years the mine produced more than £40,000 annually, and it still continues to yield largely. M. Porte, who had witnessed the heartrending spectacle of the immediate success of others where he had laboured in vain for years, did not see this undertaking in its zenith. He soon died of grief, and poorer than ever.

The three fortunate proprietors of Monte Catini compose a triumvirate which has become celebrated in Tuscany. One, an Englishman, formerly major-domo of a great lady, and chamberlain to the ex-grand duke, continued faithful to the aged Leopold in the days of his adversity. He acquired, amongst other things, the beautiful villa of Careggi, near Florence, in which is still echoed the name of Lorenzo the Magnificent. The second has an interest in a large Florentine banking house, whose principal, a worthy successor of the astute Italian financiers of the Middle Ages, exclaimed loudly when he heard that his partner intended to try his luck at the mines. The fortunate explorer remained what he always was, simple in manners, kind and modest, receiving with a smile all foreigners recommended to his house. The third, who merely served as intermediary in the cession of this mine, has since become administrator for the first railways in Italy. Happy trio! There is nothing to deplore in the whole affair, except the fate of the luckless discoverer. His marble bust adorns the entrance of the principal gallery of Monte Catini, but his heirs, who still live in Florence, have not become millionaires.

CHAPTER VII.

HIDDEN TREASURES.

Peasants and shepherds.—Rebuscadores and Cateadores.—Diviners.—The divining
wand or dowsing rod.—The adepts.—The alchemists.—The Egyptian and Etruscan
diviners.—Discovery of the iron mines of Elba by ancient Etruscan diviners.—
Mode of discovering lodes practised by Madame Rey.—Cosmo I. and Benvenuto
Cellini.—The gold-mines of Chrysonése.

ALTHOUGH it is true that chance or the teachings of geology
have alone led to the discovery of lodes, the search for metallic
veins has not the less occupied the minds of numerous persons,
and turned the heads of many in all ages. This search is akin
to that for hidden treasures. The enthusiast giving himself
up to the pursuit of the great work is actuated by the spirit
which in ancient times gave birth to the Hermetic school, and
to the alchemists of the Middle Ages. This delusion reap-
pears, in our days, in the desire of man to enrich himself
suddenly at any cost. Indeed, this disposition characterizes
the nineteenth century, and leads to the organization of the
wildest and most visionary schemes ; it is, in a word, part
of the natural course of events.

In some countries the peasants search for mines in their
own fashion. A piece of yellow mica, or a glittering crystal
of iron or copper pyrites, is gold in their eyes. They go and
consult in secret the curate, the overseer, and the mayor ;
saying they have discovered a mine, but will not reveal the
locality, wishing first to ascertain whether they have really
laid their hands on the precious metal.

The shepherds are like the peasants; that which the latter
collect in the fields and on the banks of streams, the former

find on the mountain-sides, in the beds of the torrents, and on the highest mountain-peaks. Chance sometimes favours them, as has been shown by the instances which have been already given. It is not the less true, however, that for the most part they seek without finding. These shepherds would do better, like their Chaldean ancestors, to endeavour to study the stars and to read the heavens, than to fix their eyes on the ground; but they might urge that astronomy is an advanced science, and besides, that many lodes have been found by shepherds. And then there is the fable of the astrologer who fell into a well, a catastrophe which he who seeks for mines at least avoids.

The peasants and shepherds compose what may be termed the forlorn hope of the battalion of mine-seekers, of which the poachers and hunters also form a part. These latter, who often climb the highest summits. and tread volcanic rocks under their feet, are not so deeply absorbed in the pursuit of game but that they can bestow some attention to the subject of mines. In some countries, miners themselves join this species of free-lances; such are the *rebuscadores* and *cateadores*; Spanish mine-seekers, energetic adventurers, who have more than once laid their hands on the richest lodes. It is now necessary to mention a regular body, a true regiment, which has its discipline, laws, and manœuvres—that of the users of the divining-rod. They profess to be endowed by nature with a peculiar sense or perception by which they are enabled to reveal the spots where mines are hidden, and also to discover underground water-courses, springs, and even buried treasures, by means of the divining-rod.

In 1853 M. Simonin resided at Laffrey, near Grenoble, where he had the superintendence of the metallic mines. Veins of lead and zinc traversed the neighbourhood in different directions, and all these deposits united together and formed what are called bunches. By inductions authorized by geology, the directions followed by veins to places not yet worked could be determined with tolerable approximation. One of these users of the divining-rod visited the place, coming from a village near l'Oisans, where everybody from time immemorial is born a diviner (*sourcier*).* M. Simonin had never seen any of these professors, and all he knew of their *modus operandi* was from books. He proposed that the diviner should exercise his art, warning him that he would not be paid unless he divined rightly. The man accepted the terms, demanding a freshly-cut hazel-wand as indispensable; and having obtained one from the verge of a neighbouring torrent, which, escaping from the lakes of Laffrey, descends in boiling fury to Vizille, he cut the wood in the requisite manner, held it with both hands at the extremities, and then advanced hither and thither. The scene of operations was a table-land, covered with verdure; on the left, looking towards the lakes, was a forest of firs; on the right, a perpendicular rampart of calcareous rocks crowned with beeches; in front, the village, with the lakes beyond; behind, the steep declivity of a hill, at the foot of which is Vizille. It was just the landscape for a comic opera, being suited to please the eye, but not adapted for the edification of the geologist or the miner, still less of a seeker of lodes. The diviner walked backwards and forwards, uneasy and hesitating.

" Come, my friend, what do you say? Is there not a mine beneath your feet? You bu. '"

" Here?"

" No, farther on."

" There, perhaps ? "

" You are not there yet."

" I have found it !"

" Not yet."

---

* These adepts, or users of the divining or dowsing rod, are called *sourciers* in France, or spring-finders. Sir Walter Scott gives an amusing account of this mode of divination by a rod or wand in connection with the search for hidden treasure in his novel " The Antiquary."

The wand twisted in his hands, which were seized with a convulsive movement. A physiological phenomenon undeniably took place, apparently of the nature of animal magnetism; but although science may have hitherto failed to give a satisfactory explanation of these mysteries, no one would be so bold as to pretend that they imply a certain knowledge of the future and the unknown—in a word, what is termed second sight. This diviner only furnished another proof that he and his compeers are merely sensitive and nervous subjects, availing themselves of the circumstance to become clever impostors.

The art of Rhabdomancy, or the use of the divining-rod, the *dowsing-rod* of the Cornish miner, which is still occasionally practised in Italy and the south of France under the names of metalloscopy, hydroscopy, &c., as well as in some other countries, as much for the discovery of mines as for that of water and hidden treasures, has always had numerous adepts. Before geology was raised to the rank of an exact science, before the laws of the formation and modes of occurrence of deposits were studied, the rod was in universal use everywhere. In the sixteenth century, at the time when Biringuccio in Italy, Agricola in Germany, and Bernard Palissy in France, established the study of subterranean formations on the precise basis of experiment, and introduced observation into mining, special treatises were written upon the divining-rod and the manner of using it. Agricola, in his work *De Re Metallicâ*, strongly combats this mode of procedure; but there is always a tendency on the part of mankind to believe in the marvellous and the supernatural. In 1678 a manual of mining operations, printed at Bologna, and entitled *Pratica Minerale*, minutely treated of the discovery of lodes by this means, and described the two methods of operation. By the first, a piece of the metal which was being sought for was attached to one end of the wand, and the disengaged extremity then inclined towards the vein. By the second method, the operator holds both ends of the wand (which is a branch of a tree, generally hazel, forked at the end like an inverted λ) in a particular way between his hands, the fingers being bent back, and the arms pressed close to his sides. The end of the wand is then supposed to indicate the lode or other object sought by bending forcibly towards it with a slow rotatory motion, the direction of the point indicating the required spot. According to modern practice the adept is placed in contact with some metallic or other magnetic substance, and everybody who is considered to be antipathetic to the operator must be removed, as their presence would interfere with the results of the experiment.

It is not surprising that the use of the divining-rod should have ranked, even in the seventeenth century, among the acquirements indispensable to the mining engineer. Even in the present day there are some few men of experience who attach a certain value to this mode of finding lodes; and when there exist believers in spirit-rapping and table-turning, it is scarcely surprising that there should also be practisers of divination by the magic wand. It is little more than twenty years (1844) since diviners operated officially in Saxony, where it was believed that the lodes of the east, or the rising sun, were richer and finer than those of the west, or setting sun. The Saxon engineers did not abandon the use of the wand in the search for mines until it was clearly demonstrated to be utterly powerless. In France, the Abbé Paramelle, by reducing to its true principles the art of discovering springs of water, has also, in our own day, done full justice to the diviners.

The practice of interrogating the earth by means of magnetism came from the ancients, and the Egyptians, of whom the alchemists and all the disciples of Hermes Trismegistus (who was regarded by the Egyptian alchemists as the founder of their imaginary art) are lineal descendants, transmitted their knowledge to the Etruscans, from whom it again passed to the Romans. The Etruscans were considered masters of these mysteries, and combined with the art of reading the heavens, and of interpreting the voice of thunder, that of reading the secrets of the earth. Indeed the discovery of the celebrated iron-mines in the Island of Elba is attributed to the Tuscan diviners.

Historians relate that, under the reign of Ancus Martius, some Etruscan diviners landed on the island, at the spot where now stands Rio. These priests of the Unknown had not forgotten their wand. They had scarcely disembarked when, without giving themselves time to rest from their fatigue, they addressed themselves to their favourite occupation. Fortune favoured them, and they discovered the iron-mines; there was, however, no great merit in this achievement, for these lodes crop out at the surface of the ground, and the quantity of ore is so enormous that it might be said of this part of the Isle of Elba, that it is literally nothing but a mountain of iron now, requiring simply the removal of a little soil to develope it. The mines being discovered, they began to think of working them. Ancus Martius, himself an Etruscan, ceded the mines in perpetuity to his countrymen the diviners, who certainly had some right to this reward for their discovery. They established themselves at Rio, worked the ore there, and founded the colony of miners which is still flourishing in the Island of Elba. Such is the tradition, or legend, if the term be preferred.

The iron-mines of Elba have always been a strictly preserved property, and were worked by the late government of Tuscany on their own account up to 1851; but the inhabitants of the island have frequently rebelled against the prohibition to work these mines freely; and at the time of the annexation to the kingdom of Italy, in 1860, they complained more loudly than ever. Their cause was carried before the council of the Italian States. It is not perhaps known that the lawyer of the commune of Rio relied on a pretended passage of Livy to defend the cause of his clients, a passage which is at the least apocryphal, and relates to the Etruscan diviners. "We have been," said this modern Cicero, "the legal proprietors of the mines since the time of Ancus Martius, from whom our fathers received them in due form. Unjustly dispossessed, we claim our rights."

The divining rod, so highly valued by the Egyptians and Etruscans, is not the only means employed by indefatigable seekers. In the mines of the French Alps a very curious mode of procedure is sometimes adopted. It was invented by a lady (Madame Rey) and it was near La Motte and Bourg d'Oisans, a country esteemed by mineralogists, that she carried on her operations. She went alone to the mountain, holding a piece of string to which was attached a five-franc piece, a piece of lead, or a large copper coin, and pretended that her pendulum vibrated on approaching the vicinity of a lode. Silver attracted silver, and copper or lead metals of the same nature. This was the reverse of the laws observed by electricity and magnetism in physics, when bodies of the same nature mutually repel each other. Madame Rey marked with stones the places where the pendulum vibrated, and then connected these points by an imaginary line, saying, "That is the direction of the lode."

M. Simonin says, "At the commencement of my career as a miner I have seen, during my excursions among the picturesque mountains of Dauphiny, these novel land-marks placed by this intelligent seeker in the midst of abrupt rocks. The country is intersected everywhere by lodes and outcrops of metallic veins, and at certain points copper deposits abound, the soil being marked as with a pencil with green and blue stains, denoting the presence of the metal of Venus.* Madame Rey had therefore no difficulty in finding the lodes; but truth compels me to avow that she also discovered unknown mines. She has done much for the development of mining industry in these savage countries, and it was she who found again the deposits of mercury of Saint-Arey. She aroused some of the miners of the French Alps from a sleep which had lasted more than a century, into which condition, unhappily, they were soon to relapse. Several times, during my sojourn at Laffrey, I have seen her passing with her cart full of ores, which she was taking to the copper, lead, and silver foundries at Vizille. Tall and vigorous, her very looks bespoke strength; she was something more than a woman, she was a valiant miner."

In all ages, then, the search for mines has largely occupied the attention of mankind. Princes, and even kings, have not been exempt from this passion for the marvellous and the unknown, although it would not be supposed that they were subject to its influence, since they might reasonably be expected to find in their exalted station the satisfaction of all their desires. The Grand-duke of Tuscany, Cosmo I., who founded the princely house of Medici, and who was an equally astute politician as an able administrator, was besides a great seeker of mines. In order to re-open some of those ancient workings in Etruria, which have already been frequently mentioned, he sent to Germany for engineers, from some large proprietors of mines, the Fuggers, rich bankers of Augsburg, who were the Rothschilds of their time. These were at that period, as others have been since, the kings of bankers, and the bankers of kings. They had lent large sums to Charles V., and one day when the great monarch, journeying to the Low Countries, stopped to rest at one of their châteaux, the story goes that the fire at which the emperor warmed himself was

* The symbol ♀, used to denote the metal copper, is supposed to represent the looking-glass of Venus, after whom the metal was named by the alchemists, in consequence of its easy union with other metals, and the change which ensues in its nature and appearance.

fed with cinnamon-wood, and that the paper which served to kindle it was a bond inscribed by Charles V. to the Fuggers.*

The German engineer demanded by Cosmo I. of his bankers at Augsburg hastened to place himself in communication with the Grand-duke.   In the archives of the offices at Florence, which M. Simonin was kindly permitted to examine by M. Bonaini, the superintendent of the department, are the autograph letters of the Grand-duke, with the answers of the Fuggers and the engineer.

After a long correspondence, in which all the minutest details of the voyage, the sojourn, and emoluments were negotiated by the German with a methodical precision peculiar to his country-men, the engineer arrived and set to work.   He was sent to the ancient mines of the Tuscan Maremma; and Cosmo, who despatched an overseer to the spot, was kept daily informed of the progress of operations.   His correspondence in this affair is most curious, and ought, no less than that already mentioned, to be published entire.

The search for metallic mines so absorbed the attention of the Grand-duke, that he established a laboratory in his palace, where he worked with Benvenuto Cellini, before he quarrelled with him.   In this laboratory they smelted metals, heated the forge, and brought the retort and alembic into play; those were the palmy days of alchemy, when one might compare Cosmo I. to King Dagobert, and Benvenuto to Saint Eloi.

However, they did not succeed in finding gold at the bot-tom of the mines and crucibles; and one day when a Bolognese galleon returned from America freighted with ingots of silver, Cosmo sent to Lisbon and bought up the entire cargo.   At the same time he also monopolized, by means of his correspondents in England, the tin of Cornwall, a metal which was then indis-

---

* These same bankers, who were called Fuccares in Spain, for a long time worked the quicksilver-mines of Almaden, which they held from the crown, as the Messrs. Rothschild do in the present day The name of Fugger is still popular among the Spaniards, and in the provinces of Granada and Anda-lusia, in speaking of a man of very large fortune, they say, "He is as rich as a Fuccares."

pensable. All the banks, all the money-changers, and all the smelters of Italy were obliged to apply to Cosmo, who had become possessed of all the silver and tin in the country, and he resold his merchandise retail and at very high prices. These furnished his largest profits, and were his most successful speculations in metals. In spite of all his efforts, he could not restore to the mines of Tuscany the flourishing activity of the past; and his two sons Francesco and Ferdinand, who in turn succeeded to the throne,* were not more fortunate than himself in their attempts. Still Cosmo does not the less take his place amongst the original family of lode-seekers and disciples of the great Hermes.

It would occupy too much space to describe every type of mine-seekers, and especially to pass on by degrees to the fabricators of lodes. M. Simonin says that he has himself been appealed to for his advice, if not in the formation, at least in the carrying into execution of some of these subterranean enterprises, and was struck with the readiness with which men of the greatest intelligence suffered themselves to be caught by the bait. He adduces the following as a curious example:—

In an island of the East Indian Ocean, which shall be nameless, volcanic lava, which was one day erupted from below the bed of the ocean by submarine fires, extends along the coast in the form of sand and shingle. The waves beating against the rock disintegrate it; the winds contribute to the effect of the water, and thus dunes, or shifting sand-hills, are formed along the sea-shore. The black, porous, and heavy substance of which the sands and shingle are composed is the volcanic rock called basalt. In the midst of the grains or blocks of basalt may be distinguished acicular crystals of magnetic iron, attractable by the magnet; then gold-coloured crystals, sometimes of a slightly greenish tinge,

* This Francesco was the father of Marie de Medicis and the lover of Bianca Capello.

transparent, and hard enough to cut glass; this is chrysolite, a variety of olivine or peridot, employed as a precious stone of low standard in jewellery.

The chrysolite or gold-stone, so aptly named by the ancient mineralogists,* produces a striking effect in the midst of the volcanic sands. It glitters in the sun, and mingles the golden tint of its crystals with the sombre black of the basalt.

Some years before the French revolution of 1789, the Chevalier de P——, a relative of a French poet, who was born in this island, went mad. Struck with the brilliancy of these little pebbles, he mistook them for gold, picked them up on the shore, heated them in a crucible, and fancied he produced ingots from them. His family allowed him to amuse himself in this manner; and as his pursuit was not dangerous, they even provided him with a laboratory. At the end of a garden, shaded by mangoes, those peach-trees of the tropics, was a pretty kiosk, or pavilion, as it is called in those parts; and it was here that this last representative of alchemy carried on his operations, at the very moment when Lavoisier, by his memorable discoveries, gave a death-blow to the feeble art of the blowers.

All this was nearly forgotten when, about ten years ago, some persons who had inhabited this distant island revived in a most unexpected manner the experiments of the Chevalier de P——, and even addressed a request to the French government for the concession of the gold-mines of Chrysonèse. Shareholders flocked together; great personages interested themselves in the affair. Tons of sand were carried away, and gold was always found in it. With the produce of analyses it was proposed to make sleeve-buttons for the members of the committee of management. Meanwhile the mines prospered and multiplied, silver and platinum being found associated with gold, and in the interior of the island

* The name is derived from χρυσος "gold," and λιθος, "stone."

were discovered, not only placers, but true auriferous veins.
An eye-witness, who had contributed to the discovery, exclaimed with naïve enthusiasm that one hundred thousand
men, working for the same number of years, would scarcely
exhaust even the surface of the mine, and that henceforth
Chrysonèse would produce gold instead of sugar and coffee;
and this man spoke with good faith! The imaginary placers
are the shingle and sand of the sea-coast; the veins are bluish
volcanic clays, sprinkled with brilliant and crystallized yellow
iron pyrites, which might readily be mistaken for gold, but
which do not contain an atom of that metal; besides which
they are situated in the midst of almost inaccessible mountains
nearly five thousand feet above the level of the sea.

In spite of failures without number, the mines of Chrysonèse
still survive, and the shareholders have not lost their confidence. Although they have never received any other dividends
than those paid out of the capital—as is often the case in such
undertakings—the greater part of them are still sanguine.
Some months since it was said that the precious metals could
not be worked, on account of their existing only in a microscopic state; the Company were about to change the character
of their operations, and collect the magnetic iron-ore from the
sands. Unfortunately, the steel-ore is not sufficiently valuable
to be transported from the Indian Ocean to Europe. It is
therefore necessary to smelt the ore on the spot. From
whence are the coal and the smelters to be procured? In the
meanwhile the Chrysonèse people laugh at the credulous
Parisians. From the lowest of the blacks to the highest
inhabitants of the island, no one has taken shares in the Gold
Company. The country, prudent, economical, and industrious,
continues to produce sugar, coffee, and spices. The cultivation
of cotton has recently been tried with success, and everybody
there laughs at the fabricators of lodes, who selected so remote
a spot for the theatre of their exploits.

CHAPTER VIII.

THE ATTACK ON THE GROUND.

Difficulties of the work.—Submarine workings; Botallack Mine in Cornwall; Wherry Mine near Penzance.—Mines of the Upper Harz, in Hanover.—Great galleries, Georg-Stollen and Ernst-August.—Five leagues of tunnel.—Navigable canals in mines.—Mine-shafts at Andreasberg.—Saxony, Hungary, and Rhenish Prussia.— Methods of working the ores.—Underground transport.—Moving ladders, or man-engines.—A false step.—Mining implements.—Attacking the ground by fire.— Boring rocks by machinery; Low's machine.—Accidents in metal-mines.—Ores of arsenic and quicksilver.—Eight models of miners' lamps.—The Cornish miners' tallow candle.—The mines of Spanish America.

It has been told, in treating of coal-mines, what difficulties the miner had to surmount in making shafts and galleries. In metallic mines these difficulties are equally great. Even if, from the nature of the ground he is engaged upon, falls, slides, and inundations oppose themselves less frequently to the progress of the miner, on the other hand, rocks of exceptional hardness, such as quartz, crystalline schists, granite, and porphyry, have to be patiently attacked, which the best-tempered steel can scarcely penetrate; and the work in some of these rocks only advances at the rate of a few feet in a week.

In Cornwall shafts have been sunk in such rocks to the depth of two thousand feet. At this depth the working places have been extended right and left upon the line of the lode. In that country the metalliferous deposits are frequently worked on the sea-coast, and sometimes even under the bed of the sea. It is necessary to protect the works from the irruptions of the waves, and against the rising sea itself; but these obstacles have not deterred the workers. In the mine, galleries have been driven for more than a mile beneath the waters; in some places the thickness of the rocks which protect the miners is very slight. The workings at Botallack Mine are extended 225 fathoms under the Atlantic Ocean, and are carried beneath the sea for a distance of 712 yards (nearly half a mile) from the point where the diagonal shaft leaves the shore. In one part of this mine the miners, following the small strings of tin which are disseminated through the schistose rocks, actually broke through into the Atlantic; and up to the present day the waters are kept out of the mine by a wooden wedge and some greased oakum. In stormy weather they hear the ocean roaring above their heads; the boulders at the bottom of the sea, rolling over each other, make a noise like thunder. A terrific rumbling reverberates from gallery to gallery, but the workmen tranquilly work on in these gloomy abysses, in which the ocean seems to threaten to engulf them.

In 1862 M. Simonin tells us he himself visited the mines at the extreme point of Cornwall, near the Land's End. "I had descended into the deepest levels, and had traversed many galleries, penetrating to the interior shafts. I ascended to daylight by the fatiguing ladders, a mode of traversing the shafts which is of the rudest nature. Protected by the woollen garments in use among the English miners, with heavy miner's boots on my legs, and a candle stuck on my hat, I was in a stifling atmosphere, with hardly room to move. At one stage of the ascent a gallery opened into the shaft, and the captain of the mines requested me to taste the water which oozed through the rocks; it was salt water. We climbed a few more steps of the ladder, and entered another gallery, which opened to the level of the sea (fig. 102). It was thus placed beyond a doubt that the workings I had just visited were partly carried on beneath the very waters of the ocean."

The most extraordinary attempt that has ever been made to work a mine under the sea was one made on the rocks at the Wherry, near Penzance.

At low water at spring tides a gravelly space was left bare beyond a ridge of rocks, in which were discovered some tin lodes, which crossed each other in every direction through the Elvan rocks.   The distance of the shoal from the beach at high water is about 120 fathoms, and this, in consequence of the shallowness of the beach, is not materially lessened at low water.   It is calculated that the surface of the rock is covered about ten months of the year, and that the depth of water on it at spring tides is nineteen feet.   The prevailing winds occasion a very great surf here even in summer, but in winter the sea bursts over the rock in such a manner as to render all attempts to carry on mining operations apparently hopeless. Such were the difficulties which a poor miner of Breage, whose name was Thomas Curtis, had to surmount, whose capital was not ten pounds.

As the work could be prosecuted only during the short period of time when the rock appeared above water (a period which was still further abridged by the necessity of previously emptying the excavation), three summers were consumed in sinking the pump-shaft, a work of mere bodily labour.   The use of machinery then became practicable, and a frame of boards being applied to the mouth of the shaft, it was cemented to the rock by pitch and oakum, made water-tight in the same way, and carried up to a sufficient height above the highest spring-tide.   To support this boarded turret, which was twenty feet high above the rock, and two feet one inch square, against the violence of the surge, eight stout bars of iron were applied in an inclined direction to its sides, four of them below, and four of extraordinary length and thickness above.   A platform of boards was then lashed round the top of the turret, supported by four poles, which were firmly connected with those rods.   Lastly, upon this platform was fixed a winze for four men.   Of course while all this was in progress, Thomas Curtis was supporting himself and paying for his materials, out of the tin which he separated from the rock through which he was working. This enterprising miner thought, when his structure was complete, that he should be able to work during the winter months. This was, however, found to be impossible. The water percolated into the shaft in large quantities, and on account of the swell and the surf, the tin-ore raised could not be carried from the rock to the beach. The whole winter was therefore a period of inaction, and it was not before April that the regular working of the mine could be resumed. In 1791 Huel Wherry was in the following state:—Depth of the pump-shaft and workings below the bed of the ocean was four fathoms and two feet. The roof was worked away in some places to the thickness of three feet. Twelve men were employed for two hours at the winze, in hauling the water; while six men were teeming from the bottom into the pump. The men worked for six hours afterwards; in all, eight hours. Thirty sacks of tin-stone were broken on an average every tide, and ten men in the space of six months, working about one-tenth of that time, broke about £600 worth. Such was the character of the works carried out by this enterprising Cornish miner. In 1792 the mine assumed a much more important position. Thomas Curtis was joined by men of capital, and no less than £3000 worth of tin was raised in the summer of that year. A steam-engine was then fixed on the mainland, and a drift worked out to the mine; and a wooden bridge was constructed from the beach to the shaft, along which the ore and material were carried. This mine is said to have produced ore to the value of £70,000. Nor were its treasures exhausted at its close, which was as romantic as its commencement. An American vessel broke from its anchorage in Gwavas Lake, and striking against the stage, totally demolished the machinery, and thus put an end to an adventure, which both in ingenuity and success was probably never equalled in any country.

Fig. 102.—Botallack Copper and Tin Mine at St. Just in Cornwall; worked under the sea for more than half a mile.

In Germany, specially in the Harz and Saxony, the miners are always distinguished by the patient, as well as important nature of some of their works.   The mines of the

Fig. 108.—Winze-shaft, or "sinking," in a Cornish mine.

Upper or Ober Harz, in Hanover, which have been in work since the end of the twelfth century, and uninterruptedly since the beginning of the sixteenth century, * are remarkable for the

* These mines, twenty-four in number, concentrated round Zellerfeld and Clausthal, give employment to five thousand workmen, miners, smelters, woodmen, &c.; they produce annually a hundred thousand tons of various ores, chiefly lead and silver, of the value of £200,000. The entire country round is supported by this industry.

enormous development of their galleries and for the good
management of the underground workings (figs. 109 and 110).
In those mines the access of water had always to be contended
against; and it has been removed either by means of ingenious

Fig. 109.—Ladder-shaft in the Harz mines.

hydraulic machinery erected over the shaft, or by underground
channels, adit-levels, or soughs, adapted for enormous drainage.
As the works have been carried deeper, and a lower depth
than the level of the drainage-galleries or adits has been

reached, new shafts have been sunk at a considerable outlay
of money, time, and labour. Four of these tunnels, which are
still in work, were opened in the sixteenth century; and the
fifth in 1777. It is called Georg-Stollen (George's Level) after

Fig. 110.—Harz miners working on the lode.

George III., who was then king of England and Hanover.
Until 1851, that is, for three-quarters of a century, that gallery
was sufficient for all the purposes for which it was required;
but at the end of that interval, the depth and extension of the

works rendered the addition of another underground channel
necessary, and the gallery named after Ernest Augustus, in
honour of the late king of Hanover, the father of the present
king, was planned and very soon afterwards opened.

The Ernst-August gallery is perhaps the most remarkable
work which has ever been executed in connection with metal
mines. The mouth of this long tunnel is at Gittelde in the
Duchy of Brunswick : it is about ten feet high, six feet and a
half in width, and has a fall of six-tenths of an inch in a yard.
Like a railway tunnel, it was begun simultaneously at several
different points, as many as ten having been opened at the
same time. The work, which was carried on with the utmost
vigour, extended over thirteen years, and was finished and
solemnly inaugurated in 1864. The gallery, which is lined
for a fourth part of its length with a solid vaulting of masonry,
is six and three-quarter miles long, that is, twice the length of
the longest railway tunnel ; but if the lateral branches which
open into it are taken into account, and a subterranean
gallery, navigable for boats, into which it opens, the Ernst-
August gallery is not less than fifteen miles, or five leagues,
in length.

Nothing has been left undone to render this magnificent
work the finest example of the perfection to which the art
of mining has been carried by the Germans. The survey was
so skilfully made, and the plans so carefully drawn, that all the
junctions of the different sections of the gallery fitted accurately into each other ; the admirable precision of the results
obtained having been partly insured by the use of a magnet
weighing two hundred pounds, which influenced the compass
through the solid rock sixty-five feet thick, and which was
put in one of the working places, while the compass was held
in the other. All the difficulties connected with ventilation
were overcome by the use of pipes of sheet-zinc, by means of
which a supply of fresh air was conveyed into the labyrinths

Fig. 111.—Working by direct or descending Steps at Stolberg, in Rhenish Prussia.

for a distance of several thousand yards. In a word, in a subterranean work attended with so many difficulties, the engineer was never at fault, but was always prepared to provide against every contingency.

The cost of the Ernst-August gallery and its annexes, including the navigable gallery, amounted to nearly 3,500,000 francs (£140,000). It is the last great adit-level which the conformation of the surface of the ground of the Ober Harz will allow. The greatest depth drained by it is about 400 mètres (437 yards); but a new underground gallery is already in contemplation, without any opening to the surface, at a depth of 240 mètres (262 yards) below the last, the water of which is to be raised to the Ernst-August level by a special hydraulic engine, to be placed in an underground vertical shaft or sinking, sunk expressly to hold it; and which will serve, at the same time, as a drawing-shaft for raising ore. The expense of these new works will amount to nearly 1,500,000 francs (£60,000). The Harz miners, as will be seen, have confidence in the duration and the richness of their lodes, and look forward to the workings extending over many centuries to come.

The works in the Upper Harz are not the only ones deserving of notice. In the same parts of Germany, at Andreasberg in the Unter-Harz, there are mine-shafts sunk to a depth of more than 800 mètres (874 yards), the deepest hitherto known. These shafts are rectangular, eight mètres (twenty-six and a quarter feet) wide, and three mètres (nine and three-quarter feet) broad, giving a sectional area of twenty-four mètres (twenty-eight yards). They are lined throughout the whole depth with a solid timbering of charred fir-wood, and they serve all purposes—for the extraction of the ores, pumping the water, ventilation, ladders for the descent and egress of the men, &c.

In Saxony, Hungary, and Rhenish Prussia, magnificent

2 D

works connected with mines may also be mentioned, of no less importance than those in the Harz, and which like them have been attended with a great outlay of money. In this respect all metal-mines resemble each other more or less, and are on a par with collieries; but that is not the only resemblance between these two classes of mines.

The mode of working metallic substances is not conducted

Fig. 112.—Working by descending levels.

in a way with which we are already acquainted. When a bed or stratum is worked, as is the case in many iron-mines, operations are carried on by first of all dividing the ore into pillars, as in some coal-mines. In the case of lodes, which are sometimes vertical and always have a high inclination to the horizon, the system of working is by means of steps; either in *direct* or descending steps, which consists of attacking the ore from above; or in *reverse* or ascending steps, in which the ore

Fig. 113.—Working by the Excavation of a Mass, at Campiglia, in Tuscany.

is attacked from below.   In either case the excavations are disposed in steps like a flight of stairs upon its upper or under side (figs. 111 and 112), the first of which is not in use in collieries, because the hewer would have to stand upon the coal which he has already disengaged with his pick.   This method is, however, frequently adopted in getting metalliferous ores, which have generally to be pulverized and dressed in order to be made fit for sale and the furnace.   Lastly, in the system of

Fig. 114.—Open workings at Rammelsberg in the Harz.

working in direct or descending steps, the miner easily makes a preliminary sorting of the ore and the rock, and collects the metallic dust; while the reverse steps not only do not offer the same advantages, but he even loses, by this system of work, part of the ore in the rubbish or waste.

In rich lodes, and in the case of thick masses, recourse is had to the system of working in large chambers or extensive excavations (*par grandes tailles*), fig. 113, or by transverse attacks.   Lastly, in superficial deposits like those of alluvial ores, which spread out at the surface in gigantic outcrops, like the iron-ore of Elba, or the copper-beds of Rammelsberg in the Harz, the work is carried on in a special system (fig. 114). These methods are called open workings, because they are not carried on underground, but open to daylight, in which respect

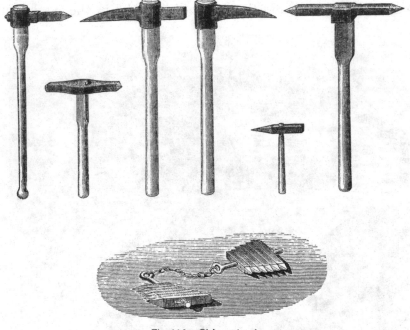

Fig. 116.—Picks and guds.
N.B.—All the tools are on a scale of $\frac{1}{8}$.

they resemble the mode of working employed in quarrying, or in large earthworks.

In one of the methods of working about to be described, only rubbish or waste is left in the mine in place of the ore which has been removed.   Nevertheless in this instance, as in most other things, there is a difference between theory and practice, and the principle adopted has often to undergo modification.   The commonest causes which compel the miner to infringe the rule, are accidents which derange the lode—-such as faults, fractures, and throws, the law in reference to which

Fig. 115.—Dialling the Plan in the La Vieille-Montagne Mines.   After F. Bonhommé.

is generally known ; thickening out and thinning away of the
lode; and sometimes its sudden disappearance altogether,
baffling all the principles of practical geology.   But the miner
is not disheartened, and by the exercise of
patience and reflection, in conjunction with
a certain sort of instinct, he nearly always
succeeds in finding the lode again.

The methods and apparatus which are
used in the underground conveyance and
winding of coal, are equally applicable to
ores.   The railways, waggons, and horses
are met with again in the levels and

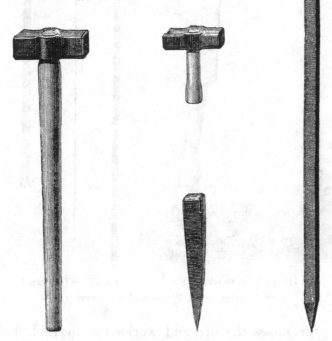

Fig. 117.—Double-hand boring hammer, single-hand   Fig. 118.— Pointed bar (Barrena
hammer, and wedge.   of the Spanish-American miners).

galleries.   The workings are conducted upon the same principles ; the ventilation is effected by similar means, and
the delicate art of drawing the plans (fig. 115) is held in
equal respect.   The arrangement of the shafts is also the

same, and the very surface buildings bear a certain sort of.
family likeness. Still collieries in the neighbourhood of the
shafts present an appearance of greater animation, it being
seldom that metal-mines, except in particular instances, raise
in the course of a month as much as some collieries do in a
single day; that is, from 500 to 1000 tons. In many mines a

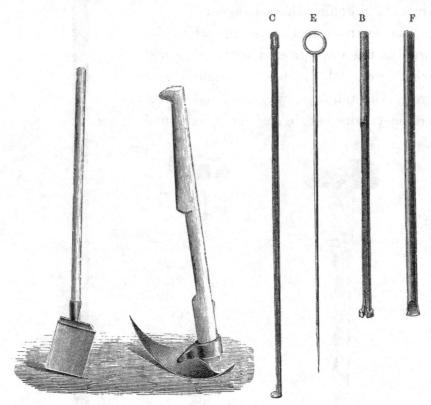

Fig. 119.—German shovels.　　　　Fig. 120.—Tools used for blasting.

C, scraper; E, copper needle; B, tamping bar; F, borer or drill.

steam-engine raises the ore and works the movable ladders,
or man-engine. The pumps are also worked by a steam-engine,
some of which, like those in Cornwall, are of formidable
dimensions. In mountainous countries the drawing and
winding engines are often worked by water-power. In the
case of small undertakings, at the outset the water and the
ore are both raised by means of a contrivance driven by horses

(fig. 61), or rather by a whim worked by manual labour. In that case the miners enter and leave the mine by galleries or levels opening to the day, or by means of ladders placed in the shafts, and sometimes by means of a rope, with the leg placed in a tub.

In very deep mines, movable ladders are almost universally in use. These man-engines (or *Fahrkunst*), which originated in the metal-mines of the Harz in 1833, have passed thence into Cornwall, where in some six or eight of the deep mines they are now used. A certain sort of skill and great vigilance and care are required in going up and down by means of them. At first one is confused by the reciprocating motion of the rods, which takes place at intervals as regular as clock-work : if the miner is nervous or afraid, if he hesitates or if he does not step from one platform to another at the proper time, a frightful accident may be the consequence. M. Simonin relates how on one occasion, when he was going down the shaft of a Cornish mine more than 300 fathoms deep, accompanied by a friend, the latter missed the platform and would have been crushed at the return-stroke of the enormous rods. Fortunately the captain, by whom they were accompanied, foresaw the danger, and saved the imprudent novice by catching hold of his dress.* Some of the tools employed in metal mines are like those used by colliers. The pick, which is of various patterns (fig. 116), is used for working in soft or fissured ground, in which wedges or gads (*pointerolles*), and the crowbar (bar or borer) are also employed (figs. 117 and 118). The ore and the rock which are disengaged are placed in baskets or bags, buckets (called in Cornwall kibbles), and waggons or trams. Some mining shovels are of very curious shapes (fig. 119).

In consequence of the extreme hardness of the rocks, nearly

---

* The man-engines have been already described, and an account of them may be found at pages 216 to 219.

all the work is effected by blasting with gunpowder. The borer (or drill), scraper, tamping-bar, and accessory tools are represented in fig. 120. The object sought is to bore holes of large diameter and to blast away very great masses. A miner

Fig. 121.—Boring a blast-hole by three men.

and two strikers almost always work in company, or at the least one miner and a striker; seldom a man by himself. Three men at work are a sight worth seeing. The two strikers, standing upright, let fall their heavy hammers in regular cadence on the head of the borer, which gives out a longing, metallic sound, whilst the other squeaking away,

Fig. 122.—Attack of rocks by fire.

a ringing metallic sound, whilst the miner squatting down, and holding the bar between his hands, gives it a slight circular turn at every stroke, so that the cutting edge of the tool may come in contact with a fresh face of the rock (fig. 121). In delicate operations, where it is necessary to keep the hole of a regular shape, picks of special shape are employed, or the gad (*pointerolle*, or *eisen*), which was used in all mines before the introduction of powder. The Saxon gad or eisen is a sort of steel chisel, resembling a long slender hammer furnished with a rectangular eye. When in use it is mounted on a handle (fig. 116), inserted into the socket, and is driven by the miner striking it on the head with an iron hammer. He carries a dozen or fifteen for his day's work, united into two sets by iron pins passing through the eyes, and connected by a chain for carrying across the shoulder. When the point becomes blunted it is sent to the smith's forge.*

Before the application of powder to underground operations, a very curious method of attacking rocks with fire was formerly practised in mines, and is still retained in some countries, amongst others in Saxony, Hungary, and the Harz. Horizontal layers of billets of firewood disposed crosswise above one another, are piled up in a nearly vertical position so as to present four free vertical faces (whence the pile has been called a *chest* by the miners), and set fire to (fig. 122). The flame plays on the face of the ore, which becomes shattered and traversed by cracks, and when cooled is very easily detached with the pick, or long iron forks. Before the use of the hardest steel tools, very strong charges of powder only could produce the effect.†

At the Rammelsberg Mine, in the Harz, to the south of

---

* The pointerolle or gad is not merely in use in metal-mines; it is likewise used in marble and granite quarries to disengage masses that are to be brought down, and which might be injured by blasting with powder.

† The splitting of the rock is increased by pouring cold water on it while in a still heated state, and advantage has been taken of this phenomenon in quartz-crushing. May not this process furnish a plausible clue to the way in which Hannibal is said to have attacked the Alps with vinegar?

Goslar, this old method is still practised, and the fires are lighted on Saturday night, and kept burning till Monday morning, when the miners return to the mine to resume their usual labours, after the fireman and his assistants have extinguished the remains of the bonfires. While the fires are burning it is impossible to remain in the galleries, for a thick smoke fills all the working places of the mine.

In consequence of the difficulties connected with the prosecution of underground operations when carried on by manual labour in a hot and vitiated atmosphere, the attention of inventors has of late years been directed to the subject; and various attempts have been made to contrive a machine by which rocks may be bored or worked away, and coal hewed.

The principle on which these machines have been devised is either that of boring a hole by the continuous motion of a rotating drill, or by means of intermittent blows delivered with a pointed tool, striking the rock after the manner of blows delivered from the elbow and shoulder of a man. In the rock-boring machine invented by Mr. George Low, and manufactured by the Messrs. Turner, of Ipswich, the work is effected by a combination of both these operations.

The machine represented in fig. 123 consists of a boring cylinder, into which the tool (very similar to the ordinary hand-tools) is inserted. This cylinder moves within another cylinder, in which it is made to rotate slightly but continually between each blow of the drill, which strikes the rock upon one spot at the rate of from 300 to 500 blows per minute. Holes of two inches in diameter can be made in the hardest granite at the rate of two inches per minute, while in softer stone double that rate has been attained. That portion of the machine which bears the drill, and to which, of course, all the rest is accessory, is borne upon an iron frame, running upon wheels, on rails about two feet apart, and is moved backwards and forwards by a small engine, which is encased within the iron column supporting the drill.

The reciprocating motion of the boring-tool is produced by compressed air or steam. The former is preferred in underground work, as it serves to ventilate the workings, whilst the steam produces an atmosphere unsuited for respiration. When air is used, it is compressed into a large receiver similar to a steam-boiler, by Low's hydro-air-compressing engine, which

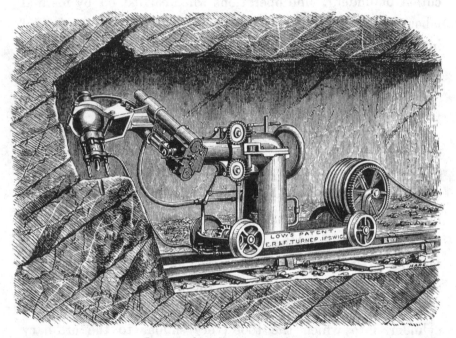

Fig. 123.—Low's Rock-boring Machine.

consists of two powerful air-pumps worked by a steam-engine. In Low's process the compressed air is forced through water, which deprives it of all its heat, and enables a higher pressure to be attained than is practicable by any other process. For working the boring-machine a pressure of from seventy lbs. to ninety lbs. per square inch is necessary. This may be conveyed through india-rubber tubing to any distance with little loss of pressure, so that the engine and air-compresser may be at the mouth of the workings, whilst the boring-machine may be at work underground at a distance of a mile or more. The air is conveyed to the boring-cylinder through tubing which is coiled upon a drum at the back of the machine, which uncoils itself as the machine is advanced, and re-coils itself when the machine is drawn back from its work. When the air is not used by the boring-cylinder it is applicable for propelling the machine to and fro by the small engine encased in the colunm, which by a suitable arrangement of gearing is made to propel the machine, or to raise and lower the boring-cylinder, and to turn it to any angle it may be required to work in.

Owing to an ingenious arrangement there is an air-space at the top of the boring-cylinder, which is always filled with compressed air. This forms an air-cushion, which receives the concussion of the blows, and prevents the crystallization of the working parts, and enables the machine to stand as steady when delivering 400 blows per minute as when it is not at work.

The fragments of stone are removed from the bore-hole by a powerful jet of water obtained from the waste through which the air is compressed, so that the tool has never to be withdrawn except for sharpening, even with a perpendicular hole.

Those who work in metal-mines are liable to fewer casualties than coal-hewers. Accidents resulting from the mere working of the mines, such as falls of ground, inundations, and those met with in traversing the shafts, are almost the only ones which they have to guard against. They escape, except in quite exceptional cases, fire-damp and carbonic acid; and, lastly, fires, which rarely happen except in seams of coal.* As to foul air, and suffocation by carbonic acid, they have still less to fear, as much owing to the different nature of the rocks

---

* Nevertheless, there are some instances of fires having taken place in metal-mines. The mine of Almaden, in Spain, was on fire for two years and a half, about the middle of the last century (1755). Silver-mines have been destroyed by fire in Mexico. At Encino, near Pachuca, the timbering which supported the roof of the galleries was consumed by fire, and most of the miners were suffocated before they could reach the shaft. A similar fire led to the abandonment of the Bolanos mines in 1787, when the works became flooded after the fire, and were not reopened till five years afterwards.

they work in as the generally less extent of the open spaces in metal-mines—both which tend to produce a more perfect ventilation of the galleries. They are likewise nearly exempt from a cause of danger which is very commonly met with in collieries, arising from extensive old workings, mostly of unknown age, and filled with water and mephitic or explosive gases. Lastly, the compact nature of the rocks places them out of the way of any danger arising from the sudden falls of bell-moulds (*cloches* or *fonds de chaudron*), or lumps of ironstone, which suddenly become detached from the carboniferous schists, or by the fall of masses of shale or rock from the roof on to the floor of the galleries—a frequent source of accident in coal-mines. But such casualties as flooding a level by the outburst of water from an old working; known by the miners as a " house of water," do occasionally occur.

The metal-miner is, however, far less fortunate than the collier, if the underground work be regarded from a hygienic point of view. From the effects of climbing, of working in air deficient in oxygen, and having an excess of carbonic acid, the average duration of a miner's life, in the deep mines of Cornwall, and in the lead-mines of the Mountain Limestone districts of the north of England, is reduced to about thirty-two years. Then certain metal-mines, such as those containing arsenical pyrites, or quicksilver-ores, entail real danger on the miner from other causes. In the quicksilver-mines the miners after a certain time become salivated, and are seized with convulsive tremblings. In the smelting-houses, where ore is reduced, it is still worse; and the poor workmen soon become pallid and frightfully thin, and look more like corpses than living men. It is, however, curious that the men employed in the arsenic manufacture live long and are generally healthy.

Having no fire-damp to fear, nor any explosive gas, the miners of metalliferous lodes make use of open lights of a particular form. First there is the iron oval-shaped lamp, movable about a stirrup-shaped handle, to which is attached a carrying-hook (fig. 124). This is the most common miner's lamp in France. A very convenient kind is that (fig. 125) which admits of its being easily fixed against the timbering by means of the iron point in which the stirrup terminates.

The lamp (fig. 126), or more correctly lantern (or *blende*), in general use both in collieries and metal-mines in Saxony, consists of a small globular brass lamp inclosed in a square-based chestnut-wood box, which is sometimes furnished with a glass in front to protect the flame. A long hook is fixed to the back of the box, by means of which it is passed through a leather strap round the miner's waist.

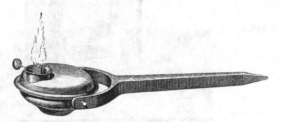

Fig. 124.—Modification of the French lamp.

Fig. 125.—French lamp.

N.B.—All the lamps are drawn to a scale of ⅓.

The Cornish miners, as we have already said, merely attach a tallow candle to their hats by means of a lump of moist clay (fig. 156), thus calling to mind the Cyclops—the ancient Sicilian miners and founders, the lieutenants of Vulcan—who are described by good old Homer as having only one eye, which was in the middle of the forehead.

In Rhenish Prussia, Mansfeld, the Harz, and the Tyrol, the lamps are made after a special model, which is almost always of elegant design, and sometimes reminds one of the ancient Greek and Roman lamps (figs. 127 and 129).

The miner's lamp sometimes characterizes an entire country; besides which, when the miners emigrate to foreign places they almost invariably take their lamps with them. It is for this reason that, in Italy, the lamps used in Saxony or the Tyrol are met with at the mines to which Saxon and Tyrolese caporals have come, with men from their own countries, to start the preliminary works.

In most of the mines of Spanish America the lamps are of

Fig. 126.—Saxon miner's lamp, or Blende.

a very original description. One kind is precisely the same as that used by wine-merchants in their cellars, and consists merely of a thin piece of wood split at one of its ends to clasp the candle, which is fixed in it upright, the other end being carried in the hand (fig. 130). But it is not only the lamps, but the workings themselves, which everywhere, in these mines, are of a strange and primitive stamp.

It is necessary to go back to the aboriginal Indians, or at any rate to the time of Isabella the Great (1474–1504), to understand everything that is to be seen there. Rude ladders are fixed in the shaft, cut out of the trunk of a tree, like enor-

Fig. 127.—Prussian miner's lamp.

Fig. 128.—Mansfeld miner's lamp.

mous notches. It is dangerous for persons who are not accustomed to them to climb these curious ladders, but the native miner goes up and down as if by magic (fig. 130).

When the shafts are deep this system cannot be adopted. Then the men suspend themselves by a rope, in a manner like to that mentioned as being practised at the salt-mines of Galicia (fig. 94), or formerly in some English collieries. Supported by a leather strap, upon which they sit, and furnished with a staff, which enables them to keep clear of the side of the shaft, the miners yield

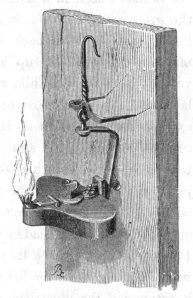

Fig. 129.—Harz miner's lamp.

themselves to the movement of ascent or descent, while the

2 F

man with a light who precedes this human cluster, provided with a torch, opens the way. The pit-mouth looks like a misty distance. Sometimes when the wearied horse stops for a moment, the rope stops also and stretches as though it were going to break, or else it turns on its axis. At length the pit-bottom or day is reached, and generally without any accident.

Equally primitive systems are practised in the various branches of the workings at almost all the mines of South America. Thus in some localities powder is not used for attacking the rocks, and the long bar of steel, the *barrena*, is merely employed (figs. 118 and 130). The Nicaraguan miner does wonders with that tool, and like the Mexican miner he is an accomplished *barretero*.

For draining off the water a contrivance is made use of, turned by three or four pairs of horses, which brings the water up in bags holding more than 1000 litres (222 gallons). The ore is also raised in a similar way, but more frequently it is the *apire* or porter who, in a stifling atmosphere, carries the heavy load upon his back in a leather sack, along the interminable galleries, and up a steep incline fitted with inconvenient ladders. In Chile and Peru the *apire* is the miner's labourer. In Mexico, where he is called the *tenatero*, he excited the admiration of Humboldt at the beginning of the present century. The animation and vigour with which the tenatero accomplishes his laborious task, making eight or ten journeys every day, without resting, up and down ladders with eighteen hundred rounds, naked to the waist, leaning on a little staff, and loaded with a burden which may exceed 100 kilogrammes (220½ lbs.), was a kind of labour of so barbarous a kind, and attended with such astonishing results, that it astounded the illustrious traveller. There were men there of sixty years of age, and boys of scarcely twelve.

If the mines of South America are, in most instances, car-

Fig. 130.—Miners of Chihuahua, Mexico.

ried on in an altogether primitive way, it must be remembered that great and gigantic works are not wanting in some localities, especially in Mexico, as at the *Socabon del Rey* ("the King's gallery"), at the mines of Guanaxuato, in which a person may ride on horseback for a distance of 100 mètres (109 yards), and which is driven for a distance of more than 600 mètres (656 yards); also the octagonal shaft of the Valenciana mine, the beginning of which was witnessed by Humboldt in 1803, which is 11 mètres (12 yards) wide, and descends to a depth of 400 mètres (437 yards) beneath the surface. The adit-level of *La Biscaina*, at Real-del-Monte, is 2½ kilomètres (1½ miles) in length. It must be acknowledged, however, that these are brilliant exceptions, which only serve to show that South America, if she really wished, could vie with Germany in the art of mining; whereas, in most cases, matters are carried on in that part of the world after a somewhat old-fashioned and primitive manner.

CHAPTER IX.

THE SISTERS OF THE CATACOMBS.

The mines of antiquity.—Convicts as labouring miners.—Opinion of Tacitus with regard to underground work.—Works of the Middle Ages.—Origin of Practical Geology.—First use of the compass.—Tuscany; ancient workings at Campiglia; formation of *Buratite* in the old galleries.—Iron-mines of Monte Valerio.—The Cento Camarelle —Bats.—Volterra; ancient Etruscan Coins.—Formation of stalac tite in the old Tuscan mines.—Old mines of Massa Maritima.—Spain; mines worked successively by the Phœnicians, Carthaginians, Romans, and Arabs.— Stone-hammers and human skulls found in ancient Spanish mines; in the copper-mines of Lake Superior; the Asturias, and in the province of Cordova.— Stone and bronze implements found in France.—The mummified miner of Fahlun. —Fossil soap.—Mines of Gar-Rouban in Algeria; lamps.—The aboriginal gold-miner and goldsmith of Chiriqui.—Mineral Archæology.

BEFORE geology became an exact science, and when the mode of working metallic mines was still handed down as a sort of mysterious tradition, mining was carried on in a hesitating, timid way peculiar to each separate locality. It was not known whether the lode lasted in depth; it was taken, so to say, by the hair of the head instead of being attacked in the very heart. So, too, if it were uncertain whether it had a horizontal exten-sion, pits were dug upon it at all points where it was visible, and never elsewhere. The workings were close together, and very narrow and shallow besides. A man could scarcely crawl through the low and tortuous galleries, and the working-places were not less confined. It was the time when persons were condemned to the mines, and it may be understood that such labour would only be handed over to slaves, or to prisoners of war. Moreover, working in the mines was regarded by the ancients as infamous, and Tacitus held the Goths in infinite contempt because they not only devoted themselves willingly to this kind of labour,* but having iron they did not make use of it to assert their liberty.

In the Middle Ages, when a slight calm followed the tumult of invasion, and allowed the mines to be reopened, subterranean industry began to improve. Amongst the ancients there were not, strictly speaking, either engines or machinery, and everything was done by manual labour, the ore even being raised on the backs of men. Moreover, they beat a retreat on the occurrence of the slightest fall of ground, or at the least appearance of water: and movements of ground and inundations often take place in works carried on very near the surface. In the Middle Ages machinery made its appearance. The water-wheel and the pump were employed for raising both the ore and the water; timber stays and props were put up with a certain sort of order and calculation, to resist the pressure of the rocks. Moreover, practical geology makes its first appearance, and it was in fact with the operations of mining,.which proves the ground in depth, that it really originated. The direction and inclination of mineral veins was noticed, and it began to be understood that these last were not distributed by chance, but that their mode of occurrence was regulated by certain laws. From the sixteenth century, the use of the compass extended in Italy to the working of mines, almost simultaneously with its extended use in navigation. The old mining code of Massa-Maritima, which makes special mention of the *calamita* or magnet in surveying underground plans, leaves no doubt on this point. Finally, the labourers were mostly freemen, working for the most part of their own free will; while in many countries, as in Tuscany and Bohemia, the works were regulated by special laws, dictated by an enlightened and liberal spirit. Still a certain sort of superstition continued to prevail, and the divining-rod asserted its mastery, and will long

---

* " Gothini, quò magis pudeat, et ferrum effodiunt."
" The Gothinians, to their additional shame, also dig for iron."
Tacitus, De Mor. Germ , lib. xliii.

continue to make its signs, for there are some countries where
it is even now implicitly believed in. A kind of disorder also
prevailed in underground works. The proprietorship of mineral
veins, when independent of the ownership of the surface,
became infinitely divided. Every discoverer appropriated the
part which he originally discovered, and bravely descended
beneath the surface; and terrible fights not unfrequently
ensued in the galleries, when two veins united.

It may, perhaps, be interesting to visit some of these ancient
workings, executed in metal-mines at periods of which, in many
instances, history has not preserved the precise date. These
are the Sisters of the Catacombs, less known but certainly
more curious than the vaunted quarries of Paris or Rome.
When the rock is firm enough the excavations are sometimes
enormous, as may be seen in the ancient copper-mines of Campiglia, in Tuscany, which have been open since the Etruscan
times; there are excavations there large enough to hold a six-
storied house with ease. These vast chambers communicate
with each other by means of narrow galleries, or rather passages, in which a person can scarcely crawl. The barren rocks,
left as rubbish or waste in the excavations, have hardened and
become cemented together under the pressure of the overlying
beds, and by the earthy debris of the mine. These artificial
masses can only be broken by blasting, like those blocks of
concrete which are thrown into the sea in the construction of
breakwaters. The wooden props are still in place, rotted or
rather carbonized by a sort of slow decomposition of the vegetable tissue; all the smell they give out may be recognized
as that of the evergreen-oak and the chestnut, which are always
grown in the country. Fragments of vases, lamps, and amphoræ, which are found in the rubbish, are connected with
Etruscan art. Wedges and bronze picks have also been met
with in the mines, affording proofs that these works date
from a period when iron was not commonly used for ordinary

purposes. Outside the mine Etruscan money has been found, mostly bearing the head of Vulcan, and on the reverse a hammer and tongs, the emblems of the god of miners, smiths, and founders (fig. 131). Agates and carnelians engraved in intaglio, or carved into the shape of the scarabæus, clearly proving the Egyptian or Asiatic origin of the Tyrrhenians, have also been dug up here and there. Enormous masses of

Fig. 131.—Triens, or the third part of an As: Etruscan coin of Pupluna (Populonia), having the effigy of Vulcan stamped on copper from the mines of Campiglia. Natural size.

rubbish cover the flanks of the mountains where the ancient pits have been opened, and over an extent of several miles follow two parallel courses marking the outcrops of the veins. In the valleys there are still enormous heaps of cinders, on the very sites of the ancient foundries; and there, also, various objects have been found which enable us to assign these works to the Etruscans.

Three thousand years have passed since these mines were first opened; when the dawn of metallurgy, and with it of civilization, took place in the Italian peninsula. And yet, so complete has the quietude of these sombre depths remained, that the trace of the tool remains still as visible on the rock as though it were only made yesterday. The gad has marked its passage by a series of oblique and parallel grooves of small extent; below there is another series of grooves, then another, and so on. When the rock had been dressed on three faces, those above and below being naturally free, it was raised on its base by means of wedges, crow-bars, and levers; and then the same hard and tedious labour began again. The vein was poor; but copper was more valuable at that time than it is now, and labour was, beyond a doubt, cheaper. On the other hand, machinery can scarcely be said to have had existence, any more than steam, gunpowder, or steel. Lastly, physics and chemistry were so imperfectly studied that they lent but little aid to the miner.

In the old galleries at Campiglia new ores have formed, and nature has used this mine as one of her laboratories, the profound calm being found there which is necessary for the production of crystallization: under these favourable conditions, with the aid of water and air, yellow copper pyrites, pitchy blendes, and the silicious rock itself, have become metamor-

Fig. 132.—Earthen lamp found in Etruscan workings at Monte Valerio in Tuscany. Scale ¾.

phosed into blue or green stalactites. One of the mineral species contained in these specimens, and recognized twenty years ago, has been named by mineralogists *Buratite*, in honour of Burat, the engineer who first had the management of the re-opening of the ancient workings at Campiglia. Chemically this accidental mineral is a carbonate of copper and zinc, the occurrence of which has been since made known in other mines.

At the iron-mines of Monte Valerio, in this same district of Campiglia—opposite to the Isle of Elba, and worked also by the Etruscans simultaneously with the Isle of Elba itself— several interesting objects have been found in the rubbish.

An earthen lamp is represented in fig. 132, the shape of

which may be compared to that of some modern mining-lamps (figs. 127 and 129). This type is also met with in Tuscany in the lamps used by the peasantry throughout the Maremma.

Near the same mine of Monte Valerio there is a series of ancient caverns, in which the Tyrrhenians also mined for iron. They are designated by poetical names, such as the *Cento Camerelle*, or the hundred chambers. Bats are now their only inhabitants, who deposit beds of guano, which the people of the locality might perhaps find it to their advantage to work in default of iron-ore.

In another part of Tuscany, at Volterra, Etruscan coins

Fig. 133.—Semissis or half-as: Etruscan coin, cast of copper from the mines of Monte Catini. Obverse; the effigy of Janus or Hermes (both young faces) capped by a petasus; reverse, *Felathri* with a club and a crescent-moon. Nat. size.

have been found bearing the effigy of Janus bifrons, or rather Hermes, the patron of the Pelasgian miners (fig. 133). The double-headed Hermes was acquainted with the secrets of futurity and those of geology as well. The copper out of which these coins were made was certainly got from the neighbouring mines, which are those of the present Monte Catini.

In most of the ancient Tuscan mines chemical deposits are found analogous to those which are produced at Campiglia. True calcareous stalactites sometimes completely stop up certain galleries, and these columns of crystalline marble are disposed here and there precisely as they are in natural caverns.

The water which trickles through the roof of the galleries percolates through rocks, from which it takes up small quantities of carbonate of lime. When the water evaporates the solid deposit is left behind; or else the drop in falling carries with it a portion of lime which, in its turn, is deposited on the ground; and the column goes on forming above and below, until the parts become united. The length of time required to produce such results is enormous, the process being essentially a slow one; but nature has no regard for years in her operations. We only, poor travellers here below, are obliged to take heed of time.

It is in the old mines of Massa-Maritima, near those of Campiglia, and dating mostly from mediæval times, that these wonderful stalactites are mostly found. Tool-marks may also

Fig. 134.—Picks from the ancient mines of Massa-Maritima, in Tuscany. Scale ⅓.

be traced on the walls of the galleries, and further and further different marks, left beyond a doubt by the managers to show the successive progress of the workings. There is still visible the niche where the miner put his smoky lamp; further on the place where wood had been lighted to break up and disintegrate the rock. The tools found in the rubbish are made of iron (fig. 134); the age of bronze having long passed by. The remains of leather-bags in which the ore was carried, and pieces of the ropes used for drawing it, have also been met with. In some places the continual friction of the ropes has produced deep grooves in the stone. In many instances the rubbish and the timber have undergone no changes; and lastly, instead of a confused labyrinth, an arrangement is apparent of galleries disposed in different stages, while a certain direction and a regular slope are perceptible in all the important parts of the workings. The sides of the shafts are truly vertical, and the ore has been cut back in regular masses for working; the open spaces having been very carefully either stayed, walled, or filled with waste.*

In the absence of written history or trustworthy oral traditions, it is difficult in most cases to fix the precise date of certain subterranean excavations. We know that in Spain the Phœnicians, the Carthaginians, Romans, and Arabs followed each other in the mines; for instance, in the lead and silver mines of the Sierra de Gador. The pits are close together, narrow, and of little depth; the galleries low and winding, having the appearance of great antiquity; and the question arises, to what extent had the people to do with them whose names have just been mentioned? In such a case, the finding of a coin, or even a lamp, must surely tend to throw some light on this archæological inquiry. In the rubbish of some Spanish mines iron picks and Carthaginian coins have also been found, showing that the works in question are at least as old as the Punic domination, and that from that epoch iron was in use in the Iberian workings. The type and character of a medal almost always furnish a date with sufficient precision, within a few years. Some ingots of lead bearing Latin †inscriptions, and unglazed amphoræ with pointed ends, such as those which are still in use all over the south of Spain, lamps, the entire skeleton of a slave in chains, a noria or chain with cups for raising water, and lastly, polygonal timbers fastened together by means of tenons have, likewise, been found in the old mines of Carthagena—and this time of Roman age.

In another Spanish mine at the foot of the Asturias, on a

---

* It should be stated that some of the Roman workings in the south of Spain are remarkable for their regularity.

† In the Museum of Practical Geology, London, there is pig of lead, of the same shape as those of the present day, inscribed with the name of the Emperor Hadrian. It came from the ancient Roman foundries in Brittany.

copper-lode, stone implements and a deer-horn formed into a
chisel have recently been found.   The primitive working of
this mine belonged to a very ancient human period, when
implements of bronze were about to supersede those of wood
or flint; but before casting the metal, the lode must have been
worked.   Hence the presence of those stone hammers and
chisels made of deer-horns, which were employed in the place
of copper, too scarce a substance to make tools of at first, and
the alloy with tin was unknown in Spain.   The upper portions of the cupreous deposits, being earthy, pulverulent, and
decomposed, yielded to wood and stone.   The objects just
mentioned have remained as irrefragable evidence in this mine,
which is perhaps the earliest in date of the localities worked
for copper in Europe; and not only have implements been
found there, but three human skulls turned green by the
carbonate of copper which in course of time is formed by the
decomposition of the ore.   Anthropologists who see the entire
man in his bony case, because it contains the brain, after a
careful examination of these crania, have declared themselves
of opinion that they were those of Basques, and moreover
brachycephalic; that is to say, with rounded skulls, the true
type of primitive European man.   We see then, in these
authentic remains, the relics of our oldest miners.

In the ancient copper-mines of the province of Cordova
a stone hammer has also been found (fig. 135).   The way in
which the handle should be fastened to it may be guessed, on
comparing it with similar hammers brought from the ancient
copper-mines of Lake Superior (fig. 136), or with those which
are still in use amongst the Indians of Texas.   These people
make use of the sinews of the bison for hammer-handles,
wrapped round with a wide band of the hide of the same
animal sewn on raw, and passed round the hammer-head over
a groove which is worked round the middle portion.   The skin
contracts in drying, and the stone, whose two ends only are

left free, is held by the handle as in a sheath that is too narrow for it, and from which it cannot escape (fig. 137).

The ancient Gallic localities offered some interesting remains of early industry. At La Villeder, near Ploërmel, department of Morbihan, at the ancient stream-tin works worked by the Kelts, a stone hatchet and a bronze hatchet have been found (figs. 138 and 139), as though the two periods characterized

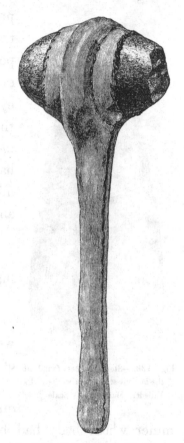

Fig. 135.—Stone hammer from the ancient copper-mines of the province of Cordova. Scale ½.

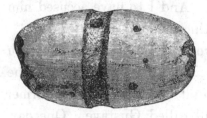

Fig. 136.—Stone hammer from the ancient copper-workings of Lake Superior. Scale ½.

Fig. 137.—Stone hammer fitted to a handle by the Kayoway Indians of North America. Scale ¼.

by those two implements went hand in hand at the very mine, the working of which would have the effect of causing the disappearance of implements of the first-mentioned kind.

At the copper-mines of Fahlun, in Sweden, it was not merely the skull of a miner, as in the Asturias, but the entire body which was found on one occasion, in 1719, in a working-

place which had long been abandoned. The body was preserved intact by the blue vitriol or sulphate of copper produced in the mine under the influence of the atmosphere and water.*
Working miners are naturally superstitious, and the body was brought out with solemnity, was shown, and even carried in

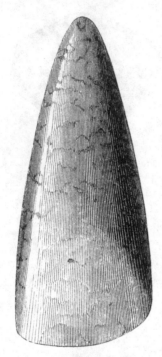

Fig. 138.—Stone-hatchet found at the ancient stream-tin works of La Villeder (Morbihan). Scale ¾.

procession. There was a great agitation throughout the country, and people came from many leagues round, everybody wishing to see with his own eyes, and to touch with his own hands, the vitriolized miner. An old woman, more than eighty years of age, had hastened like the others to witness the miracle; when suddenly she shrieked and fell to the ground senseless.

" What is it, my good woman?"

" It is he, himself!" said she, pointing to the body.

" Who, he?"

" He, Gustave! whom I have so wept for. And I to have accused him of forgetting me, and of ingratitude!"

The woman was overwhelmed with questions, and it appeared that the miner whose body had been found was actually her former affianced lover, he whom she still called Gustave. One day, sixty years previously, he had disappeared, and nobody knew where he had gone to. He was not employed at the mines, so that nobody thought of searching them underground; on the surface they are traversed by vast crevices, and the workings extend underground in gigantic hollows. Attracted by curiosity Gustave had perhaps leaned too far forward, and

* This salt possesses very curious conservative properties. Dr. Boucherie has applied these to wood used for industrial purposes, and especially for railway-sleepers, which have by this means been rendered incorruptible.

had been seized with a sudden attack of giddiness; or else, perhaps, a jealous rival had precipitated him to the bottom of this new Tenarus.

No one could clear up the mystery; science alone determined that the mummy, very different in that respect to those of Egypt, had preserved its appearance of youth and beauty, and for the first time two old lovers found themselves together, after the lapse of more than half a century—he with his twenty years of age, she with more than twenty lustres. The story might form the plot of a fantastic romance, which would recall to mind that of "l'Homme à l'oreille cassée."

The blue skull of the Asturias, the still-chained skeleton of

Fig. 139.—Bronze kelt from the ancient stream-tin works of La Villeder (Morbihan).   Scale ¾.

the mines of Carthagena, and the vitriolized miner of Fahlun, are amongst the most curious human remains that have been met with in old mines; but other interesting objects have been found of a similar kind to those already mentioned. We are told by M. Fournet that he has taken out of the argentiferous lead-mines of Pontgibaud, in the Puy-de-Dôme, which have been worked perhaps since the time of the Gauls and Romans, not only iron picks and hammers, but a lamp still containing a lump of tallow. This substance, with which miners used formerly to light themselves, and which is still used at many mines, had assumed a strange aspect, and had become in some measure saponified—the fat having become a true fossil soap.

It has been stated that the miner was sometimes able to form an opinion as to the date of a working from the form of

the tools and implements.   In the mines of Gar-Rouban, in
Algeria, the remains of earthenware lamps, of vessels for holding water or oil, with pointed ends like those of certain *alcar-
azzas*, found in the ancient excavations, enable us to assign the
origin of these workings to the Moors.   The lamps (figs. 140
and 141) have a form which is not known elsewhere, and

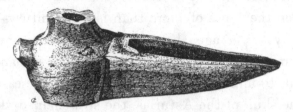

Fig. 140.—Moorish lamp found at the mines of Gar-Rouban, in Algeria.   Scale ½.

remind one of the type of the pointed shoe styled *à la poulaine*.
One of them is varnished black, and shows traces of arabesque
ornaments.   Iron hammer-heads and timber props have also
been met with in the mine.   The Arabs supported their
galleries with extreme care, and the wood used by them for
the purpose was that of the *Thuja* (arbor vitæ), which is
now found in the forests of
Algeria, and is not applied to
rough purposes, but is employed for the most delicate
ornaments of upholstery.

Fig. 141.—Fragment of a Moorish lamp found in
the mines of Gar-Rouban, in Algeria.   Scale ½.

The implements of the
Stone Age which have been
described, are not the only
ones which have come to our knowledge.   At Chiriqui, in
the province of Panama, are gold placers which have been
worked by the Indians from time immemorial.   M. J. Thévenet
(who has communicated the particulars connected with Gar-
Rouban) passed through these curious countries in 1859,
and formed the idea, as was then the custom, of working the
tombs of the ancient gold-seekers instead of the impoverished

placers. By the side of the jewellery and nuggets of the past, he found in one of these tombs a chisel to cut the metal, an awl for working it, and burnishers to polish it—all the implements being made of flint (figs. 142 to 144). The awl shows at the sides in some places traces left by the gold, as though it had also been used for a touch-stone. This aboriginal gold-hunter, whose last sleep was thus disturbed after the lapse

Fig. 142.—Chisel.   Fig. 143.—Awl.   Fig. 144.—Polishers.

Flint implements taken from the tomb of an Indian gold-finder of Chiriqui, in the province of Panama.
Natural size.

of so many centuries, was certainly a worker in gold as well as a finder of it; and he, therefore, belongs to the types of workmen we have to do with here. The period when he flourished belonged to the Stone Age of Central America; and if the epoch be not prehistoric, it is certainly long anterior to the arrival of the Spaniards in America. The soldiers of Cortez introduced the use of iron into those countries, where

2 F

not only had bronze been already known for many centuries, but also even the way to temper and forge it, so as to make it supply the place and the absence of the metal of Mars.* The stone implements which have been mentioned may then be referred to the most ancient periods of the working of mines in North America.

These researches might be extended, and the list of all the curious objects taken from the old galleries might be increased; but this is not the place to display all the remains derived from extinct civilizations which are revealed by the ancient excavations. It is sufficient for the purpose in view to point out the way, and to have shown to more able persons the road to some of the more ancient mining undertakings which are still accessible. There is more than one inspiration to make in these subterranean visits, and a novel and fruitful branch to open in the study of antiquity. The ancient mines are true Herculaneums and Pompeiis in miniature; and regarding them from this point of view, the field of archæology may be considerably extended and enriched by means of what may not inappropriately be called mineral archæology.

* The sign for iron, ♂, used by the alchemists, represents the lance and shield of Mars, the god of war.

CHAPTER X.

CRUSHING AND WASHING.

Picking and breaking of Ores.—Crushing, jigging, buddling.—Crushing-mills and washing-machinery.—Modes of washing auriferous substances described : the Indian horn or poruna, the batea, the pan process, the Chinese cradle or rocker, the Georgian tom, the Brazilian sluice, the flume.—The Chilian process.—Washing away of banks and hill-sides by water: the hydraulic process.—Treatment of auriferous quartz.—The American process of amalgamation: Mexican, Chilian, Hungarian, Tyrolese, and Russian amalgamating machines.—Mode of obtaining gold by amalgamation described.—Proportion of pure gold at the mines of Siberia Australia, and California.—The standard for gold in England and France.—Amalgamation of silver-ores: the Mexican or patio process.—German method of amalgamation; cupellation; liquation.—Pattinson's desilvering process.—Table showing the minimum amount which metallic ores should yield.

THE ores, as they are extracted from the mine, always contain a certain portion of the barren rock or gangue in which they are inclosed. With very rare exceptions, as in the case of certain homogeneous and compact iron-ores, the metallic substances are irregularly disseminated through the gangue, and sometimes in particles scarcely visible to the eye. Very often the produce of the ores does not amount to a millionth or even to a ten-millionth part for the precious metals, and to a hundredth for common metals. The first process the ores have to undergo, therefore, on being taken from the mine, is to make them richer; and this is the object of what is called the mechanical dressing. Then begins a whole series of delicate operations, in which mills, on the one hand, and water, on the other, play the principal part: whence the names of mill and washing applied to the work-places where this concentration of the ores is effected.

The first operation to be performed is that of picking or sorting, and breaking or cobbing by hand, which is done with a

hammer by boys or young girls, for there is not sufficient light
for this purpose in the mine; and besides it does not form a part
of the miner's business to do it.   By daylight, under a shed or in
the open air, boys do this work, singing the while, and at many
of the Italian mines these lively workers mark the cadences of
their songs with the strokes of their hammers.   Each one joins
in these choruses, in which the voices of the boys blend agree-
ably with those of the young girls; solo-singers are not wanting, and every one instinctively observes a wonderful accord.
It is clear that one is, in Italy, amongst a people born with a
taste for music.   In England, under similar circumstances, the
women and girls either do not sing, or if they do, often sing
out of tune.   To make up for this, however, they are dressed
coquettishly like regular ladies, wearing on their heads or in
their hair the *cornette*, or the net, and elegant and well-fitting
boots on their feet (fig. 145).

The ore, after it has been broken and sorted by hand, is
divided into three lots: the poor or barren ore, which is
rejected; the rich, which is laid aside for sale or for smelting;
and the ordinary or middling, which is to undergo other pre-
parations.   At this stage the breaking and mechanical sorting
begins, under upright or flat stones, between rollers turning
towards each other; or lastly, by the shock of heavy pillars of
cast-iron the ore is pulverized, and then separated by means
of a sieve of very ingenious contrivance into pieces of equal
sizes.   The washing is the next operation, which is performed
on jigging machines (*cribles*), or on inclined tables, called
frames in Cornwall, and on *buddles*, either fixed or movable,
over which a stream of water passes, agitating and carrying
away the sands or slimes, and leaving the richer and heavier
metallic powder behind.   The principle of this operation is
extremely simple.   The pieces of ore in each special machine
being nearly of the same size, the heavier ones having a less
velocity imparted to them, by the motion to which they are

subjected, than the lighter ones, gradually become separated from the latter, until at last there are three kinds—the barren slimes, the rich ore, and the middling, which will undergo a repetition of the operations of *buddling* and *framing*.

Metal-mines being generally situated in mountainous dis-

Fig. 145.—Cornish ore-dressers.

tricts, the neighbouring stream or torrent is used for washing the ores, and all the machinery is set in motion by a water-wheel, which is often replaced by a steam-engine. When water is scarce, the works stand still for a part of the year during summer; in which case, as it is impossible to supply all the washing machinery with water, each of which consumes a large quantity, an attempt has been made to use wind as a motive power in the concentration of the ores—in many instances with success.   The air is injected into a pipe, the arrangements of which have been calculated for the peculiar work to be done, and in its course carries away the finely pulverized particles, carrying the lightest the greatest distance.   A sharp separation is thus established between the metallic powder and the barren sand; that is to say, between the ore and the gangue. In the same way vanes are made use of in agriculture to separate the chaff from the grain, and in grinding to sort the bran from the flour.

The crushing-mill and washing-machinery are also as near as possible to the mine.   They present an animated sight; women, young girls, and boys are mostly employed in the washing, the men rarely working except at the mill.   The machines do nearly all the work automatically; and it is a pleasant sight to see them carry and distribute the ore wherever it is necessary, and in the conical turning-tables even sort the washed sands (slimes) by sweeping, with a movable brush, each different sort into the proper box for receiving it.

Washing the ores is the great preliminary operation of all metallurgy.   It is in this process that the hard work at the placers consists, which become then transformed into true washings in full work, so much so that the Spaniards of America call them in this case *lavaderos*.   The operations are always based on the same principles—the separation by means of water of substances of different weight and nearly of the same size, and the concentration of the metalliferous substance.   At the placers in California, Australia, and the Ural, washing the sand or "pay dirt" forms the concluding operation.   Before the arrival of the Europeans, the Indians employed the horn or *poruna,* which is made of an ox-horn cut lengthwise through the middle, and fashioned in hot water into the·shape of an oblong bowl (fig. 146). The natives made use also of the wooden bowl (*batea*), cut out of the trunk of a tree, and hollowed out like a large cup, as much as two feet in diameter, but very shallow. In California the Europeans have substituted for the horn, or the bowl, the pan made of wrought iron, and more recently of thin sheet-iron or tinned-iron—similar, except in size, to the article used in our households. The placer-miner's pan has a flat bottom about a foot in diameter, and sides six inches high, inclining outwards at an angle of thirty or forty degrees.

But whether the horn, the bowl, or the pan is used, the operation is still the same. A handful of dirt is placed in the

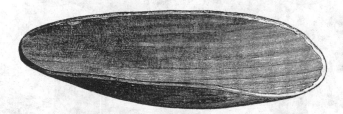

Fig. 146.—Horn or spoon for washing gold at the American placers. Scale ⅓.

apparatus, and washed with plenty of water so as to get the mass thoroughly soaked, the instrument being shaken horizontally and turned the while, and then tilted up a little to let the thin mud and light sand run out ; by this means the heaviest particles go to the bottom, and the lightest are carried off with the water, all the large pebbles and fragments being picked out by hand. At last the gold is found at the bottom of the bowl, with some substances which are almost invariably associated with it, such as platinum, which is collected with it, magnetic iron-ore, which is removed with a magnet, and certain gems, amongst which the diamond may occur.

Pan-washing, as it has been described, is sufficient for a trial and even for a small working, if the placer should happen to be very rich ; but in most cases it would not be suitable for an undertaking of importance. The Chinese have invented the cradle or rocker (*berceau*) in the shape of an oblong box, somewhat like a child's cradle and open in front, to which an oscillating or rocking motion is given (fig. 147). A hopper or riddle is placed at the upper end, an inclined frame-work made with a bottom of sheet-iron punched with holes, and covered with a canvas apron or woollen blanket beneath the riddle.

Fig. 147.—Chinese washing auriferous sands with the cradle, at the Australian placers.

The sands, gravel, and earth to be washed are thrown into the hopper, and the machine, the bottom of which is perforated with holes half an inch in diameter, is rocked with one hand while water is poured out of a dipper over the dirt with the other. The fine and light substances, the sands, the specks and spangles of gold, and the small lumps or nuggets are carried by the water through the openings of the hopper and descend to the inclined blanket, and thence on to the bottom of the cradle, from which the mud, water, and sand run off at the lower end of the rocker, which is left open. In this process the heaviest bodies travel the least distances, and nearly all the gold is found at the head of the blanket under the hopper, and behind two bars (riffle-bars) which are nailed across the bottom of the cradle to prevent its escape.

The tom or *long-tom*, a sort of large and fixed cradle called by the miners a jack-in-the-box, is a trough about twelve feet long, eight inches deep, fifteen inches wide at the head and thirty at the foot. A stream of water runs constantly through the tom, into the head of which the pay-dirt is thrown by several men, while one throws out the stones too large to pass through the half-inch holes punched in the sheet-iron riddle forming the bottom of the tom at its lower end, and throws back to the head of the tom the lumps which reach the foot without being dissolved. The tom, a great improvement on the rocker, was soon superseded by a greater, the *sluice*, which is a broad trough, from a hundred to a thousand feet long, with transverse cleets at the lower end to catch the gold. All these inventions, the tom, the *sluice* or long wooden trough (*launder*) having a considerable inclination, through which a stream of water is always flowing and washing the dirt thrown into it; and lastly, the channels of larger dimensions, or the lofty aqueducts constructed on trestles and called *flumes*—are all made on the same principle and intended to effect the same object as the cradle, only on a larger scale, and compose the great Californian processes, which have also been adopted in Australia. Formerly the Chilian process was regarded with the greatest favour, by which the auriferous dirt is washed in place by means of a stream of water (fig. 148). By this latter method the pebbles and larger lumps, which are scattered over the course of the water, have a tendency to retain the gold; and the same effect is noticed in the middle of auriferous streams, where the specks and nuggets are seen to accumulate behind any obstacle lying transversely to the direction of the flow.

In California the miners, not content with digging the placers, have likewise turned the course of the streams in order

Fig. 148.—Chilians working at the placers, after the method of their country.

to search their beds (fig. 149). They have even torn down alluvial terraces and entire banks and hillocks by means of what is called the hydraulic method—a process invented in 1852 by Edward Mattison, a native of Connecticut, and

Fig. 149.—Working auriferous sands in California by turning the water-courses.

similar to that known in Derbyshire as "hushing" (fig. 150). Provided with a flexible hose terminating in nozzles like those of fire-engines, from which the water is driven at a very high pressure, the miner washes away the face of the bank. In a little while the upper part falls; and as the gravelly ground is composed of pebbles, and is naturally incoherent and loosely consolidated, large masses of the hill come down with a violent crash. The miners avoid being buried in the wreck, and return to their work when all is quiet. Then the largest blocks are broken up with the pick, and the dirt is thrown with the shovel and barrow into a large channel (flume) put up at the foot of the hill which has been demolished. In this way ground can be worked to a profit that does not contain more than three francs (two shillings and sixpence) of gold in a thousand kilogrammes (2204·6 lbs.); that is to say, in which the proportion of gold does not exceed a millionth part.

The ores of auriferous quartz are treated in a different way from the alluvial dirt. In the first place, the rock must be pulverized after it has been extracted, which was formerly effected in the Spanish colonies of America by rude machines worked by the natives (fig. 151). In California and Australia machinery of a more improved construction is employed, which, whether crushers, rollers, or stamps, now serves for all the ores indiscriminately. The powder or mineral dust which is obtained is not washed, but placed directly in contact with mercury, which possesses the singular property of acting as a solvent to certain metals, amongst others gold and silver, just as water dissolves sugar—a combination which is called an amalgam. It is effected in America in the treatment of gold, under the stamps or in the particular machines, which, until very lately, were of an extremely primitive kind, like the Mexican or Chilian crushers, the *arrastres* and *trapiches* (figs. 152 and 153), which have been considerably modified by the Californians. The tables of amalgamated copper on which the

Fig. 150.—Hydraulic Method invented in California for working Auriferous Alluvium.

ore coming in contact with the quicksilver is stopped, are also of Californian invention. The Hungarian or Tyrolese amalgamating mills, the object of which is to agitate the pulverized ore in a bath of quicksilver, are in use in the respective countries whose names they bear, and have been adopted at almost all the mines in the world : as have also the Russian amal-

Fig. 151.—Crushing ores by the old Mexican process.

gamators, now used in the Ural. In these last machines an inclined trough is divided into compartments full of quicksilver, in which a series of forks, having an oscillating movement communicated to them, place the sands in contact with the fluid metal as they traverse the troughs.

The object of the woollen coverings and the sheep-skins employed in California and in all gold-mining countries, is to retain, in the labyrinth of their filaments, the last specks of gold or amalgam which have escaped in all the previous processes. The use of fleeces to collect the gold, which is very likely as old as the time of Jason in Colchis—the California of mythological times—allows us to substitute reality for fable, and to consider the story of the golden fleece and the expedition of the Argonauts, which has been handed down to us by

Fig. 152.—Arrastre, or Mexican mill for the treatment of gold and silver ores.

ancient authors, no longer as a story made to amuse, but rather as a true history.

It has been stated that in working the placers the miner treats with profit the dirt only containing a millionth part of gold; that is to say, half-a-crown's worth of gold in 2204·6 lbs. of ore. In the quartz-mines the proportion of gold ought naturally to be much greater, since it is necessary to execute all the work which is partly done at the placers by nature; that is, to get, transport, pulverize, enrich the ore, and finally amalgamate it.   In California the quartz-ores of least hardness only begin to be profitably treated when they yield 30 francs (24s.) worth of gold to the ton, or about one hundred thousandth, or a proportion of one part of gold in a hundred thousand parts of ore.   This proportion is ten times higher than the minimum for the sands at the placers ; but it is still so small, that the great value of gold and the simplicity of the

Fig. 153.—Trapiche, or Chilian mill for treating ores.

metallurgical methods which are employed in the treatment of its ore, only allow the extraction of such small quantities.

Quicksilver is not employed in the treatment of auriferous quartz-ores only.   When the gold of the placers is very fine, and invisible to the eye, or that the work is done by means of long channels, cups filled with quicksilver are interposed in the bed of the stream and the sands, which detain all the gold in its passage, however small the specks may be.   The scales

themselves encountering in their passage these basins of quicksilver, are caught by them, and are better secured by the cups
than by any other obstacles of more or less ingenious contrivance intended to produce the same effect.

The amalgam which is obtained is generally liquid; and to
solidify it, it is squeezed through a chamois leather, folded into
a bag and wrung between the hands like wet linen from which
it is sought to express the water. The quicksilver being separated from the gold, passes through the pores of the skin in
the form of silvery drops, affording a curious fact, which is
employed in lectures on physics to prove experimentally the
porosity of bodies. The gold has now to be separated from the
pasty amalgam left behind, which in shape and appearance
resembles a button or ball of tin. It has been already stated
that quicksilver dissolves gold just as water dissolves sugar; but
the quicksilver can be volatilized. The amalgam is, therefore,
distilled to extract the gold, in the same way that a solution of
sugar in water is evaporated to make sugar-candy. In this
instance, if the comparison may be allowed, the gold is the
sugar-candy of commerce.*

All the balls of amalgam are introduced to an iron retort,
which is heated in a furnace. At a temperature of 360° Cent.
(680° Fahr.) the quicksilver boils, and ascends in the form of
vapour towards the neck of the apparatus; and the vapour is
condensed by means of wrapping wet cloths round the descending neck of the retort, from which it drops into a vessel full of
water. When the operation is concluded, the quicksilver is
found condensed at the bottom of the water in the receiver,
and a cake of gold is left at the bottom of the retort. The
metal is porous and spongy, of a pale yellow colour, here and
there a little stained and interspersed with blackish specks.
It is remelted in black-lead crucibles with a small quantity of

---

* The ancients were acquainted with this property of quicksilver to dissolve the precious metals,
and to yield them again on being volatilized. They turned it to account to collect the gold from old
stuffs, and in gilding copper.

borax, which removes the last impurities. When thoroughly fused the gold is poured into moulds of cast-iron, and formed into ingots or bars, in which form it reaches the assayers.

The proportion of gold at the mines is generally, in California, from eight to nine hundred thousandths ($\frac{800}{1000}$ to $\frac{900}{1000}$); that is to say, in a thousand parts there are only from eight to nine hundred parts of pure gold. The rest is composed chiefly of silver, sometimes a little copper and iron, and lastly, some stony substances, accidentally present, as silica. The country from which the gold comes may frequently be known from its chemical analysis. The gold of Australia and the Ural has not the same composition as that of California. Sometimes, in the same country, the gold from the different mines varies. The purest gold now obtained is from the Altaï; after which comes that from the Ural and from Australia, both of which have sensibly the same composition. Californian gold contains a large proportion of silver, and the gold from the northern mines is purer than that from the mines in the centre of that country.

Below, as an example and epitome of all these results, are some chemical analyses of gold from these several countries:—

ANALYSES OF SOME SPECIMENS OF GOLD FROM MINES.

Composition in a thousand parts.

		Gold.		Silver.		Iron, copper, &c.
Siberia,	Altaï,	980	...	20	...	—
	Ural,	950	...	50	...	—
Australia,		940	...	60	...	—
California,	Northern mines,	900	...	100	...	—
	Central mines,	800	...	198	...	2

The above Table shows how the character of the gold varies with the country which produces it, and how it often differs from that of the gold which is coined or worked. The standard gold coin of England, by an Order in Council issued in the reign of her present Majesty in 1855, is made of an alloy consisting of 22 parts of pure gold and 2 parts of alloy, the alloy being an indefinite proportion of silver and copper. The standard for gold used by jewellers and goldsmiths (hence called jewellers' gold) is 18 parts of pure gold in 24 parts of alloyed metal (called gold of 18 carats fine); but articles of plate and jewellery may also be wrought of 9, 12, or 15 carats pure. In France the proportion is 9 parts of pure gold to 1 of copper for gold coinage, and $7\frac{1}{2}$ of pure gold to $2\frac{1}{2}$ of copper for the gold employed for jewellery and goldsmiths' work. It need scarcely be added that the object of alloying the gold with copper is to make the precious metal harder and less liable to wear.

It is not merely in the treatment of gold-ores, but likewise in the treatment of silver-ores, that quicksilver plays an indispensable part. The method of amalgamation by cold (called the *patio* or yard process) invented by a poor Mexican miner, Bartolomeo Medina, in 1557, has given a vigorous impulse to all the South American silver-mines, where fuel was wanting, and Medina may be considered a benefactor to the human race. M. Michel Chevalier does not hesitate to compare him to Triptolemus, who discovered wheat. The Mexican mode of reduction is very slow, and is unsuited to the cool climate of Nevada, at an elevation of 5000 or 6000 feet above the sea.

The productions of the American mines was still further increased when, in the last century, the method was thought of in Mexico of stamping the material by means of horses instead of manual labour. The ore having been first of all crushed under stamps, or the flying-stone of the *Arrastre* (the primitive mill with which we have already become acquainted in reference to the treatment of gold, fig. 152), and afterwards ground to the necessary degree of fineness under porphyry mullers worked by a couple of mules, the mineral is made into *tortas* or cakes. These cakes mixed with salt, quicksilver, and iron or copper pyrites (sulphate of copper being preferred to copper pyrites), which is called the *magistral*, are trodden by mules in a large court-yard or *patio*, generally paved with flagstones, carefully cemented together to prevent the escape of the quicksilver through the joints. The mixture, after exposure to the air, ferments, when complicated chemical reactions take place, the nature of which has not yet been altogether satisfactorily explained; but the ultimate object of which is the formation of an amalgam of silver. At the end of forty days the amalgam is separated, by washing, from the stony matters in which it is contained, and it is then distilled in an iron retort. In Mexico a bronze bell, called a *capellina*, is preferred for this purpose, which covers the *bollos* or balls of amalgam, and is surrounded with lighted charcoal. The quicksilver is then volatilized, and afterwards condensed into a liquid form in a current of cold water flowing at the bottom of the apparatus, forming the mode of distillation called by the alchemists *per descensum*, or by descending; that which had been previously described being *per ascensum*, or by ascent.

The method of amalgamation which has just been described has been practised in South America for the last three hundred years, without having undergone any material change. Even now, the chemists, metallurgists, and engineers, who go out from Europe to those distant countries with the intention of changing, or at least modifying everything, and improving the methods of treatment already in use, have always been obliged to go back to the primitive process of Medina.

When the distillation is complete, the porous cake of silver which remains in the bell is removed in leather bags to the smelting-house, where it is assayed, remelted, and cast into bars, which are those which we have seen under the name of *plata fina*, loaded into the Pacific steamers, for conveyance to England by way of the Isthmus of Panama.

The metallurgy of silver might, perhaps, be described here at length, and the other methods besides the American process of amalgamation by which the metal is extracted from its ores.

We will, however, be content with merely mentioning the German method of amalgamation by the wet way, which was invented in Saxony about the same time as the Spanish method; then the cupellation, which has been practised from time immemorial; the process of extracting silver from argentiferous copper by liquation or parting, practised in the Upper Harz; and lastly, Pattinson's desilvering process, for separating silver from lead by the concentration of the former metal by the crystallization of the lead in cooling.

By these latter processes, one of which has been in use in Germany since the sixteenth century, while the other has been recently (in 1833) invented by Mr. H. L. Pattinson, of Newcastle-on-Tyne, whence it has passed to the Continent—the smallest traces of silver can be profitably separated from metallic copper or lead.

Resuming these studies in metallurgy, a science of very high antiquity, certain methods of which acquired at the very outset the same degree of perfection they now possess, our admiration continually increases.

Here we should see complicated and learned inventions, in which are applied the most elevated chemical theories, come to the help of the smelter; on the other hand, we should see nothing grafted on to long experience, and the workmen handing down the secret of their art from one to another, practising by rule of thumb feats which are inexplicable theoretically. In many instances the philosopher is struck with amazement and confusion before the results which he arrives at, some of which still remain inexplicable and mysterious. But to enter into all these details would render it necessary to increase the size of this book, which should treat of the miner and not of the smelter—of mines and not of factories. The washing of ores and American amalgamation formed part of our plan—not so with regard to foundries.

We are, therefore, compelled to pass by metallurgical processes, as varied as they are ingenious, by means of which not only silver, but all the other metals besides, are separated from their ores. It falls, however, within the scope of our plan, to inquire as to the lowest limits at which the reduction of metallic ores ceases altogether to be profitable. It is clear that such a minimum does exist, and that it depends on a great many elements, such as the mineralogical variety, and the greater or less purity of the ore, the nature of the gangues, the difficulties of working, mechanical preparation or dressing, metallurgical treatment, distance of factories, cost of carriage, materials, and wages; and lastly, the commercial or current value of the metal to be smelted.

This last element sometimes takes the lead of all the others. Mines of quicksilver and antimony have been mentioned, which are perhaps for ever stopped by the fall in value of the metals, which were previously worked there with advantage. It is evident that, to work at the same profit, the richness of the ore in metal should increase as the value of the metal falls.

Taking all these elements into consideration, an attempt has been made, in the following Table, to show the minimum percentage at which the ores cease to be workable at the present time. The figures would alter if the prices of the metals underwent a sensible change.

TABLE OF THE MINIMUM PRODUCE WHICH METALLIC ORES SHOULD YIELD.

COMMON METALS.

Nature of the ores.	Lowest percentage.	Price of pure metal in 1865.		Observations.
			lbs. kilo.	
Iron,	25 to 30	16s.	per 220·46 (100)	Cast iron, 11s. 8d.
Zinc,	20	45s. 10d. to 50s.	"	Part of the metal is lost in the process of smelting, by volatilization.
Lead,	20	45s. 10d. to 50s.	"	For non-argentiferous ores.
Antimony,	20	104s. 2d. to 125s.	"	Very volatile when melted.
Copper,	2	187s. 6d. to 208s. 4d.	"	
Tin,	1 to 1½	187s. 6d. to 208s. 4d.	"	Very easily reduced.
Quicksilver,	1 to 2	£25 to £30	"	Very volatile.

PRECIOUS METALS.

Nature of the ores.	Lowest percentage.	Price of pure metal in 1865.			Observations.
		lbs	kilo.		
Silver, ·	$\frac{1}{2000}$	£8 4s.	per 2·2046	(1)	⎰ The average American ores do not exceed 2 thousandths.
Platinum, ·	$\frac{1}{10,000}$	£40	"		"
Gold, ·	$\frac{1}{100,000}$	£136	"		For quartz-mines.
" ·	$\frac{1}{1,000,000}$	"	"		⎰ For washings and diggings.

The above Table teaches us a very important fact, by demonstrating that the present methods of treating ores has arrived at such perfection, as to enable us to extract the very minutest particles of metal from them. An ore of iron that contains less than 20 per cent. of metal becomes merely a ferruginous earth. In the case of lead-ores, it is now no longer one-fifth, but less than one-twentieth part of lead that can be profitably extracted; the ancient cinders resmelted in Sardinia rarely contain more than 10 per cent. of lead. It is shown in the Table how poor in silver most of the South American ores are, which are commonly supposed to be so rich; but that is nothing to the infinitesimal limits at which gold can henceforth be worked. In that respect the Californian methods have effected wonders, and it was with propriety that, in 1862, M. Michel Chevalier, as President of the French section of the Jury at the International Exhibition in London, directed attention to the inventions of the gold-miners of the Eldorado, as one of the most brilliant conquests of industry. Nature is diversified sometimes by microscopic proportions; man has overcome nature by reuniting these invisible and scattered atoms, and in forming from them solid ingots of metal.

CHAPTER XI.

THE PHALANX OF MINERS.

The metal-miner and the coal-hewer.—The German miners.—Good and evil spirits : Nickel and Kobold.—Old customs.—Division of labour.—Military discipline.— England : Cornish miners.—Pharaoh's liard.—Miners of the Italian Alps : Tyrolese, Piedmontese, Tuscans.—Spanish miners.—Quicksilver miners.—Miners of Spanish America—Chile, Peru, Mexico—compared with Anglo-Saxon miners.— Coca.—The Mita.—The native Indians.—The Aztecs.—Mines of Central Asia : the Kirghese.—The Chinese as miners and gold-diggers.—The French in California.—Social condition of the metal-miner.

THE miner in metal, like the coal-hewer, occupies a distinct place, and is marked by peculiar characters in the great family of workers. He constitutes the true type of miner, a sort of subterranean pioneer, a type more varied, and of yet greater originality than that of the collier. In coal-mines a certain sort of family likeness prevails, the same discipline distinguishes the works, the strictness of the regulations in force in each district are the same, and these deprive the workman of a portion of his local habits and customs. In metal-mines, on the contrary, the deposits are dissimilar, and each of them has a special mode of occurrence, necessitating different systems of working ; and the discipline also is less exact. Lastly, while the ordinary way of mining for coal is scarcely a hundred years old, some of the metal-mines have been worked for ages as they are now worked. In each of them the workman has retained, as it were, the stamp of the ground—peculiar manners and customs.

Throughout the whole of Germany, in the Harz, Saxony, Prussia, Bohemia, Hungary, and the Tyrol, the miners are remarkable for a sort of invincible attachment to altogether primitive customs. The dress remains the same, while the systems of working have, for the most part, improved. A leathern apron worn behind, and long leather gaiters up to the knees to protect the limbs in the underground journey, and to save them from the vitriolic waters; a leather strap buckled round the waist to hold the hammer, pick, and lamp; a short vest, with narrow sleeves puffed towards the shoulders, like the slashed jackets of the time of Charles-Quint; a round hat made of hard felt, raised, turned up, and strong enough to withstand hard knocks—is the sort of uniform commonly met with in most of the olden German mines, and the invariable costume in which the visitor is dressed who is about to lose himself in this world of another age. The dress has not undergone any change since the earliest days of mining. Holy fidelity to ancient usages, pious memories of the past, which must meet with respect, perhaps now that everything, and amongst them ancient usages and old beliefs, is in a rapid state of change.

In some workings it is never forgotten to offer up a prayer before going down the mine. The engineer often piously recites the orison composed for the occasion, while the men are grouped around him at the opening of the mine; and the austere and solemn gathering is lighted up by the lamps suspended from the timbers, and the caporal, with his hand to his hat, who is about to superintend the change of shifts, has partly descended the ladders (fig. 154).

The German workman occupies a place of his own in the world of miners, having retained not only the manners and dress, but also the simple superstitions of his forefathers. He believes in gnomes—genii of the mines—in the good-natured and malevolent gods of the subterranean domains, which are not, in this case, Ormuzd and Ahrimanes, but Nickel and Kobold. In order to propitiate them, he has given their names to the two metals nickel and cobalt, which were originally

Fig. 154.—Miners at Prayer at La Vieille-Montagne before descending the Shaft.

discovered in the mines of Saxony. They are the gnomes who fill or empty the lode, and who reproduce the ore as fast as it is removed. They prowl about the old galleries and the abandoned working-places, where they have been met by more than one person; they blow upon the lamps in order to put them out, and drag by the nose or the hair the miner whom they encounter by himself. When he has greatly displeased them, they cast spells upon him, throw him down the ladders, or crush him under a fragment of rock. What will he not do, then, to render them propitious? Provisions are left for them in the mine; and bread, cakes, and pieces of money are put in secret niches. "Saints Nickel and Kobold, preserve us from explosions, falls, inundations, and from falling away in the shaft. Be propitious to us. Enable us to recover the vein when we have lost it, and above all lead us towards the rich parts of the lode. Finally, intercede with God or Beelzebub for us: *Sancti Nickel et Kobolt, orate pro nobis!*"

This strong attachment to the customs of ancient times has not been at the expense of any of his manly qualities in the case of the German miner, and has only added one trait more to his already attractive personality. We must not forget that it was in Hungary, Saxony, and the Harz, that the art of working mines was renewed directly after the tumult arising from the invasion of barbarians, and that the art has gradually spread from those countries into all the other mines in Europe. These last, in their turn, have made fresh progress, and given a stamp of activity if not of grandeur to the new undertakings, which is sometimes wanting in the German mines. Thus, one would like to see some of the old-fashioned customs disappear from the old Saxon or Harz establishments. The use of the divining-rod has been already proscribed; but why, when the lode is rich, wall up the working-places, to return to them in times of scarcity? A mine is so much capital underground, as the English say, and the owners should try to raise the greatest quantity in the shortest possible time, in order that the capital may yield the highest interest. It is true that in the Harz and Saxony the mines constitute a territorial property; but in all cases the economical methods dictated by the narrow spirit of avarice of the past must be cast off. One cannot help praising the division of the work which is so wisely maintained in the German mines, where there are engineers and overlookers for the surface as well as the interior of the mine. The person on whom the geological part of the management devolves has nothing to do with the washing, or dressing, or smelting of the ores. To one belong the works of the mine, strictly speaking, such as the getting, disposal of the waste, and underground transport; to another belong the works of construction, such as carpentry, timbering, and masonry: to this one, all the mechanical part, the steam-engines and waterwheels; to that one, surveying the underground plans and the maps of the surface, or else, again, the chemical assays in the laboratory. Nobody encroaches on the duties of another; engineers and workmen execute well what they have to do, because they are familiarized with it, and because their labours are not oppressive.

Not content with having established such a methodical division of labour in the works, and with having so sharply defined the duties of every man, the Germans have also introduced into their mines a regular military discipline. In this legion of labour the workmen are the private soldiers; the master-miners or overmen, the corporals; the engineers, the captains—all of whom wear uniforms that have been designed for ages. The Hungarians, Saxons, and Harzois are proud of these old costumes, which are set off by lace and gimp for the corporals, and for the chiefs by epaulets and gold embroidery (fig. 155). These chiefs themselves come from celebrated schools of mines, amongst which those of Freiberg in Saxony, and of Schemnitz in Hungary, hold the foremost rank. On Sundays, or other solemn occasions, the soldiers of the peaceful army appear with all their insignia, all their medals and decorations, with colours flying in the wind, and music at their head; and the captains with sword or sabre by their side, and the baton

Fig. 155.—Saxon mining-captain in full dress, with his baton of office.

of command in hand. Let us once more applaud this regiment of soldiers; let these brave and intrepid miners defile before us: the soldiers of war too often take precedence of the soldiers of peace.*

* In Russia the mine-parade is no longer a jest. They have improved on the Germans with regard to the higher ranks; and the government engineers-in-chief and inspectors of mines are actually colonels and generals, having officially the title and the pay of their colleagues in the army.

In England, where all industrial undertakings are sustained by private enterprise, the engineers also bear the title of captain in all metalliferous mines; but in this instance there are neither uniforms, decorations, nor swords.   There are many specimens in Cornwall of these old captains, hard in their bearing, simple and modest in manners and behaviour, but with uncouth gait, who have won all their grades in the exercise of their duty, having become corporals from being soldiers, and then captains.   They compose a type such as is not met with elsewhere, where examinations and diplomas send at once a young sucking engineer to command old and experienced hands.   All the English captains welcome you at their mines with marks of the greatest kindness.   They show you their works, their machinery, and their plans, saying: "We have nothing to conceal."   "What is hid, we look for."   "We seek hidden treasures," added one of the number, and he pointed out to me the adage in large letters in the office, where the miners change their dress before going underground.   Admirable definition of the labour in metal-mines, which are but in fact buried treasures, which it is sought to discover; they are the veritable liard of Pharaoh, that the quarrymen of Paris talk about.*

In these interesting Cornish mines (Map IX.), the soldiers are worthy of their chiefs, and in every part of this country, at Saint Just, Saint Ives (fig. 156), Redruth, and Camborne, and at Tavistock, in Devonshire, the miner deserves the reputation which he has acquired of being one of the best miners in the world.   Like his German brother he adheres to old opinions; he has his legends, and believes with no less sincerity than his Teutonic brethren in subterranean gnomes, the genii of the lodes, and eternal guardians of the mines.

---

* There is a legend current amongst the Parisian workmen, that King Pharaoh concealed a treasure in the old quarries, and that he who discovers it will suddenly become wealthy.   *Pharaoh's liard* is the name given by the quarrymen to this legendary treasure, which nobody has hitherto succeeded in putting his hand on.

The Italian Alps furnish types of miners which, without approaching those of England or Germany, are likewise distinguished by very curious traits. The Tyrolese is both German and Latin ; cold, reserved, true to his word, a steady

Fig. 156.—Miners of St. Ives in Cornwall.

worker, and until lately a greater traveller than he is at present, he has advantageously stocked the foreign mines to which he has migrated, and planted his tent for good. The Piedmontese—also an emigrant, but with a greater spirit of adventure than the Tyrolese—also seeks employment at times in the mines of France, and throws them into disorder. The

Germans have succeeded better, and in some localities, as at Vialas and Pontgibaud, have even made durable way. Of athletic form, and a skilful handler of the borer (or needle, as he calls it), the Piedmontese has few rivals in breaking a hole for blasting ; but he is terrible when he has drunk, and he also makes a free use of the knife, in the Italian fashion.

In Modena, to the north of Tuscany, in the midst of the Apuan Alps, the character of the Italian miner seems to borrow something from the boldness of the scenes by which he is surrounded. Nearer the south, at Monte-Catini and in the Maremma, the working man is more slenderly built—in a word, he is less heavy and more light and lively—than the miners of the north. All these districts furnish good examples of able and brave mining-captains, and the captains of the mines of Massa, some of whom are Germans, also deserve to be honourably mentioned.

In the iron, zinc, lead, silver, quicksilver, and copper mines of Spain, the miners of the Sierra Morena in the Asturias, and of the Sierra Nevada in the Pyrenees, are remarkable for their personal peculiarities, more than is the case in Italy. The Asturian is hardy, somewhat grave, neither familiar nor a talker, like all the Spaniards of the north. The inhabitants of the sierras of Gador, Almagrera, and of the Alpujarras, spurs of the Sierra Nevada, which produce inexhaustible quantities of lead and silver; the miners of Carthagena and Almeria; but above all, the Andalusian miners of Huelva or Seville—are more gay and genial than the Asturian. Their character changes with the happy influence of the magnificent sky of their country that land of brooms, of rose-laurels, that *beau pays des Espagnes* celebrated in romance. The miner of Almaden alone, poisoned with the vapours of mercury, feels its slow but fatal effects, and goes sadly to death, the victim to underground labour.*

* The work in quicksilver-mines is so dangerous, that up to the beginning of the present century only criminals were employed in them in Spain. Almaden was the site of a *presidio*, or correctional prison, and a gallery communicated from the prison with the mine. Now all the labourers are free.

In the Asturias and at the foot of the Pyrenees the Spanish miners and founders are a sober and frugal race. A little bread; thick wine kept in the leather bottle, and smelling of pitch; *garbanzos* or sorry peas, those mealy beans despised by Charon himself in the "Dialogues" of Lucian—compose the national fare, everything being dressed with rancid oil such as we would not burn in our lamps. This frugal repast is only made once in the four-and-twenty hours, at noon: morning and evening they usually eat whatever comes to hand. The cigarette is smoked several times during the day, even while getting the ore. On leaving the mine, the miner dons the snuff-coloured cloak and the broad-brimmed sombrero, and silently returns to his hut. Galicians, Castilians, Basques, Navarrese, Catalans, and Arragonese, may all be recognized by these principal traits.

In the Sierra Almagrera or Sierra Gador, in the south of the Peninsula, far from any inhabited centre, the life is harder still, and more full of privations; but who thinks of comfort in Spain? A wide pair of cloth breeches ending at the knee; a belt fastened round the waist, in which tobacco, knife, and money are put; a handkerchief folded round the head by way of a hat; together with a shirt—constitute all the accoutrement of the miner. Sandals made of esparto,* and the cloak, *la manta,* a covering of many bright colours, and no longer chocolate-coloured—an article of dress which is for the labour-

---

They are attracted by numerous immunities, grants of land, and exemption from military service. They only work on alternate days, and receive medical advice and medicine gratuitously. Very few escape the effects of the mercury. Emaciated and wan, their gums become salivated, and they lose all their teeth. They are subject to tremblings, convulsions, and finally die of consumption or become idiotic. In that country of desolation the very vegetables are poisoned by the metal, the land is sterile, and the only industry pursued at Almaden is that of the quicksilver-mines, which supports nearly four thousand workmen.

The quicksilver, on being separated from its ore by distillation, is put in iron bottles, the cork of which is screwed down. Muleteers convey the precious flasks to Cordova, under the escort of a guard of soldiers to prevent brigands, on the watch for the metal, laying hands on the convoy. From Cordova the quicksilver is carried to Seville by railway.

* Esparto is a wild grass which grows all over the south of Spain, the tough and flexible fibres of which are plaited for making not only sandals, but also mats, panniers, ropes, pack-saddles for beasts of burthen, and it is now used largely in the manufacture of paper.

ing man of the south of Spain what the *poncho* is to the Chilian miner, or the *serape* for the Mexican : at once a cloak, a blanket, and a bed at need. A good miner of Almeria lives and dies in his manta, and transmits it to his descendants. In Andalusia, besides the manta, the hat with tufts also forms part of the miner's dress. On Sundays he puts on the tight embroidered velvet jacket and knee-breeches, and he has almost the air of a *mayo*, or eldest son of a good family.

Fig. 157.—Common room of Spanish miners of the Alpujarras (province of Granada).

The dwellings of these hardy mountaineers of the Sierra Nevada and Sierra Morena are on a par with their dress, and consist of a bad cabin built of stones and mud. Here and there some culinary utensils of iron or copper, the water-bottle or *alcarazza* of porous clay for keeping the water cool, and baskets made of esparto or wicker. The fire-place is in the middle of the floor; the bed nowhere : they lie on the ground, anywhere, rolled in the woollen manta (fig 157).

The fare is on a par with the house and dress. For break-
fast, soup made of cold water and oil (and such oil!) swimming
with garlic, tomatos, and slices of bread : for dinner, a bowl of
rice with salt-cod, lupins, haricots or peas, the whole dressed

Fig. 158.—Barretero, or miner, of Cerro de Pasco, in Peru.

with oil and pimento: for supper, the same medley; the formula
being as unchanging as those of the Codex. The three meals
cost $6\frac{1}{2}d$. a day, inclusive of bread. With all that, they are
a warlike, orderly, brave, and intelligent race, all of whom

whether miners, muleteers, or founders, do their duty without
noise, and obey the orders of their chiefs.

The miners of Spanish America have preserved somewhat
of the character of their Andalusian or Castilian ancestors;

Fig. 159.—Apire, or ore-bearer, of Cerro de Pasco, in Peru.

retaining their sobriety, pride, and a seeming indolence. In
Chile, Peru, and Mexico, they form an interesting and vigorous
race, inured to fatigue : the *barretero*, or miner, and the *apire*,
or ore-carrier, are represented in figs. 158, 159.

It has been found that these Spanish Americans do little work compared with the Anglo-Saxons, who have often been employed in the South American mines; and they eat still less. What a striking difference between the men of the two

Fig. 160.—Miners of Spanish America in full dress.

races! The Cornishmen have not their equals in blasting a vein of quartz, and earn by the work from ten to fifteen shillings a day, while the earnings of the Spanish, Chilian, or Mexican miners scarcely exceed four to eight shillings. The former are hardy, robust men, great consumers of roast-beef, gin, and whiskey; fond of drink after their work is over, when they become quarrelsome and given to fighting: the latter always calm, temperate, and impassive, whether over their work or in their recreation; great smokers of *papelitos*, inveterate gamblers, but subsisting on little—dried figs and a little *charqui*, or flesh cut into thin strips and dried in the sun; only working enough to earn this meagre fare (*la comida*, as they call it), doing nothing as long as they have any piastres, and playing for them to their last halfpenny at *monte*, the American baccara. They love everything that produces effect; pistols, the revolver thrust into their belt, the knife or *machete* stuck in the boot, the *serape*, the *poncho*, bright-coloured mantles cast over the shoulder, rows of silver buttons on their jackets and down their breeches, large vicuña hat with gold lace, embroidery, and ornaments everywhere (fig. 160).

However sober the miner of the colonies may be, the charqui and dried figs are indispensable to him all over South America. Among the Chilian miners this food is obligatory, and the mine-owner has to provide it. There is also a shrub of the tea-plant genus, the *coca*, which is grown in those countries, especially in Peru. The miner mixes the leaves of this plant with quick-lime, and chews them incessantly, so that they constitute an integral part of his habitual nutriment, and the Indian of South America can no more dispense with his coca than the Hindoo can do without betel, the Chinese without opium, or the sailor without tobacco to chew. The Chilian and Peruvian miners have learned this practice from the natives; and they, also, chew the leaf of the American tea-plant. They pretend that it makes them work better, and that they experience neither fatigue, nor hunger, nor thirst. Curious examples of this are cited. The Indian who traverses the Andes, in many instances loaded like a mule, can pass several days without food, provided he chews the coca. Is

Fig. 161.—Chilian Miners.

this an effect of the alkaline matters contained in the vegetable, which tan and, as it were, deaden the coats of the stomach? Physiologists are far from having properly explained this phenomenon; but its existence is undeniable, however surprising it may seem to us.[*]

The American Indian also worked at the mines before the War of Independence, and the victors compelled him to do so, with a cruelty and an avarice of which the ancients never perhaps have furnished an example, even in connection with their slaves or prisoners of war condemned to similar labour. This compulsory drafting of the savage into the mines was called the *mita*. The Indians revolted several times in Peru and Bolivia, in order to throw off this kind of forced enrolment but the risings were always suppressed,[†] and troops of slaves were despatched every year to the mines of Pasco, Potosi, Huancavelica, and Huantacajo, where they perished by thousands of excessive labour and barbarous treatment. The mita was scarcely less severe in Mexico than in South America. It was not entirely suppressed until the colonies seceded from the mother-country; since which time the Indians over the whole of America have no longer worked in the mines, but have returned to their homes in the woods and plains, the only kind of life agreeable to their nature. Sometimes Indians may be met with working underground, but then it is of their own accord. In case of war they are also made to work in the mines by compulsion. They are especially employed in the dressing of silver-ore in the open air.

[*] The consumption of this leaf is so great in Peru, even amongst the Indians only, that the growth and commerce of the coca form a very important branch of industry. The Indian who chews coca carries in his belt a calabash filled with lime, which he takes out with a little stick. The habit becomes so great with some persons, that they do not leave off chewing the leaf even when asleep. This practice eventually affects the brain, and impairs the mental faculties, just as tobacco affects those who smoke too much.

[†] The most celebrated insurrection of this kind is that of 1780. The Cacique Tupac-Amaru, a descendant of the Incas, armed the Indians, and held the viceroy of Peru in check for two years, when he was taken prisoner and put to death, together with his wife Micaela, with the most horrible tortures. The limbs of the two victims were cut in pieces and exposed in different cities; their bodies were burned, and the ashes thrown to the winds.

The Indians of Sonora and Chihuahua in Mexico, who revolted and were conquered, have also been condemned to break up the ore. The mouth of a cannon turned against them is the means adopted to suppress all complaints (fig. 106).

In Nevada and California the Indian no longer works in the mines; according to his notions it is the work of a barbarian. In Mariposa Co. the Indians who passed the mines were observed by M. Simonin to regard the amalgamation-house with a sort of disdain : they had no idea of giving themselves so much trouble. A bow and arrows, a plume stuck in their hair, a bone passed through the ear or nose,* a hut beneath the trees, were their arms, their ornaments, and their dwelling. What need of more ? What were nuggets to them ? What use could they. be of ?

Towards Lake Superior the indigenous tribes, the Chippeways, display as profound an indifference for work connected with the copper-mines, as do the Californian Indians for the gold-mines. It was, nevertheless, the ancestors of these same Chippeways who first opened up the deposits of Lake Superior, and who worked them with those stone-hammers which are met with still in old excavations (figs. 136 and 162). The Aztecs of Mexico were likewise acquainted with copper, and, moreover, alloyed it with tin to make bronze, which they even knew how to temper and harden. The Indians are now everywhere degenerating, and are no longer capable of deriving any advantage from the natural wealth of their land. Civilized man has come to colonize the country, as was predicted in all their legends; and it is not to be wondered at if the aboriginal races should disappear before the European. It is but just that he should depart and die off, who did not know how to profit by the treasures which nature had dealt out to him with so liberal a hand.

* It was on that account that ancient travellers gave the name of *Nez-percés* to these tribes of North American Indians.

Passing from the mines of North America to those in Central Asia, we see in the Ural and Siberia indigenous races submitting, with somewhat greater difficulty than in America,

Fig. 162.—Indian miner of Lake Superior (tribe of Chippeways or Sauteurs), with his hammer, and in mining costume.

to the painful toil of the mines. The Kirghese, especially those of the south and east of the Ural, are both shepherds and miners, occupied with the care of their flocks and herds, and in working the placers. Nomadic miners, washing the sands at

intervals, and camping under tents, they contrast not only with
our severely disciplined European miners, but also with the
Russians condemned to the mines, who form a disciplined
battalion, made to work as a punishment. There is no special
type amongst these last. Nobody thinks of founding a family
in Siberia; the only object in view is to get away from it
as soon as possible. In the placers of the Altaï, and in the
graphite mines of Mount Sayan in Eastern Siberia, some few
of the nomadic natives of the country give themselves up willingly, like the Kirghese, to the practice of underground labour.

These instances of indigenous populations remaining in a
semi-barbarous state, and occupying themselves with mining,
are unfortunately very rare. Unless it is a question of deposits
that can be easily worked, especially placers, it is always the
European, with his skill and energy, and never the native of
the country, who enriches it by developing the wealth of the
subsoil.

The placers lead us to speak of an altogether new class of
miners, who have sprung up in our times in California and
Australia. Recruited from every country in the world, the
gold-diggers soon became compacted into one family, every
branch of which has, nevertheless, preserved its distinctive
characters. At the head of them the Chinese present themselves, as steady and industrious workmen, who imported the
rocker to the placers, and who, contented with the smallest of
profits, wash the sands that it will not pay the other miners to
work. But the Chinese are a yellow race, who live together,
and scarcely consume anything except the products of their
own country. They have been expelled from California and
Australia on account of their colour, and they have also been
accused of causing a fall of wages wherever they are. It was
also fancied, from their ever-increasing numbers, that they would
come and invade the country. From their pious custom of
sending their dead to their native land, the *Alta California*,

a daily newspaper published at San Francisco, takes up their defence in these terms :—"We do wrong to drive away the Chinese, for they are our best merchandise. We import them in the raw state, living, and export them, manufactured and refined when they are dead." At the present time there are forty thousand Chinese in California.

In Australia they have defended themselves through the eloquent mouth of one of their number, "Quang-Chew, a man of sense, fifth cousin of the Mandarin Ta-Quang-Tsing-Loo, who owns many gardens near Macao," and they have also gained the day.

It is not only the Chinese, but also the Spanish Americans, especially the Mexicans and Chilians, besides the host of gold-finders from Europe—Italians, Germans, and French—who attract the attention at the placers. The number of French in California amounted to fifteen thousand in 1859. All the miners have preserved a special physiognomy: the Mexican and Chilian are fond of gambling and the cigarette, are sober, and often taciturn; the Italians gradually become traders, sailors, and fishermen; the Germans are united to each other by the ties of fatherland and the love of music; the French are active and restless, running from one placer to another, always grumbling, for ever looking back upon that belle France which they would love to see again, and to which they hope to return at last; the best fellows, however, in the world, joyous livers, spirited workers, having no equal on the *claim* in handling the pick and shovel, or in cooking or dressing a dish; unrivalled also in striking up a song, or in making a joke, amidst the resounding echoes of the placer.

All the miners reassemble in the hut in the evening, if the camp be not in the neighbourhood. From time to time they change the scene of their labours, and carry their tools else-where in search of richer ground. (Fig. 163.)

The moralist who interests himself with regard to the social

position of the miner, especially of the gold-finder, soon sees
in him the first colonist of the age.   He is the pioneer of all
others, whom nature employs to fertilize the virgin countries
which she wishes to deliver over to the industry of civilized
man.   In California and Australia the family of gold-miners,
collected from every quarter of the globe, composed of elements

Fig. 163.—Californian gold-finder prospecting the ground.

that are partly bad and even vicious, becomes purified by
degrees, and regenerated by labour, without which nothing is
consolidated.   In our survey of the metalliferous world we
see in these colonies prosperous and powerful states, who have
little cause to envy the best governed countries of Europe, and
which have been founded by enchantment, as it were, or by
the stroke of the fairy's wand.   That fairy was labour—above

all, labour connected with mines—which was the primary means of raising these distant countries from their savage condition. All honour then to the metal-miner, to whom this happy transformation is due; and let us recognize in this energetic pioneer, not less than in the coal-miner, one of those hidden instruments of which Providence makes use sometimes for the development, the progress, and the prosperity of humanity. But it is not only in the foundation of colonies that the miner plays a most important part here below; he also exercises a very considerable influence, as we have already shown, even on the progress of civilization. No industry is possible without the metals which the miner provides; without gold and silver part of the arts connected with decoration would disappear, and the basis of value, from which commerce really takes its rise, would no longer exist. From this point of view the miner is the first and most meritorious of workers, as he is the most ancient in date. Without him neither agriculture nor shipping could have been developed; for did not the pick precede the plough and the ship?

CHAPTER XII.

THE WEALTH OF NATIONS.

Continual progress.—The metal of peace and war.—Industrial feudality.—Production of iron and steel in 1865.—Production of gold and silver.—Statistical Tables.—Quantity and value of the different metals extracted from the Globe in 1865.—Economic and social laws.—Exhaustion of the metals.—Discovery of aluminium and of aluminium bronze by Dr. J. Percy.—Spectrum analysis.—Stones fallen from the sky.—The metal of the future.

THE production of metals affects material prosperity in so high a degree, and one might also say the artistic, intellectual, and moral prosperity of nations, that the subject is worthy of our careful consideration. This production, it will be understood, determines the true riches of a country, and like that of coal, it progresses so rapidly year by year, as to astonish even those who are most conversant with the industrial and economical phenomena of our times.

The universal introduction into use of steam-engines and railways has involved so large a consumption of wrought and cast iron, steel, and even copper, zinc, tin, and lead,* that in many countries the production of these metals has, like that of coal, doubled itself in less than fifteen years. The application of the combustible mineral instead of wood has singularly facilitated metallurgical manufactures. The making of implements and instruments for all professions, for the humblest as well as the most refined, civil and naval constructions, and in a host of other works, have absorbed considerable quantities of iron. In London iron was selected for the Great Exhibition Building of 1851, which has since with additions become

---

* Most of the pipes and joints of machines are made of lead; copper, either pure, or alloyed with zinc or tin, and then forming yellow-copper or brass, and bronze, is employed in machinery in a variety of ways.

the Crystal Palace at Sydenham. In Paris, cast and wrought iron have also been used for the construction of the Halles Centrales, and again for the framing of the International Exposition of 1867. A new style of architecture, unknown to our ancestors, and suited to the exigencies of modern times, owes its origin to iron, especially in combination with glass.

This is by no means the extent of the services destined to be rendered by the metal of Mars; it will be found in every way worthy of its name. The art of war, by land as well as by sea, with a boldness justified by the results of recent conflicts, has possessed itself of iron and steel, and uses enormous quantities of both. Iron plays a most important part in the military defences of a country. In the last American war, the Northern States ended by bringing it to bear on the South. In the recent wars of Germany, Prussia so quickly defeated Austria, and both, at the commencement of a struggle in which they were at first united, overcame Denmark so easily, partly because the conquerors had the advantage of metallurgical superiority. In the contests which will unhappily long continue to take place, victory will henceforth generally remain with those who produce steel in the largest quantity and of the finest quality. If proofs are necessary for the support of this assertion, which may at first appear paradoxical, the following may perhaps suffice: An armour-plated vessel possesses an engine and boilers of a thousand horsepower, a plating six inches thick, forty steel guns, with a magazine of projectiles, a metal ram of two hundred tons, an enormous amount of pig-iron ballast; in short, a weight of at least three thousand tons of pig-iron, wrought iron, and steel.

Merchant vessels in turn have also as great a demand for metal as ships of war. The hull alone of the *Great Eastern* required ten thousand tons of iron. To give an idea of what these figures represent: twenty thousand tons of ore, containing fifty per cent. of metal, and from forty to fifty thousand tons of coal; that is to say, the whole year's produce of a very rich iron and coal mine.

Before metallurgy underwent a great change by the employment of the mineral fuel, the same figures would have represented the annual produce of several mines and foundries. Now the small smelting-houses, with their somewhat antiquated processes, but whose products used to enjoy a well-merited reputation, and whose brands were even celebrated, disappear by degrees, and in their places have sprung up immense works, furnishing about two hundred thousand tons of cast and wrought iron per annum, as those of Dowlais and Cyfarthfa, in South Wales; Low Moor and Bowling, and the newer works in Cleveland, in Yorkshire; or the half of this formidable amount, as the establishments of Creuzot and Moselle in France. This production of a hundred thousand tons per annum, at which some of our foundries have arrived, would not have been believed possible some fifteen years ago. The concentration and centralization, which governments do not appear to wish for themselves, pass into the hands of individuals, and metallurgical and mining affairs are carried on much in the same way as were formerly political and administrative matters. A kind of industrial feudality has commenced, by which a few influential persons will monopolize the production of coal and iron of a whole country.

Although by the concentration of several furnaces at one point the yield of individual works may be increased almost without limit, the chief economic effect of such working is to cheapen the produce at the expense of its quality. Wrought iron, especially of the kind used for the finer classes of cutlery-steel, is still produced in forges scattered over wooded mountain districts in Sweden, Russia, the Eastern Alps, and the Pyrenees. Most of the famous cutlery of Sheffield, as is well known, is fabricated with Swedish iron.

France, possessing numerous and very rich mines of iron,

occupies a good position with regard to that metal, as in coal-mining.   In half a century, from 1815 to 1865, the amount of her production in iron increased ten-fold; and in ten years, from 1851 to 1861, it doubled.   At the present time France consumes annually five millions of tons of ore, of which nine-tenths are extracted from her own soil, the rest being derived from Belgium, Africa, and the Island of Elba.   This quantity of ore yields twelve hundred thousand tons of metallic iron, being as much as the entire produce of the United States. Great Britain produces nearly five millions of tons, that is to say, twice as much as France and North America together. Belgium and Prussia each produce about half a million of tons, or a tenth part of the British Isles; Austria, three hundred and fifty thousand tons; the German Confederacy, excepting Prussia, two hundred and fifty thousand; Russia and Sweden, each two hundred thousand; Italy and Spain, fifty thousand each; and all the other producing countries, India, Africa, &c., a hundred thousand tons at the most.   The total amounts to nine millions and a half of tons, as will be seen by the following Table :—

TABLE OF THE PRODUCTION OF THE GLOBE IN WROUGHT IRON, CAST IRON, AND STEEL, IN 1865.

Names of the countries.	Production in tons of 2000 lbs.
British Isles,	4,900,000
France,	1,200,000
United States,	1,200,000
Belgium,	500,000
Prussia,	500,000
Austria,	350,000
Other German States (Bavaria, Saxony, Hanover, Nassau, Wurtemburg, &c.),	250,000
Russia,	200,000
Sweden,	200,000
Italy,	50,000
Spain,	50,000
Other producing countries (India, China, Africa, Central Asia, &c.),	100,000
Total,	9,500,000

These nine million five hundred thousand tons of wrought iron, cast iron, and steel, reckoned at the mean price of £8 per ton, would give a value of £76,000,000.

The common metals have for the most part, in the rate of production, kept pace with that of iron. But there is no parallel for the marvellous production of the precious metals, especially gold. California and Australia, since 1848 and 1851—that is to say, in an interval of not more than eighteen years—have alone produced a quantity of gold equal to that which the whole of America had produced during three centuries and a half, from the period of its discovery to the year 1848.* Nothing is wanting to add to facts already so startling. At the very moment when it was imagined that gold was about to take the place of silver, at the moment when some of the European governments, by the advice of economists, thought of lowering the standard of their silver money, in order to restore in an indirect manner the uniform relation of a fifteenth between the value of the two precious metals, which had existed for centuries—silver in its turn was suddenly discovered (1859) by the people who could secure the best part, the North Americans, in the territory of Utah. Scarcely seven years had elapsed when the region occupied by mines formed a State, which, under the name of Nevada, adorns the star-spangled banner of the Union with another star; and already as much silver is annually extracted from these mines as of gold from California (see page 346). In 1864 President Lincoln, in his last message, valued the produce in precious metals of California and Nevada at a hundred mil-

* The total quantity of gold, declared or not, extracted from California and Australia alone, may be estimated at twelve million sterling annually, on an average; or for fifteen years only, and at the rate of twenty-four million sterling annually, three hundred and sixty millions of pounds sterling. In 1848 the whole amount of gold furnished by America from the year 1500, was estimated at four hundred million sterling. At the present time, the aggregate quantity of precious metals in existence throughout the world is estimated at two thousand four hundred million sterling, which is supposed to be about equally divided between gold and silver. Of the amount of the precious metals now in existence, £1,400,000,000 are estimated to have been obtained from the continent of America, £400,000,000 from Europe, £400,000,000 from Asia, and the rest from Africa and elsewhere.

lion dollars (£24,000,000). The equilibrium between the two standard metals, which it had appeared impossible to preserve, was now instantaneously restored. Strange oscillation in the value of gold and silver, which nature herself appears to have ordained! The history of the precious metals is indeed a curious one. The mines of Mexico, Peru, Chile, and Bolivia, for the most part falling off or exhausted, threatened the drying up of the source whence the world acquired its silver. We were about to be inundated with gold, and economists began to apprehend the most serious financial disasters, when the silver-mines of Nevada were discovered, and the balance between the two metals was restored, contrary to all human calculations.

Let us follow the precious metals in their voyage through the world, by the isthmuses of Panama and Suez, which, although not yet actually cut, are beacons placed between the oceans, pointing out the two greatest routes in the world. The silver of the American mines goes to London by way of the Isthmus of Panama. The gold of California follows the same route, the greater part going to New York; that of Australia reaches Europe chiefly by sea, a little of its gold may, however, pass by the Isthmus of Suez, and continue its way intact towards the great metropolis of the industrial and maritime world. Silver completes its journey in going from Europe, by the Mediterranean and Egypt, to the Indies and China, and never returns. These two countries, to which might be added Arabia, Persia, and Japan, have been compared to gigantic sponges which absorb European silver. The more white metal is produced by metallurgy, the more is sucked in and buried by the East. Silver is, besides, the chief coin current in Asiatic countries, and has been so from time immemorial; while silver even more commonly than gold is used among them in jewellery, and bank-notes are unknown to them. In default of coined money, English, French, or American ingots, bearing the stamp of commercial assayers, are accepted.

2 I

Having shown the continued progression which character-ized the extraction of metals from the commencement of this century, it would perhaps be interesting to present a complete statistical account of the metallic production of the globe in 1865, as has already been done with regard to iron. But this series of long and interminable tables, full of figures, would be somewhat out of place. It will be better to give a summary comparing with each other, in a single outline, the total quan-tities and value of the metals extracted, beginning with iron.

TABLE OF THE QUANTITY AND VALUE OF THE DIFFERENT METALS EXTRACTED
FROM THE GLOBE IN 1865.

FAMILY OF IRON.

Name of the metal.	Total weight in tons.	Price per ton. £	Total value in pounds sterling. £
Wrought iron, cast iron, steel,	9,500,000 ...	8 ...	76,000,000

COMMON METALS OTHER THAN IRON.

Lead, ...	250,000 ...	24 ...	6,000,000
Zinc, ...	115,000 ...	24 ...	2,760,000
Copper, ...	65,000 ...	100 ...	6,480,000
Tin, ...	22,000 ...	100 ...	2,200,000
Quicksilver, . . .	3,000 ...	240 ...	720,000

MINOR METALS.

Antimony, nickel, cobalt, bismuth, aluminium, platinum, &c.,	1,640,000
	19,800,000

PRECIOUS METALS.

Gold, ...	32,000,000
Silver, ...	18,000,000
	50,000,000

In comparing the different figures of this Table, for the pur-pose of deducing some economic laws from them, it will be found, first, that the quantity of zinc is half, and the quantity of copper a quarter, of that of the lead produced annually; secondly, that the quantity of tin is a third of that of copper; thirdly, that the value of the minor metals is scarcely a tenth part of that of the useful common metals; fourthly, that all the common metals do not represent in value the half of the precious metals, or the third of the iron extracted annually, and in weight a twentieth part of the quantity of iron. There is a special law for the precious metals, as the value of the gold annually produced is nearly double that of silver; but the contrary was the case before the discovery of the Californian and Australian mines.

If we had given minutely detailed statistical tables, other economic and even social consequences might have been drawn from them. For example, it would be seen that Europe still holds the highest rank for the production of the common metals, which insures its superiority, as well from a commercial and industrial, as from an intellectual and moral point of view; and in their turn, the American colonies and Australia continue to be the great purveyors of gold and silver, thus attracting constantly towards them the floating population of the colonies.

In the manufacture of the common metals, the British isles have the pre-eminence, which is assured to them by their enormous production of coal, and by the economical application which they make of the combustible mineral to all metallurgical operations. Quicksilver and platinum are the only metals they do not produce, and only a few countries surpass them in the production of certain metals—Chile for copper, the East Indies for tin, Spain for lead, Prussia and Belgium for zinc. California and Spain are the great purveyors of quicksilver, and Russia furnishes eight-tenths of the platinum annually produced on the globe.

The Asiatic States, comprised in our Tables, make a very poor figure. They do not even produce enough for their own consumption, not only of gold and silver, but of the baser metals. The copper of Chile goes to China, and competes with the copper of Japan. The iron of England holds its own in the Himalaya against the Indian steel.

When we consider that the extraction of metals is constantly progressing in an increased ratio, we are naturally led to inquire, as in the case of coal, whether and when the supply will come to an end, and what metal will be employed by man when those which nature appears to have held in reserve for him shall be completely exhausted.

The question is less pressing with regard to metal than coal, and offers more than one solution.

In the first place, the entire exhaustion of the mines cannot be looked forward to; for if some become barren at a certain depth, the productiveness of others never diminishes, however deeply they may be excavated. In the Harz, at the present time, there are mines regularly worked at a depth of more than two thousand five hundred feet; and in Saxony and Cornwall two thousand feet have been reached without losing the ore. Here there are no limited and circumscribed basins, as with coal, but indefinite deposits, so to speak, as they are imbedded perpendicularly in the ground. Metalliferous veins, moreover, are widely distributed, and those which remain undiscovered are possibly as numerous as those which are already known. These facts cannot be adduced in speaking of coal. Besides, the quantity of metal annually consumed does not amount to ten millions of tons, being less than the coal produced by France alone. Again, metals do not vanish away in smoke like coal; they no doubt waste and wear out,* but these elements of disappearance may be compared to the infinitesimal quantities of algebra. All these reasons, and others which might be adduced, combine to assure the future. It may be said that metals do not become exhausted, because new discoveries of them are constantly being made, although they are not like coal, only worked yesterday, as it were, and of which the eventual disappearance is calculated with

* The average loss of coin by abrasion is estimated to be a tenth of one per cent. per annum; and the average loss by consumption in the arts and destruction by fire and shipwreck, at from £80,000 to £3,200,000 sterling per annum.

mathematical certainty. Metals, on the contrary, have already presented many brilliant stages. Assyria, Phœnicia, Egypt, Judea, Greece, and Etruria, without mentioning the extreme East, have each in turn owed a part of their refined civilization to their metals; and the mines, often the same, continue, after three and four thousand years' working, to reward the labours of the miners. Some of the veins are as yet scarcely more than touched even at the surface.

Let us speculate on what might happen to our little globe in the course of ages, in the event of the failure of metal.

Modern chemistry, so exact in its operations, has recently demonstrated, by the beautiful experiments of M. Henri Sainte-Claire Deville, that one of the most unchangeable metals can be extracted from argillaceous substances, such as clay; and there are works established for this purpose at Nanterre near Paris, at Salindres near Alais (Gard), and at Newcastle in England. Aluminium appears to be destined for many important uses; its lightness, its inalterability, and susceptibility of being applied to many useful purposes, declare its value. Ten per cent. of aluminium, combined with ninety per cent. of copper, produces aluminium-bronze, an alloy discovered by Dr. J. Percy, having the colour of gold, which is but slightly affected by exposure, and which is harder than gun-metal, and is found to be well adapted for the bearings of heavy machinery. Magnesium is possessed of many curious properties, and calcium, sodium, and potassium may be separated from their combinations. These metals present an aspect hitherto unknown, instead of retaining the form believed by chemists until recently to be their only one. They are metals with so great an affinity for oxygen, as to burn spontaneously in air and water. As the ores of these newly-discovered metals are to be met with everywhere, composing the class of minerals commonly called earths, and are not among those rarer substances usually found in veins only, it may readily be seen how the field of our metallic resources is enlarged, and that the exhaustion of metals cannot possibly be foreseen.

The existence of the heavy metals as vapour in the sun's atmosphere has been recently demonstrated by one of the most wonderful discoveries ever made by man, that of spectrum analysis. Although these metals of the sun are far beyond human reach, yet their counterparts are found in our terrestrial ores. It has been shown, by the study of meteorites, that those bodies are of the same character as our earth. These masses sometimes consist of nearly pure iron, but they oftener are compounds of earthy matter with a little iron. Such are the minerals which our brother worlds send us by way of samples, to initiate geologists in the great phenomena going on in the remote regions of space. If, according to a dramatic poet, gold should ever become a chimera, when other metals shall have disappeared, iron will replace them all. A new age of iron, more complete than the preceding, may then dawn on mankind, and again, as in ancient times already remote, this metal may even be employed in exchange, under the name of money. It will lose none of its present uses; but if war should at length happily disappear from the world, and the metal of Mars be made subservient only to the arts of peace, then this latter age of iron would be a real counterpart of the fabulous golden era of the ancient poets.

There is thus every prospect of a brilliant future for metallurgy, and our posterity, if condemned to seek for the source of force in the solar energy, will not be compelled to have recourse to that luminary for the working of metals, as the earth will ever be in a position to supply them.

PART III.

—

MINES OF PRECIOUS STONES.

MINES OF PRECIOUS STONES.

CHAPTER I.

THE FAMILY OF GEMS.

Gems or Precious Stones.—Origin and mode of occurrence of gems.—Crystallization and cleavage.—Colour, brilliancy, and transparence.—The Diamond : physical and optical properties ; forms into which it is cut ; the carat ; value of diamonds ; the crown-jewels of England ; the jewels in the Imperial crown ; the French crown-diamonds.—The most celebrated known Diamonds: the Regent or Pitt ; the Sancy ; the Star of the South ; the Grand Duke of Tuscany, or Austrian ; the Koh-i-noor ; the Great Mogul ; the Rajah of Borneo ; the Shah ; the Orloff ; the Moon of the Mountains ; the Pole Star ; the Portuguese crown-diamond.—Composition and origin of Diamond.—The Sapphire and Ruby.—The largest known Sapphire, the Ruspoli.—The Emerald : colouring matter ; the largest known stone.—The Topaz.—The Garnet, and its varieties.—Lapis Lazuli ; Ultramarine.—The Turquois ; Odontolite or Bone Turquois.—The Group of Silicates : Rock Crystal, Morion and Smoky Quartz, False Topaz, Cairngorm, White Stone ; Irish, Cornish, and Bristol Diamonds.—The Amethyst.—Opal.—Nephrite or Jade.—Chalcedony, Aventurine, Agate, Onyx, Onicolo.—Jasper, Bloodstone, Carnelian.—The Group of Irregulars ; Malachite, Marcasite, Miroir des Incas, Amber.—Ambergris.—Jet or Black Amber.—Pearl.—Coral.

AMONGST the numerous and interesting class of what are commonly called stones, the family of gems or precious stones is the most remarkable. These are called *pierres fines*, or fine stones, by the French collectors and jewellers, most likely on account of the fine polish which can be given to them; and they are also called hard stones, because they are generally hard enough to scratch the most resisting substances, such as glass and steel. The diamond, which stands at the head of the precious stones (Plate X., figs. 1 and 2), is the hardest substance in nature; it scratches all other substances, and cannot itself be scratched by any.

Like most metalliferous substances, precious stones are generally met with in veins or in simple fissures, and even in cavities, which either traverse or occur near eruptive rocks. Sometimes they are found disseminated through the rock itself. Most gems have been produced, like metallic ores, from aqueous solutions and hot vapours. Time, repose, and the means being favourable, sparkling crystallizations are the result, true tears of nature, and the gem slowly appears, crystallizing itself out from the surrounding rock in which it originated. Volcanic lavas have given birth to some gems, but how different from their elders, if not in composition, at any rate in beauty and brilliancy!

In the cavities and geodes in which they are slowly formed, precious stones not only display the richest colours, but also forms which obey the law of beauty—being both simple and symmetrical. The elementary or primary form is always the same for the same species. That form is the mineralogical individual, the very body of the atom or ultimate molecule to which it would be reduced by cleavage,* if it could be carried so far. The fixity of the primary form does not exclude variety, and the derivative forms resulting from symmetrical modifications on the edges and angles are very numerous, giving to the crystals those facets on which the rays of light play. Lastly, gems are distinguished by a particular transparence. The result of the aggregation of so many qualities places them at the head of all inorganic bodies. If we have styled the metallic ores the princes of the mineral world, may it not be said with more reason that precious stones are the queens of it?

The Diamond, with which it is convenient to begin the

---

* By this term is meant the property possessed by natural crystals, of splitting along planes which always intersect each other at the same angle for the same species, and which are parallel to the primary crystalline form. This property of crystallized mineral substances has given rise to cutting especially the diamond. The operation of cleavage had been practised from time immemorial by lapidaries when the Abbé Haüy deduced from it the beautiful laws of crystallography, which has placed mineralogy amongst both the exact and the natural sciences. Cleavage, it may be said without exaggeration, enables us to dissect the bodies of minerals.

review of the principal gems, was called by the Greeks adamant (αδαμας), signifying unconquerable—a name which it deserves in every respect. It is infusible at any but the very highest temperatures, is insoluble in all acids, and lastly, it is the hardest of all substances, and can only be acted on by its own dust—a faculty which has been profitably turned to account by the cutters of this and other jewels. The numerous facets which are obtained in this way on the mill give the stone its greatest value, by enabling it to reflect, or rather refract or bend the light which passes through it, at all angles and in all directions, giving rise to those radiant reflections in which all the colours of the prism, especially the primary colours blue, red, and yellow, are thrown off in dazzling rays. Candle-light is the most favourable for producing this optical effect, and it even heightens the faculty of luminous reflection which the diamond possesses.

The Diamond has a peculiar lustre, which is known as *ada-mantine*. It is generally colourless and transparent, pure and limpid, and has the *water* of the finest crystal; but diamonds are sometimes, though rarely, found of a blue colour, which may be mistaken for Sapphire; rose-coloured, like some kinds of Ruby; red, like Garnet; green, like Emerald; yellow, like Topaz; and violet, like Amethyst. There are even black and opaque diamonds, which the Brazilians call *carbonado*, on account of their coal-like appearance; these, though seldom employed in jewellery, are used with advantage for several industrial purposes. Amongst others they have been recently used for arming the end of the borer in a new rock-boring machine, for making holes in the hardest rocks, such as granite and porphyry.

We are familiar with the forms which can be given to the diamond by cutting. The double-pyramidal form, or *brilliant*, is the most highly esteemed; the *rose*, or *rosette*, flat on the under surface, is the next valuable; and the *table* diamond, with broad flat under and upper surfaces, is the least valuable. The stone is also cut in the form of a pear covered with small facets.*

The diamond is sold by weight, by the carat,† a conventional weight, equal to $3\frac{1}{4}$ grains Troy, and subdivided into half, quarter or grain, and eighth, and so on. It is almost impossible to name a value for uncut diamonds, even approximately, since it fluctuates considerably according to size, quality, and also their abundance or scarcity, and the demand. The price of polished diamonds at the present time is £12 per carat for brilliants, the same weight of gold being worth five hundred times less, or sixpence. Twenty years ago the carat was only worth £8; but the precious gem, like everything else, has become dearer. The diamond cut in rose fashion is worth two-thirds of a brilliant of the same weight, and it is calculated in both cases that cutting doubles the price of the rough stone.

The price of the diamond exceeding a carat increases, not in proportion to the weight, but to the square of the weight; that is to say, to the weight multiplied by itself. A diamond of one carat, cut as a brilliant, being worth £12, that of a diamond of two carats is worth 2×2, or four times as much, or £48; that of three carats, 3×3, or nine times more, or £108, and so on. The value, therefore, of a polished diamond is found by multiplying the square of the weight by the price of a stone of one carat: thus, for a diamond of two carats, first find the square of the weight, viz., $2 \times 2 = 4$, then $4 \times 12 = 48$; so that the true value of the wrought diamond would be £48.

Although the preceding rule is that generally given for calculating the value of polished diamonds, it must be remem-

---

* For figures of cut diamonds, &c., see Bristow's "Glossary of Mineralogy," p. 112.

† The word *carat* comes from the Hindoo *kuara*, the name of a bean formerly used for weighing diamonds. It is also applied to the small Indian shell the cowry, which is used for money, and for making necklaces of. On the West Coast of Africa, in Dahomey, a slave may be bought for so many thousand cowries, varying with the country and the rate: for this curious money experiences great fluctuations in value, varying from 2s. 6d. to 8s. 4d.

bered that the value of each stone varies according to its more or less approach towards absolute perfection; so that there may be a difference of as much as £1000 between the price of a perfect stone of fifteen carats, and that of another stone of the same weight but faulty.

It is as difficult to give a constant price for cut stones as it is for stones in the rough; but it may be stated, that the value of a fine polished diamond of one carat varies from £12 to £12 16s. It must, however, be borne in mind that the value increases considerably with the weight; thus, fifteen carats of diamonds, weighing one carat each, will be worth about £180, while a single diamond of fifteen carats weight, in perfect condition, will be worth from £2000 to £2800.

This being well understood, the following Table (taken from Barbot*) will be found to give the average value of diamonds of good quality, according to their weight and form.

UNCUT DIAMONDS.

		£	s.	
A diamond of 1 carat is worth .	. . . . . .	6	0	
Uncut diamonds of 8 to the carat are worth	. . .	5	12	per carat.
" " 16 "	" . . .	6	0	"
" " 20 "	" . . .	6	8	"

BRILLIANT-CUT DIAMONDS.

		£	s.	
A diamond of 1 carat is worth	. . . . .	12	0	
Diamonds of 3 grains "	. . . .	9	12	per carat.
" 2 " "	. . . .	8	8	"
" 1 " "	. . . .	7	4	"
Cut diamonds of 8 to the carat are worth	. . .	7	4	"
" " 16 " "	. . .	7	12	"

ROSES DE HOLLANDE.

		£	s.	
A rose of 1 carat is worth	. . . . .	8	0	
Roses of 3 grains "	. . . .	6	16	per carat.
" 2 " "	. . . .	6	8	"
" 1 " "	. . . .	5	12	"
" 8 to the carat are worth	. . .	6	8	"
" 16 " "	. . .	6	12	"
" 50 " "	. . .	7	4	"
" 100 " "	. . .	8	0	"

Then from 200 to 500 to the carat they are sold, one with another, in lots of 2500 at a shilling a-piece.

1000 to the carat are worth about 5d. a-piece.

* "Traité Complet des Pierres Précieuses," par Charles Barbot, p. 237.

ROSES D'ANVERS.

Large and well-cut stones are worth £4 per carat; and from two grains they are sold in mixed lots at from £2 8s. to £3 4s. per carat.

Diamond dust, made from refuse diamonds, and used for polishing purposes, is worth £50 per ounce.

The geometrical progression in the price of diamonds furnishes an explanation of the great value of diamonds, and of the estimation in which those of very large size are held, and also for the minute precautions which are taken against any chance of losing articles, which are not only so precious but so small in size. The crown-jewels of England are kept in the Tower of London, in an iron cage surrounded with glass. The jewels comprised in the Imperial Crown consist of 5 Rubies, 17 Sapphires, 11 Emeralds, 273 Pearls, 4 Drop-shaped Pearls, 147 Table Diamonds, 1273 Rose Diamonds, and 1363 Brilliant Diamonds. The famous heart-shaped Ruby in the front of the crown, in the centre of a diamond Maltese cross, is said to have been given to Edward the Black Prince by Don Pedro, king of Castile, after the battle of Najera, A.D. 1367. It was afterwards worn in the helmet of Henry V. at the battle of Agincourt in 1415. In the centre of the cross, on the summit, is a splendid rose-cut Sapphire. In France, during the troubles of the first republic, in 1792, the crown-diamonds disappeared, although they were kept under seal, and in the treasury itself. Some of them were found buried in the Allées-des-Veuves, in the Champs Elysées, in a place that was named in an anonymous letter. The famous Regent diamond was in the casket. At a later period Napoleon had all the other Crown diamonds repurchased of which he could obtain possession.

Perhaps the finest and best-cut diamond, unrivalled for shape and water, though by no means the largest in point of size, is the Regent, named after the duke of Orleans, who was regent during the minority of Louis XV. The regent bought it in 1717 for £135,000 of Governor Pitt, who paid £12,500 for it to Jamohund, the most famous native dealer in India.

During the five years the stone remained in his possession, Pitt lived in such constant dread of having it stolen, that he never made known beforehand the day of his coming to town, nor slept two nights following in the same house. Cut as a brilliant it weighs 136⅞ carats, and has been valued at £480,000; by some it has only been valued at £16,000; taking the mean of the two valuations, the sum is still very large. The Regent was found at Parteall, in the province of Golconda, where it received its first cutting; the weight in the rough was 410 carats.

After the "Regent" we will mention the Sancy Diamond, which fell from the helmet of Charles the Bold at the battle of Granson, and was sold for two francs by a Swiss soldier, and bought in 1589 by Nicolas Harlay de Sancy, treasurer to Henri IV. of France, after whom it was named. The Sancy Diamond was stolen in 1792, together with all the other crown-diamonds, from the *Garde Meuble*, and was subsequently purchased in the year 1838 by Prince Demidoff for 500,000 roubles (£75,000). In 1865 it was bought of the Demidoff family by Sir Jamsetjee Jeejeebhoy for £20,000. The Star of the South, a facsimile of which in its original rough state is given in Plate VIII. (fig. 1), was found by a negress in Brazil in 1853, and bought by M. Halphen, who had it cut. The weight in the rough was 254½ carats, and admirably cut as a brilliant at Amsterdam it now weighs 124¼ carats. The "Grand Duke of Tuscany," or "the Austrian," is of a yellow colour and weighs 139½ carats, cut as a double rose. It is said to have been lost by Charles the Bold at the battle of Morat, and to have been subsequently bought for a mere trifle as a coloured crystal out of a jeweller's shop in Florence. The Koh-i-noor, or Mountain of Light, the oldest known diamond, is said to have been worn by the Indian Rajah Karnah, 3001 years B.C. In our own times the Koh-i-noor was taken from the king of Kabul by

Runjeet Singh, "the Lion of Lahore," and now belongs to her Majesty the Queen of England. The story of the Koh-i-noor is a strange one. According to Hindu legend it was found in the mines in the south of India during the great war which forms the subject of the heroic poem the *Maha'bha'rata*, and was worn by one of the warriors slain on that occasion, Karnah, king of Anga. This would be about 3001 B.C. It is then said to have become the property of Vikramaditya, rajah of Mjayin about 56 B.C., and to have remained in his family until the principality was subverted by Mohammedan conquerors, into whose hands the jewel fell with other spoils.

The Koh-i-noor was first accurately described by Jean Baptiste Tavernier, who in 1665 was permitted by Aurungzebe to examine this treasure of the Delhi cabinet. Nadir Shah, on his occupation of Delhi in 1739, compelled Mohammed Shah, the great-grandson of Aurungzebe, to give up everything of value, amongst other jewels this magnificent diamond. Mohammed Shah is said to have worn the Koh-i-noor in front of his turban at his interview with his conqueror, who compelled him to exchange turbans in proof of his friendship. After the death of Nadir Shah, this diamond became the property of Ahmed Shah, the founder of the Abdali dynasty of Kabul. The jewel descended to the successors of Ahmed Shah, and when Mr. Elphinstone was at Peshawur it was worn by Shah Shuja on his arm. Shah Shuja, when driven from Kabul, became the nominal guest, but actual prisoner, of Runjeet Singh, and in 1813 he was compelled to resign the Koh-i-noor to the latter for the revenues of three villages, not one rupee of which he ever realized. This gem was preserved in the treasury of Lahore for many years, being occasionally worn by Rhurreuk Singh and Shu Singh, and after the murder of the latter, by Dhuleep Singh. On the annexation of the Punjaub the civil authorities took possession of the Lahore treasury in part payment of a debt due from the

A. Faquet pinx.   Regamey Chromolith

1 Turquois. *Persia*.   4 Amethyst. *Saxony*.

2. Lapiz-lazuli. *Bokhara*.   5 Chalcedony. *Ireland*.

3 Opal. *Hungary*.   6. Agate-Pebble. *the Palatinate*.

7. Yellow Amber. *Prussia*.

CHAPMAN & HALL. London.   Imp. Lemercier & Cie Paris.

Lahore government for the expenses of the war, and it was stipulated that the Koh-i-noor should be presented to the Queen of England, who now wears the jewel on all state occasions.

It would be out of place more than to mention here very briefly either the Great Mogul Diamond, found at Coulour, and cut as a rose weighing 280 carats; that of the Rajah of Borneo, and the Shah, a facetted prism weighing 95 carats; the Orloff, a rose-diamond weighing 193 carats, and set in the top of the imperial Russian sceptre; the Russian "Moon of the Mountains," * and the "Pole Star" of the Princess Yousoupouff, brilliant-cut and weighing 40 carats; and some others—full accounts of which are given in books, with all the histories, stories, and legends relating to them. The number of these priceless stones, and of such sovereign or princely diamonds, is very limited. The famous diamond of the crown of Portugal must, however, not be forgotten, the largest known diamond, but which it is said nobody has ever seen. The Portuguese value it at £3,200,000, but malicious people pretend that it is only an old topaz. White Brazilian Topaz is the nearest approach to the diamond, being extremely hard and very brilliant, but without its adamantine lustre and iridescent play of colours.

The rough diamond in its gangue, or if it be preferred, in the cemented pebbles which frequently contain it (Plate X., fig. 1) is not easily discernible. In the midst of those sands, gravels, and alluvions which the Brazilians call *cascalho*, ("rolled pebbles") or *feijao* ("beans"), either on account of the form or the appearance of some of the pebbles, the stone *par excellence* does not challenge attention; and in order that it may be more easily perceived, the sands have to be enriched by washing (Plate X., fig. 2) If it were not for the practice

* Stolen in the last century by a French grenadier from an idol of Bramah, one of the eyes of which was formed of this diamond. The soldier could not manage to steal the other eye.

and the skill acquired by long experience in the diamond-washings, the divine gem would in most instances be passed over without being even noticed. And what after all is the diamond? Nothing more than pure crystallized carbon, supposed to have been formed by the slow decomposition of some vegetable or bituminous substance.

If the sun's rays be conveyed to a focus by means of a lens, and directed upon a crystal of diamond placed beneath a bell-glass filled with oxygen gas, it will burn, and the result will be that merely carbonic acid will be found beneath the glass, an invisible gaseous combination of oxygen and carbon. This experiment of Lavoisier's shows that everything here below, glory and the most beautiful jewels, all end in smoke: *Sic transit gloria mundi.*

The Sapphire and Ruby, one of an azure-blue, and the other fiery or rose red, are the two most beautiful of coloured gems. They hold nearly the same rank as the Diamond (Plate X., figs. 3, 4, 5, and 6), and are almost as highly esteemed by the lapidaries when they are of good water; indeed, perfect rubies of one carat and upwards are more valuable than diamonds of equal weight, and fetch higher prices. In the rolled and rounded form, with a natural polish, as they are found in the placers, rubies and sapphires form very beautiful *cabochons.* Both these stones are comprehended by mineralogists under the same denomination of Corundum, which is said to be a Chinese word. One form of common Corundum is Emery, the powder of which is applied to the same uses as diamond-dust; that is to say, for seal-engraving, grinding, cutting, and polishing glass, metals, and other hard substances. Chemically Corundum is pure alumina, or clay without its silicious ingredient, one of the commonest substances, but when crystallized what fire does it not display? The Sapphire and Emerald sell for £3 per carat when fine. The largest known Sapphire, called the Ruspoli, is in the Musée de Mineralogie in Paris.

A. Faguet pinx!   C. Regamey Chromolith.

1. Diamond, *Cascalho (Brazil)*.   4. Sand with Sapphires, *Ceylon*.   7. Emerald, *New-Grenada*.

2. Diamond Sand, *Borneo*   5. Rubies, *United-States*.   8. Topaz and Rock Cristal, *Brazil*

3. Sapphire, *Ural*.   6. Sand with Rubies *Burmah*.   9. Garnet *Bohemia*.

CHAPMAN & HALL, London   Imp. Lemercier & Cie, Paris.

It is of a lozenge-shape with six faces, and weighs 132 carats. It was bought by Perret, a French jeweller, for £6800. The finest varieties of Corundum, used in jewellery, are brought from Ceylon; but very fine stones are also found in Pegu and Ava.

The green stone represented in Plate X., fig. 7, is the Emerald, of which the Aqua-marine is a pale bluish or sea-green variety. According to chemists the Emerald is a double silicate of alumina and glucina, the latter being an oxide of glucinum, a metal of which little is known, since it occurs but rarely in nature. As a silicate it is found in Phenacite, and associated with other silicates it is detected in Beryl, Euclase, Leucophane, Helvite, and some varieties of Gadolinite. It is also found as an aluminate in Chrysoberyl or Cymophane. The beautiful green colour of the Emerald has been attributed to oxide of chromium; but the more recent researches of Lewy have raised the belief that the colour may be due to organic matter, the existence of which he believes he has proved by experiment. The largest known Emerald is that in the possession of the Duke of Devonshire. It is an uncut six-sided prism, 2 inches in length, $2\frac{1}{4}$, $2\frac{1}{5}$, and $1\frac{7}{8}$ inches across, and weighs 8 oz. 18 dwts. It is also of the finest colour, and free from flaws. This Emerald was found in New Granada, at Muzo, near Santa-Fé-de-Bogotá.

The Topaz, with its characteristic wine-yellow colour (Plate X., fig. 8), is a silicate of alumina, with a small quantity of fluorine; sometimes it is blue, green, pink, and even colourless. Most gems are merely silicates of alumina, mixed with some foreign substance, which in this instance is fluorine, an element of uncommon occurrence. In combination with calcium, fluorine forms almost a gem—fluor-spar, or " Blue John," which has been mentioned amongst the gangues of metallic ores. Combined with hydrogen, fluorine furnishes one of the most powerful acids, hydrofluoric acid, which corrodes glass, and is used instead of the diamond for etching on that hard substance. The best Topazes come from Brazil, and are sold in the rough for £4 per kilo (2·2 lbs.). White Topazes are obtained from New South Wales, Ceylon, and Brazil. Large yellow Topazes are also found on the Cairngorm mountains in Scotland, and are often mistaken for the yellow varieties of Rock Crystal, named after that locality Cairngorm Stones. An inferior kind is brought from Saxony.

Garnet is another silicate of alumina in combination with (according to circumstances) oxide of iron, manganese, chromium, lime, or magnesia. The usual colour is deep red; and the variety which displays the characteristic garnet-red colour is the Iron Garnet, called Precious Garnet, Almandine, or Oriental Garnet (Plate X., fig. 9), also the Pyrope, characterized by containing chromium, or Syrian Garnet (after Syrian, the capital of Pegu), or Carbuncle; but there are also black varieties of Garnet (Melanite), as well as yellow, green, rose-coloured, and white or colourless. Garnets, like the Diamond and the Topaz, present a great variety of appearances, according to the metallic oxides which they contain.

Lapis Lazuli (Plate IX., fig. 2), on the contrary, is always of the same colour. It is a silicate of alumina, lime, and soda, without transparency, and without much lustre, but of a very uniform and soft tone. The stone may take all tints of azure-blue, whether light or dark; and it is often interspersed with yellow specks and veins of iron pyrites, which from their brilliant appearance in the comparatively dull blue stone, might easily be mistaken for gold. The finest quality of Lapis Lazuli sells in the mass for £30 per lb. The most richly coloured varieties are used for jewellery, such as rings, pins, crosses, ear-rings, and for making costly vases and ornamental furniture; it is also employed for architectural ornaments, inlaying, and in the manufacture of mosaics. When powdered and purified by a peculiar and a tedious process, it furnishes the celebrated and beautiful pigment called *ultramarine*, used by painters in oil, and said never to fade. The Chinese have employed Lapis Lazuli for a long time in porcelain painting.

Lapis Lazuli is chiefly brought from Persia, China, Siberia, Great Bucharia, Thibet, and Chile. It is also found in California, but of comparatively indifferent colour and quality.

The Turquois (Plate IX., fig. 1) is allied to Lapis Lazuli in appearance and slightly in colour. It is not, however, a silicate of alumina and other bases, simple and compound, but merely phosphate of alumina coloured by oxide of copper. Odontolite, or Bone Turquois, is merely petrified teeth or ivory, stained green or blue by phosphate of iron, like the bones of the fossil miners alluded to at p. 428. The name Turquois is derived from that of the country (Turkey) from which it was first brought, and it is from Turkey in Asia and from Persia that the most beautiful specimens are still procured. In 1849 Major Macdonald found turquoises of fine colour in the Sinaian range of mountains in Arabia Petræa. These were collected largely by him. Sometimes they are found in nodules, varying in size from a pin's head to that of a hazel-nut, and when in this state they are usually of the finest quality and colour. Another mode of occurrence is where they appear in veins, and sometimes of such a size as to be of great value. They are also met with in a soft yellow sandstone, inclosed in the centre, and of great brilliancy of colour. Another curious formation is where they are combined with innumerable quartz-crystals, the whole presenting the appearance of a mass of sand, small pebbles, and turquoises all firmly cemented together.

Silica, which next claims our attention, is an important and highly useful substance; for it furnished man with his first implements, then with a mill to grind his corn, and afterwards was used for gun-flints. Now-a-days it is used for making macadamized roads and for paving streets; but these useful purposes being beyond our domain, we will confine our notice to silica as a gem, and not as a rock-mass or as flint. The first

in the numerous and varied forms of silica is pure crystallized
Quartz (Plate X., fig. 8), or Rock Crystal, possessed of double
refraction, or the property of giving two images of any object
viewed through it.   All colourless gems except the Diamond
possess this property to some extent.

The black and smoke-brown varieties of Rock Crystal are
called Morion, and Smoky Quartz; when yellow, like topaz,
False Topaz, and Cairngorm; the limpid and colourless kinds
are occasionally used in jewellery under the name of " white
stone," and are often called, after the localities where they are
found, Irish, Cornish, Bristol, Brighton, and other diamonds.

Perhaps the most useful application of Rock Crystal is for
the " pebble" lenses of spectacles.

The commonest form of Rock Crystal is that of a six-sided
prism, terminated by a six-sided pyramid.   The crystals frequently intersect and penetrate each other, and thus form the
richest and most beautiful groups.   Some crystals inclose
internally fine needle-shaped crystals of Asbestos, and Rutile,
or red oxide of titanium, called Fléches d'Amour, Venus'
Pencils, &c., &c.   Others contain cavities containing water and
other liquids, which move as the crystal is inclined in different positions.   Quartz is one of the components of granite,
together with crystalline felspar and lamellar mica.   In its
compact and massive form it is no longer a gem; but in this
case it forms veins which are often of great size, and serve
as the matrix of gold, silver, platinum, tin, and almost all the
metals and their ores.

Amethyst, or Violet Rock Crystal (Plate IX., fig. 4), is
supposed to be coloured by exceedingly minute quantities of
oxide of manganese.   The Amethyst, like colourless rock
crystal, often forms very beautiful groups, lining cavities or
the hollow spherical masses which are called geodes, from
which the pyramidal terminations of the six-sided crystals
shoot out boldly.   In consequence of its colour this is the

episcopal stone, bishops being restricted to the use of violet as a colour. By the ancients it was regarded as an antidote against intoxication, the Greek name αμεθυστος (amethystos) having that meaning. Amethysts are imported from the East Indies and Brazil, and in large quantities from Hungary, Bohemia, and especially from Oberstein in Saxony, where they are cut and polished by steam and water power.

The Opal, with its changing and flashing tints (Plate IX., fig. 3), one variety of which is called Harlequin Opal, is a hydrated kind of silica. The Precious or Noble Opal is generally milk-white or colourless, and exhibits a rich play of colours—green, red, blue, and yellow—of various tints. It occurs in porphyry at Czernawitza in Hungary, at Frankfort, and at Gracias á Dios in Honduras. When large, and exhibiting its iridescence in perfection, it is a beautiful and very valuable gem.

The list of fine stones or true gems ends with the Amethyst; but green Jade* may, nevertheless, be mentioned, which is cut and polished by the Chinese with such labour and patience by means of the diamond; milk-white Chalcedony (Plate IX., fig. 5); Aventurine with its glittering spangles; Agate and Onyx (Plate IX., fig. 6), a variety of which is also called *Onicolo*, with parallel and concentric bands of different colours,† and carved into cameos, especially by the ancients; the variety of dark green Jasper, interspersed with bright red specks, commonly called Blood-stone, and made into ring-stones, brooches, bracelets, and other ornaments; red Jasper or Carnelian, out of which the Egyptians carved many of their scarabæi, and in which they, as well as the Assyrians and the ancient Etruscans, Greeks, and Romans, have cut the most magnificent intaglios. Our forefathers contented themselves with mounting their watch-keys and seals with oval-cut pieces of Carnelian, and

* Jade or Nephrite does not belong to the family of vitreous quartz or pure silicates, being a silicate of magnesia and lime.

† Whence the name onyx (ονυξ), "nail," given to it by the Greeks.

the Germans of the present day make it into cheap ornaments, with which all their fairs and bazaars are inundated.

We will close the list of precious stones with the group which may be termed the "irregulars," from their relation both with the mineral and organic kingdoms of nature. There are still some fine and costly stones, but none that are sufficiently hard and transparent to be considered true gems; nearly all the substances belonging to this class being rather ornamental stones, suitable for inlaying, mosaic, or for personal wear, than true precious stones.

Beginning with the inorganic kingdom, we find, amongst others, Malachite, with its tints of emerald-green, every shade of which is displayed in the same specimen passing into each other by almost insensible gradations. It is essentially a copper-stalactite, a metallic mineral, being a hydrated carbonate of copper, which is described under the family of metals (Plate II. of metal-mines, fig. 4). By the side of Malachite ranks Marcasite, a variety of iron pyrites (Plate VII., fig. 2), which was formerly made into buckles, brooches, and other ornaments. When cut into plates and polished it was known by the name of "Miroir des Incas," because it reflects the images of objects like a looking-glass; and that mirrors made of it have been found in the tombs of the aboriginal Peruvians. Recently a similar mirror of Roman origin was found in a peat deposit on the site of London wall.

Yellow Amber (Plate IX., fig. 7) is of vegetable origin. The Turks make it into beautiful necklaces, bracelets, and very elegant mouth-pieces for their pipes, chibouques, and narghilehs; and devout Mahometans and Greeks convert their amber necklaces into chaplets. Amber ornaments have come into fashion at the present day; and half a century ago the yellow transparent variety was much worn, cut in facets, or simply into beads, for bracelets and necklaces, also for the heads of canes :—

"Let beaux their canes with amber tipt produce,
Be theirs for empty show, but thine for use."
"Trivia, or the Art of Walking the Streets of London," by John Gay.

This substance is merely a fossil resin of a pine-tree which flourished at a recent geological Tertiary period. The resin, when in a viscid state, often entangled insects, such as flies, gnats, ants, grasshoppers, and which, caught and embalmed in this sudden manner, have been admirably preserved. The most delicate organs of these insects may be recognized in transparent amber.

Amber is mostly found in Asia Minor, China, and Sicily; and also along the Prussian shores of the Baltic, to which latter locality probably the Phœnicians went in quest of it, three thousand years ago. It is found on the sea-coast of Eastern Prussia, and on the shores and at the bottom of the Fresh and Curish Haff.

Formerly Amber was only procured by picking it up on the sea-shore, but large deposits of amber have since been discovered from sixteen to thirty feet below the surface of the sea in a Tertiary stratum. It is fished for in the surf with nets, or dug up out of the sands, but the most successful method is to dredge for it at the bottom of the water. About fourteen machines and above four hundred workmen are employed in dredging, who work day and night when the Haff is not frozen up. In 1866 the quantity collected by this method amounted to 73,000 lbs. Amber now constitutes an important article of trade at Memel and Dantzic, whence it is exported to Austria, Turkey, France, and also in considerable quantities to England, whence it is re-exported to the East Indies, Africa, and China.

The price of rough Amber varies from 5 silver groschen (6d.) to 80 dollars (£12) per pound. The largest piece ever found at Memel weighed about 5 lbs., and was valued at about 400 Prussian dollars (£60).

Amber is also met with in the plastic clay of the Paris basin, and small specimens in the London Clay of Sheppey, and elsewhere; but it has never been worked in either of these localities, being of very indifferent quality.

Ambergris, which is found on the sea-coast or floating on the surface of certain parts of the Indian Ocean, usually in small pieces, but sometimes in masses of from 50 to 100 lbs. weight, must not be confounded with Amber. Ambergris has an agreeable odour, and is used in perfumery. It is a solid, opaque, generally gray or ash-coloured, and very light substance, and may be said to be literally worth its weight in gold, for the Arabs of Zanzibar, when they sell ambergris, place it in one of the scales of the balance and counterpoise it with gold pieces, the number of which show the price and quantity of ambergris weighed. Various opinions have been entertained respecting its origin. It has been supposed to be the concrete juice of a tree or a bitumen; but it is now believed to be a concretion formed in the stomach or intestines of the spermaceti whale.

Jet (the Gagates of the ancients), like Amber, is of vegetable origin, being merely fossil carbon, a perfect and compact form of lignite, susceptible of a brilliant polish (Plate VIII.). Jet is generally found (in connection with fossil wood) in very small patches, from an eighth of an inch to a few inches in thickness, and varying in weight from an ounce to two cwts. in the top and bottom portions of the Upper Lias. Small pieces are sometimes found washed up on the sea-beach. The best portions sometimes realize the extraordinary price of from 12s. to 14s. per pound.

Jet furnishes a very important article of trade at Whitby, in Yorkshire, where it is very extensively worked and made into a great variety of articles, such as snuff-boxes, bracelets, necklaces, crosses, and other mourning ornaments. The total annual value of the articles made at Whitby and Scarborough amounts to £125,000. About 250 men and boys are employed in searching for the Jet, and between 600 and 700 are engaged in its manufacture. In Prussia it is called Black Amber, because it is electrical when rubbed, and is often found in sand and gravel beds.

The stones now to be described are of animal origin. The most important of these is the Pearl, which is produced by a malady of the pearl-oyster. Some kinds of this gem of a bluish-gray colour and a pear-shape are called *veuves* (widows) at Panama, doubtless because owing to their rarity they seem to droop in lonely loveliness ; when two can be met with that are exactly alike, they are almost priceless. The substance of which the pearl consists is merely carbonate of lime, soluble in the weakest acids. The story of Cleopatra is well known, how she drank at dessert two priceless pearls dissolved in vinegar, in order to give a lesson to Marc Antony, who had boasted of the cost of the dinner given by him to the voluptuous princess.

Coral, like the Pearl, is a product of the marine world. The minute and sometimes microscopic animals, the Polyps, who build up whole islands beneath the waters of the sea, secrete the solid coral. The red and rose-coloured varieties of this substance are held in great estimation, the latter more especially; and since coral ornaments have come into fashion again, these fetch very high prices.

CHAPTER II.

IN THE EAST AND UNDER THE TROPICS.

The East Indies, Golconda, Visapoor, Ceylon, Borneo.—Oriental stones.—Gold and gem-producing countries.—Siberia.—Pegu.—Australia.—Africa: Madagascar, Egypt, Abyssinia, Natal.—Pearls and Coral.—America: Mexico, California, Bay of Panama, New Granada, Peru, Chile, Brazil.—Gem-producing countries of Europe: Bohemia, Saxony, Hungary, Elba, Limoges, Oberstein, Prussian shores of the Baltic.—Yorkshire Jet.—Coral of Sicily and Tunis.

THE two chief gem-producing parts of the world are America and Asia, the Tropics and the East. Asia from the earliest historical times, and America since the time of its discovery, have not ceased to furnish the civilized world with the most beautiful precious stones. In India there was a stream which the Greeks called " Adamas," or the Diamond. The chain of Ghauts which runs parallel with the Coromandel coast, is one of the great sources whence gems are procured. Who does not know of Golconda, Raolconda, Visapoor, and the island of Ceylon? Borneo, Java, Sumatra, and the Celebes, are also amongst the localities where precious stones most occur, and especially the diamond. Birma and China, whence rubies and sapphires are obtained; Persia, Bokhara, the countries of fine turquois and lapis lazuli; the whole of Central Asia and the East—are the parts of the world most abounding in gems. When a lapidary wishes to express a stone without a rival, he styles it *Oriental, Indian, Syrian;* or the stone is *noble, precious,* or of the *vieille roche.* It is, in fact, in the Asiatic localities that the most beautiful types are found; and up to the present time it is only Tropical America that has to any extent divided this favour with Asia. Hardness, lustre, transparency, and colour are only found combined in the highest degree in the stones that come from those two sources. Some lapidaries believe, with the older mineralogists, that the sun which fertilizes those climates, and that the warm light with which these countries are flooded, have something to do with this effect; but it is more likely that the result may be more correctly referred to the cause which has been already mentioned in connection with the deposits of the precious metals. Probably a sort of attraction has been designedly given by Providence to those regions which are also so remarkable for the luxuriance of their vegetable productions, in order to draw thither the men of civilized but temperate climates, and tempt them to settle there, since without some such sort of inducement they would not migrate towards the tropics.

Whatever may have been the cause of the phenomenon, it holds good as a fact, that the gold-producing countries are also those which produce the gems in the largest quantity and of the most beautiful kind. In Asia, in addition to the countries which have been already named, Siberia must be added, as well as the gem-yielding placers dependent from the Altaï and the Ural, and lastly, the localities in Asia, especially those of Pegu, whence the finest (Syrian) garnets are brought.

In Australia, diamonds, sapphires, rubies, and topazes have been found in the gold-fields of Victoria.

In Africa, the great island of Madagascar, the country of rock crystal, should be especially mentioned; then Upper Egypt and Abyssinia, where emeralds are found and have been worked from very remote times. Diamonds of very fine water have also been lately found in South Africa, on the other side of the mountains which divide Natal from the interior of the country, about two hundred miles from Pietermaritzburg.

The finest pearls and the most beautiful coral are fished for in the Red Sea, the Persian Gulf, and in the Indian Ocean.

In America, Mexico is celebrated for its opals, and the Cali-

fornian gold-diggings also yield gems.   In the Bay of Panama
there are pearl-fisheries rivalling those of the Red Sea and
Ceylon.   New Granada and Peru are especially distinguished
for their emeralds, and Chile for Lapis Lazuli.   But no country
can compare with Brazil, which is the land, *par excellence*,
of the Topaz, Amethyst, Rock Crystal, and Diamond.   The
district of Minas Geraës, besides being rich in gold and all
kinds of metals, affords emeralds as fine as those of Bogotá.
In that respect Brazil vies even with India; but this richness
in gems is of comparatively recent date, the diamond having
been first discovered in South America less than a century
and a half ago.   The finest stones are now brought from this
ancient Portuguese colony, chiefly from a tract adjacent to the
head-waters of the Rio San Francisco, and the Rio Grande
del Belmonte.

In Europe, the rock-crystal and amethyst of the Alps must
be mentioned, also the garnets of Bohemia, the opals and
topazes of Saxony and Hungary, the emeralds and aqua-
marines of Elba and Limoges, and lastly the agates of Oberstein on the borders of the Rhine, though they are not able
to bear comparison with the marvellous jewels of Asia, Africa,
and America.   The Prussian shores of the Baltic have already
been mentioned in connection with its amber, which is collected between Dantzic and Konigsberg; and England for
its Yorkshire jet: to which may be added the Mediterranean,
especially the coasts of Sicily and Tunis, in connection with
coral.

CHAPTER III.

THE SEEKERS.

Disciplined seekers.—The Californian diamond washings.—Diamond-washings of Brazil, India, Borneo.—The Diamond trade.—Individual seekers.—Pietro Pinotti of Elba.—Coral and Pearl fisheries: Genoa and Naples, Bay of Panama, Aden, Colombo.—Anecdote of an Indian pearl-diver.

OF the miners who systematically search for precious stones, the only ones who form, as it were, a disciplined body, are those who work the diamond-mines. The alluvia which contain the gems are worked in the same way as the auriferous gravels, both having been produced by the same causes, that is, by the disintegration of the rocks in place, in which the precious substances were originally contained. The analogy is striking with regard to the deposits containing gold and platinum. The impoverished placers are now being reworked in California, not only for the sake of extracting any small specks of gold that may have been left, but also for diamonds. The gold-finder washes the dirt in his pan (*batea*), and no longer finds in it merely the yellow metal, but likewise small hard pebbles, which he tests upon a stone that he carries about with him.

The gemmiferous sands are also worked in Brazil by washing them in a bowl, or in a running stream of water, by slave-labour, carried on under the eyes of vigilant overseers. The pick and shovel are used for breaking up the diamond-bearing gravels (fig. 164), almost in the same way as they are employed by the Chilians in washing for gold. The water carries away the clay, fine dirt, and sand, and the larger lumps are thrown aside by hand; the middling-sized gravel that remains being

picked over in the open air, where the sunshine playing upon
the faces of the gem readily leads to its detection by experienced
eyes.　　The searchers have a particular knack in doing this,
and it is wonderful to see how they perceive the smallest particles of diamond amongst the largest heaps of pebbles.

Fig. 164.—Working diamond-washings.

As the gem is found it is placed in a little bamboo case
(fig. 165), ornamented more or less outside, and regarded with
a certain sort of superstitious feeling.　　A case which has
already held diamonds will lead to the finding of more, and it
is owing to this belief that the negroes are so unwilling to part
with one of these little tubes of reed.

The search for diamonds is confined to the provinces of Bahia and Minas-Geraës, where it is carried on by companies employing numerous hands, or by individual settlers. The precious stone is not only contained in the sands and cemented gravels, but also in the talcose sandstones, the itabirite and itacolumite rocks, the quartz grains of which are interspersed with crystals of magnetic oxide of iron. These rocks, whose names remind us of the chref gemmiferous Brazilian localities, contain the diamond in place, not in a rolled or partially rounded state, as it occurs in the *cascalhos*, but in the same crystallized state in which it was originally formed. It is asserted by M. Liais and others, that the green porphyritic rocks and serpentines are, on the contrary, the principal diamond-bearing rocks. This point has not yet been satisfactorily settled, and like the original auriferous and platiniferous rocks, furnishes a problem that has still to be solved.

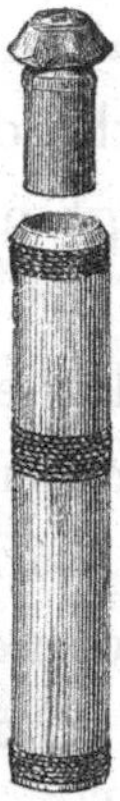

Fig. 165. — Diamond-case of the Brazilian washers. Scale, ½.

The miners, without taking any interest in geological speculations, only attend to their own operations. Discoveries of any importance are publicly announced, and when a slave finds a fine diamond of a certain weight he is immediately set at liberty. At the end of the day the workpeople are made to undergo a most rigorous examination, in spite of which thefts are of very frequent occurrence. The province of Tejuco, the capital of which is called *Villa-Diamantino* by the Brazilians, is the central depot for the washings of the province of Minas-Geraës, and diamonds are current there like money. The river Diamantino, with the town of the same name, in the province of Mato-Grosso, was also so called from the valuable diamonds found in its bed. When the great diamond-washing companies, or the agents of the firms at Bahia and Rio Janeiro, who carry on this important trade, have collected a certain quan-

2 L

tity of diamonds, a convoy is despatched to those capitals, accompanied by an escort of soldiers (frontispiece). The distance to be traversed is a hundred and forty leagues to Rio Janeiro, and two hundr and fifty to Bahia. There are no railways, and the Brazilian robbers, who are inventive and ingenious in their dispositions, know that a rich booty awaits their attack; yet it would seem that the convoys rarely meet with any accident of this nature. A new diamond-producing locality has lately been discovered at Goyaz in Brazil. Many of the diamonds found in these deposits are of the best class, weighing from three to four grammes (15 to 20 carats).

The working of the diamond-mines is carried on in India and the far East nearly in the same way as it is in Brazil. Curious and quaint accounts of former workings are given in the works of Tavernier, Chardin, and other old travellers, and Madame Ida Pfeiffer describes the way in which the workings are carried on in the placers of Borneo at the present day.

The most famous of these mines are situated on the west coast of the island, in the neighbourhood of the towns of Sambas and Pontiana in the province of Landak. The diamonds of Golconda have obtained great celebrity throughout the world; but they were merely cut and polished there, having been generally found at Parteall, in a detached portion of the Nizam's dominions, near the southern frontier, in lat. 16° 40′, long. 80° 28′, a place which affords no favourable indication of the wealth to be derived from the vocation of seeking diamonds, as it is in ruins,* and the inhabitants are ill-clothed, and half-starved in appearance.

The other principal diamond-producing regions of Central India are the high valleys of the Pennar in Cuddapah, and of the Kistnah, near Ellora; also the mountainous district containing the sources of the Nerbuddah and Sone, and the mines of Pannah, in the Bundelcund table-land, south-west of Alla-

* "Journ. As. Soc.," Beng.; Voysey, "Second Report on Geology of Hyderabad."

Convoy of diamonds in Brazil.

habad. The value of precious stones exported from India in 1860–61 was £153,748, of which £9,534 was to the United Kingdom.

The Indian diamonds are sent to Amsterdam and Antwerp, where nearly the entire trade in diamonds has been in the hands of the Jews since the Dutchman Berqueen improved or discovered the art of cutting this incomparable gem in 1475. For some years important works have been carried on in London and Paris, supplied more especially from Brazil, and from the other diamond-producing provinces of America generally—as New Granada, Peru, Mexico, and California. The quantity of rough diamonds annually produced by North and South America, may be estimated at one hundred and fifty thousand carats, or about eighty pounds weight. Taking the carat at £4, the value is about 15,000,000 of francs, or £600,000.

The search for gems is not carried on in Europe in the same systematic manner as it is in Brazil, California, and India, but by simple mountain trackers, a credulous, persevering, and hardy race, who habitually give themselves up to the business of collecting. Prompted by a sort of instinct they work likely places in the granite and quartz, either by blasting or with the pick, and very often come upon magnificent geodes, and deep pockets in the rocks, lined with rock-crystal, amethyst, emerald, aquamarine, garnet, and topaz—all retaining their original lustre, and the beautiful and bright colours given them by nature thousands of years ago, when they were formed by her in these her mysterious caskets.

Some of these collectors are very original characters. The excellent Pietro Pinotti is mentioned by M. Simonin as one of this class, living in the Isle of Elba. The nickname of *Cervello fine*, or "the knowing one," was given him as much for his marvellous skill in finding gems as for the special talent he has of selling them at very high prices.

Other not less meritorious collectors are the brave coral-fishers, the rough sailors of Genoa and Naples, who for many ages have torn from the depths of the sea the stony shrubs with their white and red boughs. Neither must those bold divers be forgotten, who at the Pearl Islands in the Bay of Panama, at Aden, and Colombo, descend, at the risk of their lives, to the bottom of the sea, in search of the shell whose pearl will furnish ornaments for beautiful and fashionable ladies. The diver has been seen in the Bay of Panama, after remaining too long under water, to bleed at the mouth and ears; but that is merely one of the least dangers which he incurs. He dives with a knife in his belt to protect himself against the sharks. M. Simonin was told on the spot of an Indian having been swallowed by a shark, and that his companion who accompanied him in the boat only learned what had taken place from the blood which came bubbling up to the surface of the water. Maddened by despair, and only heeding the promptings of his courage, he dived immediately in search of the monster, and finding it in the act of finishing its hideous repast, he ripped it open and emerged in triumph, well pleased to have avenged his friend. The types of these rude coral-fishers and bold pearl-divers deserve to be placed in a parallel rank with those brave and hardy miners whose struggles are mentioned in another part of this work.

INDEX.

THE END.

PRINTED BY WILLIAM MACKENZIE, 45 & 47 HOWARD STREET, GLASGOW.

Printed in the United States
By Bookmasters